CANCER CRUSADE

RICHARD A. RETTIG

Cancer Crusade

The Story of the National Cancer Act of 1971

PRINCETON UNIVERSITY PRESS
Princeton, New Jersey

Copyright © 1977 by Princeton University Press

Published by Princeton University Press,
Princeton, New Jersey
In the United Kingdom: Princeton University Press,
Guildford, Surrey

All Rights Reserved

Library of Congress Cataloging in Publication Data will
be found on the last printed page of this book

This book has been composed in VIP Times Roman

Printed in the United States of America
by Princeton University Press, Princeton, New Jersey

To Angie,
for her patience

Contents

Tables

Chronology of Events

1912   Congress establishes the U.S. Public Health Service (PHS)

1922   Cancer research initiated within the PHS

1930   PHS Hygienic Laboratory renamed National Institute of Health (singular)

1937   Congress establishes the National Cancer Institute within PHS

1948   Congress establishes National Institutes of Health (plural), National Heart Institute, and National Institute of Dental Research

1965   Congress enacts Social Security Amendments of 1965, thus establishing Medicare and Medicaid; Congress enacts Heart Disease, Cancer, and Stroke Amendments of 1965, thereby establishing Regional Medical Program

1970   *April 27:* Senate Resolution 376 authorizes a Panel of Consultants on the Conquest of Cancer
December 4: Panel reports to Senate Committee on Labor and Public Welfare

1971   *January 22:* President Richard M. Nixon calls for additional $100 million for cancer research in State of the Union message
January 25: Senators Kennedy and Javits introduce S. 34—the Conquest of Cancer Act
March 9 and 10: Senate Subcommittee on Health holds hearings on S. 34
April 20: Ann Landers column appears in syndicated newspapers
May 11: President Nixon announces Cancer Cure Program; S. 1828 introduced by Senator Dominick
June 10: Subcommittee on Health holds hearing on S. 1828
July 7: Senate adopts S. 1828 by vote of 79 yeas, 1 nay
September 15: Representative Rogers introduces H.R. 10681—the National Cancer Attack Amendments of 1971—and four weeks of House hearings begin
October 12-15: House Subcommittee on Public Health and Environment meets, reports H.R. 11302 to Committee on Interstate and Foreign Commerce by unanimous vote

November 3 and 4: Commerce Committee considers H.R. 11302, reports bill to House by vote of 26 to 2

November 15: House of Representatives adopts H.R. 11302 by vote of 350 yeas, 5 nays

December 1 and 7: Joint Senate-House conference committee meets, reports compromise legislation to both houses of Congress

December 9: House adopts compromise legislation

December 10: Senate adopts compromise legislation

December 23: President Nixon signs the National Cancer Act of 1971

Preface

The National Cancer Act of 1971 was a political event of intrinsic significance, directed as it was to increasing our scientific understanding of cancer and to developing improved means of prevention, diagnosis and treatment of this highly feared disease. The Act was of great significance, as well, for the entire biomedical research enterprise, carrying substantial implications for the allocation of resources, the strategy of research management, and the organization of the enterprise that reached well beyond cancer.

In political terms, the Act is of interest because it indicates how a small but powerful elite composed of private citizens mobilized sufficient political resources to secure passage of legislation opposed by the National Institutes of Health and by most of the biomedical scientific community. In policy terms, the Act captures much of the current conflict between the public and its elected representatives eager to see life-saving and life-prolonging results flow from biomedical research and, on the other hand, a scientific community acutely conscious of the long time and great uncertainty characteristic of the process by which medical research is translated into clinically useful results. These reasons justify our attention to this statute.

The story of the National Cancer Act of 1971 has essentially three parts. The first consists of the antecedents to the legislation, the second is the legislative history of the Act itself, and the third—which continues to unfold—involves the implementation by the National Cancer Institute of the mandate of the 1971 legislation.

The first part of the story deals with the general antecedents to the Act and with the specific agenda-setting activities that brought cancer to a position of prominence on the national legislative agenda in 1971. It is told in chapters 2 through 5. Chapter 2 sets forth the activities of Mrs. Mary Lasker and her colleagues over the years in attempting to influence the rate and direction of cancer research. This consideration includes a historical view of the development of the underlying assumptions, strategies, and objectives behind the cancer crusaders. In chapter 3, the perspec-

tive is administrative rather than political, focused on the evolution of the National Cancer Institute and its relationships with the
National Institutes of Health. The emergence of contract-supported directed research is among the topics considered here.
Chapter 4 details the establishment of the Panel of Consultants on
the Conquest of Cancer, the selection of its members and chairman, and the process by which it developed its report. Included is
a discussion of the steps taken by the panel and its supporters to
secure a positive reception to its report in the House of Representatives as well as the Senate, and in the executive branch as well as
in the legislature. How were the recommendations of an obscure
panel, working in closed and largely private sessions, reporting in
a poorly covered hearing to a defeated Senate committee chairman transformed into a matter receiving attention in a State of the
Union message for 1971? Chapter 5 analyzes the emergence of
cancer as an important matter in the presidential politics of 1971
and 1972.

The second part of the story is the legislative history of the Act
itself. Events that began in the Senate, moved to the House, then
to the joint conference committee, and finally to the White House
are detailed in chapters 6 through 10. Here is recorded the debate
over the issues in the proposed new cancer program as it moved
through the discrete stages of the legislative process. The relative
weights attached to each issue, the degree to which each was
thoroughly or perfunctorily considered, and the degree of analysis
supporting views on each can be observed. The manner in which
the agenda-setting work of the Panel of Consultants affected the
nature and outcome of the legislative debate is indicated, as is the
importance of the provisions that emerge in the final stages of the
legislative process—the "end game" provisions.

The story of the National Cancer Act of 1971 has, in effect, no
conclusion. Rather, the third part of the story is still unfolding in
the implementation of the national cancer program. In chapter 11
we provide a general assessment of the legislative debate, touch
briefly on some of the highlights of implementation, and focus on
several issues of continuing importance in the continuing appraisal of the cancer program. This is appropriate, since the "conclusion" to this story will be found only in scientific laboratories, in

the translation of research findings into improved means of prevention, diagnosis, and treatment, and in the willingness of the
public to confront the nature of cancer and the means available to
deal with it in personal and collective terms.

This book is not simply another legislative history. It represents
an effort to place the case of the National Cancer Act in a
framework of agenda-setting and policy formation. Though it is
an addition to the small literature on the politics of biomedical research in the U.S., as well as an account of the most significant
political event in biomedicine in recent years, it is also an attempt
to place the detail of the story in a larger context of the policy
process.

There is a logic to this larger perspective that places nearly as
much emphasis on the origins of a piece of legislation as on the
legislative history itself. We may liken the legislative history to a
dramatic production—a play, and agenda-setting to the processes
which precede opening night. These processes include a writer
who conceives of the basic idea for the drama, writes a script, and
sells it to a producer; a producer who secures investment backing
for the production and locates a director; a director who selects the
actors, guides rehearsals, and modifies the script; and the actors
who learn their roles and their lines with reference to the script
and the director's interpretation of it.

In this instance, the individual most responsible for the idea of
a cancer initiative was Mrs. Mary Lasker. She and her associates
saw a need for such an endeavor and developed that perceived
need into a preliminary script. Senator Ralph W. Yarborough (D.,
Texas) played a critical role as chairman of the Senate Health
Subcommittee and of its parent, the Committee on Labor and Public Welfare, in securing passage of the resolution that enabled the
Panel of Consultants on the Conquest of Cancer to do its work in
1970. Mrs. Lasker and the senator were basically co-producers
of the cancer initiative, though the entrepreneurial skill and commitment that was the essential ingredient of success was primarily
Mrs. Lasker's. Selection of the players was dominated by Mrs.
Lasker and the choice of Mr. Benno Schmidt was a stroke of political artistry. As director, Schmidt had the task of writing the detailed script and extracting from the individual talents of the Panel

of Consultants a collectively functioning cast that would—both on and off the stage—perform their roles with effectiveness. It would be inappropriate to try to map the structure of the theater onto political events too slavishly, partly because the drama of politics emerges from the unrehearsed and unanticipated, as several important "end game" provisions make clear. But the analogy is sufficiently apt to make the point, amply supported below, that pre-production or agenda-setting activity is an important determinant of what happens on the political stage.

The literature on agenda-setting has distinguished between the public agenda and the formal agenda of government.[1] The public agenda includes issues that have achieved a high level of public interest and visibility, that require government action in the view of a sizable proportion of the public, and that are the appropriate concerns of government. The formal agenda of government includes those issues or demands that are under active and serious consideration by the government.[2] In this light, cancer has been on the public agenda since the early part of this century. It has been on the formal agenda of the government since 1937, when the legislation establishing the National Cancer Institute was enacted. And it has remained on the formal agenda since that time through the annual appropriations process. Why then was there a need for the initiative of 1971?

Jones has written, "The agenda of government is not set by new problems emerging in a state of nature. Most of what government acts on results from the continuing application and evaluation of ongoing policies."[3] Consistent with this view, the National Cancer Act of 1971 does not represent the emergence of a new policy so much as it reflects the reformulation of an existing policy. But we are still left with our question of why the need for this reformulation.

Agenda-building, or agenda-setting, has been defined as "the process by which the demands of various groups in the population are translated into items vying for the serious attention of public officials."[4] In the case of the National Cancer Act of 1971, we are concerned with the process of reformulation of existing policy. Specifically, we are interested in three questions. First, what were the sources of concern, the motivations, which led to the effort to reformulate the government's policy toward cancer? Second, what were the processes by which this desire for policy reformulation advanced the issue of government policy toward cancer to the stage of new legislation? Finally, how did the prelegislation stage of activity determine the outcome of the legislative history of the National Cancer Act?

There are three audiences to whom this book is directed. First, there is the large body of biomedical scientists, physicians, policy-makers, and others of the attentive public, who make, carry out, and are directly influenced by the policies of the U.S. government toward cancer and medical research. It is my hope that they will find in this story a useful account of some of the main forces shaping biomedical research policy today. The second audience consists of those members of the general public who are genuinely interested in how their government arrives at policy formulations regarding cancer and medical research. I hope that they find the account informative about the policy process related to this complex area of science and medicine. Finally, this book is directed to political and social scientists interested both in health policy and in the nature of the policy process. It is hoped that some will be encouraged by this account to study the many policy issues that surround biomedical research in the United States. It is also hoped that they will find this case study useful as a detailed account of the way in which the prelegislative processes of agenda-setting contribute substantially to the outcome of legislation.

I am indebted to many individuals for assistance in the writing of this book. Stephen P. Strickland read the entire first draft and made many helpful comments. And, in a fundamental sense, his book, *Science, Politics, and Dread Disease*,[5] with its broad scope and historical perspective on the politics of biomedical research, provides an important contextual orientation to an understanding of the cancer initiative.

The idea for this book originated while I was at Cornell University. There a small group of students—Peter F. Bachman, Sal Chieffo, Pierce B. MacKay, Sister M. Juliana O'Hara, and Spencer C. Johnson—speculated with me in the fall of 1970 about the work of a little-known group of consultants to the Senate

Committee on Labor and Public Welfare, a group whose activities are now spelled out in detail in chapter 4. Thanks are also due to two others at Cornell: H. Justin Davidson, dean of the Graduate School of Business and Public Administration, and Franklin A. Long, then director of the Program on Science, Technology, and Society. Each gave encouragement to a young scholar in very helpful ways.

This book was written mainly while I was at Ohio State University. William Wyman and Alan Boyd provided helpful research assistance. Special thanks go to Ronald Smith who functioned as research assistant as well as counselor on style and grammar. Dr. Clinton V. Oster, director of the School of Public Administration, provided support for these research assistants. Dr. Richard C. Snyder, director of the Mershon Center at Ohio State, was encouraging and provided critical support for travel associated with the manuscript. To all these individuals this effort owes a great deal.

This account is based upon the careful examination of legislative hearings and reports, official government agency documents, and numerous journalistic reports on the progress of the cancer legislation. In addition, over sixty individuals were interviewed, either personally or by telephone, and to them I am greatly indebted for invaluable information and perspective. Some of these individuals have been identified in the text, but others who preferred to remain anonymous have not been identified. Alan C. Davis, of the American Cancer Society, and Robert F. Sweek, formerly of the special staff of the Senate Committee on Labor and Public Welfare, were both helpful in providing access to materials in their files.

Portions of an earlier draft of this manuscript were read by Carl G. Baker, Kenneth M. Endicott, Carl Fixman, Gene Godley, Robert Harris, Stephen Lawton, Thomas J. Kennedy, G. Burroughs Midler, Edwin Mirand, Gerald P. Murphy, James A. Shannon, and Robert F. Sweek. A nearly final version of the book was read by Theodore R. Marmor and John F. Sherman. These individuals made a number of helpful comments, though none of them is responsible for any errors of omission or commission. Responsibility for interpretative judgments, of course, is wholly mine.

I am greatly indebted to Roy M. Cromer of Ohio State University for his watchful supervision of the preparation of this manuscript. Altha Shear and Mary Hixon, for their typing and retyping of several drafts, also have my very deep thanks. The fact that they found the story interesting and readable confirmed the worthwhile nature of this effort. The completion of the manuscript came after I had moved to the Washington, D. C. office of the Rand Corporation, and special thanks go to Sally Croasman and Shirley Lithgow for typing the final changes.

The members of my family, who collectively breathed a sigh of relief to see this work completed, deserve a measure of thanks that cannot be calibrated. Jerry good-naturedly saw the manuscript claim time that might have been spent watching baseball games. Kirsten, with less comprehension, watched several birthdays come and go while her father continued to write his book. Finally, my wife, Angie, endured with good humor the subtle deprivations that authorship imposes on a family and managed to bring her editorial capacities to bear upon the manuscript in a way which led to its improvement. My thanks to her are inestimable.

CANCER CRUSADE

A National Crusade
for the Conquest of Cancer

During the last Congress we appropriated $10,000,000 to eradicate the corn borer. For the present fiscal year we appropriated for the investigation of tuberculosis and paratuberculosis in animals more than $5,000,000; for meat inspection, more than $2,000,000; for the improvement of cereals, more than $700,000; for the investigation of insects affecting deciduous fruits, vineyards, and nuts, more than $130,000. I favored and supported all of these appropriations. . . .

But in view of our unequaled liberality in protecting our domestic animals against every sort of disease and pest, and in view of the vast expenditures we have made in protecting every species of food-yielding plant and tree, and in further view of the fact that the Government has never yet appropriated a dollar for the particular purpose of combatting cancer, I beg, in the name of all the vast hosts of cancer victims living and dead, for an appropriation that will make it possible for the work of rescuing suffering and perishing humanity from this frightful scourge immediately to begin.

> Senator Matthew M. Neely
> Democrat, West Virginia
> May 18, 1928

[W]hat are we doing to combat this dread disease? Our government this year spent, per person in the United States, $410 on national defense, $19 on foreign aid and $125 on the Indochina War. In the vital field of cancer research—to study a disease that will kill 330,000 of our people in one year—this government spent only 89 cents per person.

The government spent a total of $200 million for cancer research this year. People in this country spent more than that for ball point pens in the same year, and nearly twice that much on chewing gum in the same period of time.

We can no longer afford such half-hearted efforts in the field of cancer research. We need to get moving; we need to get busy with the job of saving American lives.

> Senator Ralph W. Yarborough
> Democrat, Texas
> December 4, 1970

2 · A National Crusade

In late 1970, a relatively obscure group called the National Panel of Consultants on the Conquest of Cancer presented a report entitled *National Program for the Conquest of Cancer* to a hearing of the Committee on Labor and Public Welfare of the United States Senate. The hearing had been hastily called, the committee chairman was concluding his Senate career because of an earlier defeat at the polls, and the event received little press coverage.

Yet one year later, on December 23, 1971, President Richard M. Nixon signed into law the National Cancer Act of 1971, the legislative result of the panel's report. Between the hastily called Senate hearing in December 1970 and a well-attended signing ceremony in December 1971, cancer research legislation had occupied a prominent place in the priorities of the president, the Senate, the House of Representatives, the medical-scientific community, and the interested public. The outcome of that year of legislative struggle and controversy was a much-expanded national cancer program, a program that held out the promise of major progress in the war against cancer and the concurrent possibility of failure to deliver on that promise.

The Panel of Consultants' findings were concisely stated in its report.[1] Cancer was the number one health concern of the American people—the most feared disease, even though it was only the second leading cause of death in the U.S. But funds spent on cancer research were, in the panel's view, grossly inadequate to the task of finding cures for cancer. Furthermore, the incidence of cancer was increasing, owing in part to the growing proportion of the aged in the population, and also to the effects of smoking upon lung cancer. Though the nature of cancer was not fully understood, the report indicated, the cure rate was gradually improving. Equally important fundamental advances in scientific knowledge in the past decade had opened up far more promising areas for intensive investigation than had ever existed. Consequently, the report concluded, "a national program for the conquest of cancer is now essential if we are to exploit effectively the great opportunities which are presented as a result of recent advances in our knowledge."[2]

The panel made three far-reaching recommendations, each to

remedy an existing defect it saw in the current situation. First, it recommended the establishment of a new government agency— the National Cancer Authority—whose mission would be defined by statute as "the conquest of cancer at the earliest possible time."[3] The Authority would absorb all functions of the existing National Cancer Institute (NCI) and would be completely independent of the National Institutes of Health (NIH). This new agency would remedy the lack of "effective administration with clearly defined authority and responsibility" that was alleged to exist in the relationship between NIH and NCI.

The second recommendation was that "a comprehensive national plan for the conquest of cancer" be developed.[4] Such a plan would continue the research currently being supported by the NCI, provide that initial program expansion be built upon existing facilities and manpower, strengthen existing cancer centers and create new ones, consider the manpower requirements of an expanded program, and make ample provision for fundamental research. This would fulfill what was believed to be the unmet need for "a coherent and systematic attack on the vastly complex problems of cancer."

Finally, to provide the necessary financial resources, the panel recommended a greatly expanded budget for cancer research.[5] The NCI budget for fiscal 1972 should be brought up to $400 million, it recommended, from the $230 million appropriated for fiscal 1971. Thereafter, it should increase annually by $100 million to $150 million until it reached a level of $800 million to $1 billion by fiscal 1976.

The underlying motivation for the creation of the Panel of Consultants, the recommendations of its report, and the National Cancer Act of 1971 was the very natural and quite intense desire of the American public to be free from the threat of cancer. According to estimates of the American Cancer Society, the decade of the 1970s will see 3.5 million deaths from cancer, 6.5 million new cancer cases diagnosed, and more than 10 million cancer patients under medical care.

Cancer accounts for roughly one-sixth of all deaths in the United States each year, second only to cardiovascular diseases.

But though second in mortality, cancer is consistently cited by the public as the most feared disease. The roots of that fear were captured in this way by one report:

> No affliction that man is heir to is quite so heavily freighted with dread and mystery as cancer. One reason for this is that, to many, the word itself is synonymous with death—and with protracted suffering. For no part of the human body is immune to cancer. The malignancy eats into nerve and muscle, bone and organ, blood and lymph gland alike; and it acquires an extra measure of terror because of the stealth with which it arises and because its deadly origins are inexplicably intertwined with the secret of life itself.[6]

Cancer is a disease of many forms.[7] If classified both by the organ or tissue site in which it originates and by the type of body cell involved, it is possible to identify nearly one hundred distinct varieties of cancer. Cancers can be divided into three broad groups according to organ of tissue site: carcinomas, sarcomas, and leukemias and lymphomas. Carcinomas arise in the sheets of cells covering the surface of the body and lining the various glands—the epithelia. Sarcomas, which are much rarer, arise in supporting structures like fibrous tissue and blood vessels. The leukemias and lymphomas arise in the blood-forming cells of the bone marrow and lymph nodes.

It is perhaps more useful to classify all cancers as solid tumors or hematologic malignancies. Solid tumors are initially confined to specific tissue or organ sites. Hematological disorders, on the other hand, involve the blood and lymph system and, consequently, are usually disseminated throughout the body at the time of diagnosis. Solid tumors are often described as slow-growing in relation to the fast-growing cancers of the hematological malignancies.

All cancers, however, have two common characteristics. First, unlike normal cells, cancerous cells manifest abnormal cellular growth—essentially uncontrolled cellular proliferation. Abnormal cellular growths are called tumors, though some tumors are benign and others malignant. It is the difference between benign and malignant tumors (cancers) that brings us to the second characteristic of cancers. Benign tumors remain confined to their site of origin. Malignant tumors, by contrast, can escape their site of origin, invade surrounding tissue, enter the blood stream and lymphatic system, and—having spread to distant sites of the body—set off secondary growths. These secondary growths are known as metastases. It is cancer in this metastasized or disseminated form that is so lethal.

The statistics of malignancy give the popular fear of cancer a rational basis. The estimate of newly diagnosed cases prepared by the American Cancer Society for 1972 was 650,000, or 339,800 among men and 310,200 among women.[8] The five leading types of new cancers were estimated to be: skin, 118,000; lung, 76,000; breast, 70,000; large intestine (including colon and rectum), 55,000; and uterus, 43,000. For men, the leading types of new cancers were expected to be skin, lung, prostate, large intestine, bladder, and pancreas, while for women the leading types were breast, uterus, skin, large intestine, ovary, and lung.

The mortality from cancer differs from the pattern of incidence. In particular, skin cancer occurs with great frequency in both men and women but is not a leading cause of death for either. The five-year survival rate for skin cancer is over 90 percent.[9] This is because skin cancer is easily detected, both by the individual and by a physician, and is therefore susceptible to definitive surgical treatment; in some cases it can be treated by topical application of chemotherapeutic agents.

Other forms of cancer are less manageable. Five-year survival rates for lung cancer, for instance, are 29 percent where the cancer is localized but only 9 percent when there is regional involvement. Breast cancer survival rates are 85 percent when it is localized, 53 percent when regional involvement exists.

While leukemias and lymphomas are leading causes of death among children under age 15, cancer is mainly a disease of middle and old age. The proportion of deaths in the 55-years-and-older age category in relation to the total deaths for the five leading causes of cancer mortality for both men and women is estimated to be over 80 percent.

Historically, the age-adjusted national death rate for cancer rose from 112 cancer deaths per 100,000 population (age-adjusted) in 1930, to 120 in 1940, 125 by 1950, and 130 in 1968. Behind these statistics are a greater longevity and, thus, more people at risk; a dramatic increase in lung cancer among men; a decrease in uterine cancer in women; a marked decrease in stomach cancer; steady increases of leukemia, and cancer of the pancreas and the ovary. Concurrent with this upward trend in mortality, though, has been an improvement in the survival rate. Fewer than one in five cancer patients lived five years after treatment in 1930. In the late 1940s this had changed to one in four, by the late 1950s one in three.

Can cancer be prevented? Skin cancer, which is caused mainly by over-exposure to the sun, is preventable. So is lung cancer, the leading cause of death among men, the primary cause of which is cigarette smoking. Certain other forms of cancer caused by identifiable chemical carcinogens in the industrial workplace can also be prevented through the application of the appropriate public-health safeguards. Cancer, however, takes a long time to develop from the initial cellular insult to its appearance in the diagnosis of a patient, and the nature of the insult may be in the form of sustained exposure to a low-dosage carcinogen. Lacking the immediacy between insult and disease, the incentive to adopt preventive measures appropriate to infectious disease is frequently not very strong. So the question of whether cancer can be treated is and always will be a very central one.

There are three present modes of treatment—surgery, radiotherapy, and chemotherapy. If a solid tumor is in an early stage of development, where the symptoms indicate that it is still localized or has only invaded the region near the tumor site, then the appropriate mode of therapy is surgery, perhaps in conjunction with radiotherapy. Under such circumstances, treatment can be highly effective. The key to effectiveness clearly lies in the capability for early detection.

The Pap test, for example, can detect cellular changes that may signal the appearance of cancer of the uterus and can detect presymptomatic forms of uterine cancer. In the first case, the lesions can be easily destroyed by either minor surgery or radiation; in the latter case, where the cancer has not metastasized extensively, treatment can be effective more than 90 percent of the time. The basic problem is getting women over 20 and those "at risk" women under 20 to take the examination on a regular basis. The five-year survival rate for breast cancer, to take another example, is nearly 85 percent if detected before it has spread to the lymphatic system.

The basic problem, however, is that there are many forms of cancer that are not easy to detect in their early stages, either because the diagnostic tools do not exist or because access to patients is difficult. Lung cancer is in the former category; it is very difficult to diagnose at an early stage that would permit successful therapeutic intervention. Cancer of the colon and rectum, on the other hand, is capable of early diagnosis by means of proctoscopy. But many physicians do not include a proctoscopic examination in physical check-up and many patients do not express much interest in an annual proctoscopic examination.

Radiotherapy is normally used in conjunction with surgery in the treatment of early solid tumors. It is used exclusively for such tumors when they are inaccessible to surgery. Five-year survival rates from radiotherapy have improved substantially during the past fifteen years or so, but here again the effectiveness of therapy is directly related to whether the cancer is diagnosed in an early stage before it has metastasized. Where radiotherapy is used with chemotherapy against the disseminated forms of solid tumors, it is essentially palliative in nature.

Chemotherapy is the treatment of choice for the hematological malignancies that normally present themselves in disseminated form when diagnosed. Certain leukemias and lymphomas can be treated effectively by chemotherapy. Acute childhood leukemia is the most common form of childhood cancer in the United States. Prior to chemotherapy, the median survival was four months and long-term remissions were practically unknown. The median survival in many clinics is now three years and some children have been living more than five years without evidence of disease. Though a five-year survival rate of 75 percent is possible for early Hodgkin's-disease patients treated with aggressive radiotherapy, high rates of remission have only recently been realized for the advanced stage of this disease through the use of combinations of chemotherapeutic drugs. Other cancers against which chemotherapy has proved effective in some measure include choriocarcinoma, metastatic hydatidiform mole, Burkitt's tumor, embryonal carcinoma of the testis, adenocarcinoma of the uterine corpus, and carcinomas in the superficial layers of the skin.

The scientific report of the Panel of Consultants repeated this statement of one panel member: "The long-term future may belong to the immunologist and the geneticist, the intermediate future to the chemotherapist, but the present and the immediate future belong in the main to the surgeon and to some extent to the radiologist."[10] As this medical scientist indicated, the slow-growing, solid tumors are by far the more prevalent cancers, early detection is essential to definitive therapy, and the indicated mode of therapy is still surgery.

Yet, for four decades, the promise of new treatments has helped stimulate medical research on the nature of malignant cellular growth. In 1937, the Congress passed and President Franklin D. Roosevelt signed legislation that established the National Cancer Institute. This organization, the first "categorical" disease institute of what was to become the National Institutes of Health, has been the principal administrative agency through which the United States government has supported cancer research. Slightly more than $2 *billion* had been appropriated to the National Cancer Institute between its establishment and the time when the Panel of Consultants began its work. The annual appropriations level was approaching $200 million as the panel wrote its report. Why then, it might reasonably be asked, were the sweeping recommendations of the panel's report necessary?

The Panel of Consultants gave three reasons for its initiative, one in each of its recommendations. These reasons were central issues in the controversy over the National Cancer Act of 1971. They were—money for cancer research, the strategy of research management, and the appropriate organizational home for cancer. We sketch in some of the background of each issue to provide the reader with a framework for following the story.

Until recent years, the politics of biomedical research have been the politics of the appropriations process.[11] Nurtured by the philanthropist Mary Lasker and her associates, sustained by powerful appropriations committee chairmen—Representative John

Fogarty in the House and Senator Lister Hill in the Senate, and assisted by the leadership of NIH, James A. Shannon, M.D., and his colleagues, the rapid and sustained growth in funds for medical research made mockery of White House and Bureau of the Budget efforts to exert budgetary control over the NIH from the late 1950s until the late 1960s.[12] A small number of highly prominent spokesmen for the biomedical research community actively participated in the annual budgetary ritual. A larger number, in the hundreds, were always actively engaged in the numerous advisory committees and study sections of the National Institutes of Health. Far more, in the thousands, actively engaged in laboratory or clinical research, tended to be passive observers of political developments as they unfolded. All, however, were beneficiaries of the largesse bestowed on medical research. Indeed, many came to regard such public largesse as their due.

During these years, Don K. Price wrote, "it had begun to seem evident to a great many administrators and politicians that science had become something very close to an *establishment*, in the old and proper sense of that word: a set of institutions supported by tax funds, but largely on faith, and without direct responsibility to political control."[13] Robert C. Wood, writing at the same time, saw the basis of the scientist's power in their status as an "apolitical elite."[14] Wallace S. Sayre, on the other hand, saw the scientists who were influential in American science policy as "scientists in politics," and subject to limits that American politics places on experts and interest groups.[15] They would be effective in a democratic order in the degree to which they understood the political process, accepted its rules, played their part "with more candor than piety," and accepted the fact that they were "in the battle rather than above it." Of these three views of scientists in politics, time would clearly favor Sayre's interpretation.

Rapid changes began to occur in the established order of biomedical research in the late 1960s. First, the rate of growth in funds for medical research slowed beginning in fiscal 1968 and never recaptured its prior momentum, thus placing a serious financial squeeze on all NIH programs. Second, NIH acquired new functions within this more slowly growing budget, further exacerbating the pressure on existing research programs. Third, priorities in the allocation of the federal health dollar began to shift from medical research toward health manpower and the delivery of health services. Fourth, major leadership changes affecting biomedical research occurred in Congress, NIH, the Department of Health, Education, and Welfare, capped by the transition in the White House from Lyndon B. Johnson to Richard M. Nixon. Finally, the new Bureau of the Budget officials in the Nixon administration increasingly demonstrated the intent and ability to impose financial discipline on appropriations to NIH. Under these circumstances, Mary Lasker and her associates began casting about for a stratagem to restore medical research, especially cancer research, to its financial *status quo ante*. But the intent on this occasion ran well beyond money.

The NIH and the political forces that sustained it in the late 1950s and early 1960s achieved a remarkable degree of autonomy from external political control. It was true that Representative L. H. Fountain (D., N.C.) had scored some telling points in 1962, but these were handled with relatively little loss of momentum. The controversies that did occur were largely in the nature of family quarrels. Positions were taken, issues fought out, compromise solutions arrived at among the parties to the debate. All participants knew each other well, all were sensitive to the power resources of the others, and all had at least grudging respect for the commitment of the others to the overall purpose of the medical research enterprise.

There was essential agreement among all on objectives. Medical research was supported to generate scientific results that would undergird the practice of medicine and lead to the improved health status of American citizens. NIH was a health agency, not a science agency, and the investment in scientific research was made because of its health mission. But agreement on ends hardly brings with it agreement on means. The large issue that divided people the most had to do with the strategy or philosophy of research management. How should scientific research be supported in order to improve the health of the American public most effectively and rapidly? On this key issue there was and is today an underlying, unresolved conflict.

It is sobering to realize that the underlying conflict in

biomedical-research strategy is an old one. The medical advisory committee that reported to Vannevar Bush in 1945 on a postwar program for scientific research put the matter succinctly: "Research in medicine may be carried out effectively in two ways: First, by a coordinated attack on a particular disease; or second, by independent studies of the fundamental nature of the human body and its physiological mechanisms, of the nature of bacteria, viruses, and other agents on disease, and of the influence of the environment on both."[16] The committee had in mind the major advances in medicine that had occurred in World War II. The successful development of antimalarial agents was an example of "a coordinated attack on a particular disease," or "categorical" research. Penicillin, on the other hand, arose from "fundamental" research.

In 1945, the prevailing view within academic medicine was that the fundamental research approach was ultimately shorter and more direct than the categorical disease approach, that advances in clinical research in the war had come at the expense of basic research, and that the situation had to be rectified in the post-war world. The medical advisory committee put it this way:

> Discoveries in medicine have often come from the most remote and unexpected fields of science in the past; and it is probable that this will be equally true in the future. It is not unlikely that significant progress in the treatment of cardiovascular disease, kidney disease, cancer, and other refractory conditions will be made, perhaps unexpectedly, as the result of fundamental discoveries in fields unrelated to these diseases. . . . Discovery cannot be achieved by directive. Further progress requires that the entire field of medicine and the underlying sciences of biochemistry, physiology, pharmacology, bacteriology, pathology, parasitology, etc., be developed impartially.[17]

These words, thirty years later, still capture the prevailing view within academic medicine.

It is also important to note that these two differing strategies toward medical research are embedded in the statutes that created the NIH and the NCI. The 1930 legislation that established the National Institute of Health stated its purpose as "study, investigation, and research in the fundamental problems of the diseases of man." Supporters of the fundamental-research view lay stress upon "fundamental problems," while advocates of the categorical-disease view emphasize "the diseases of man." The 1937 cancer act emphasized that the purpose of NCI was to be "researches, investigations, experiments, and studies *relating to the cause, prevention, and methods of diagnosis and treatment of cancer*" (emphasis added). Both points of view, it should be stressed, are in essential agreement that there should be a continuing advance of biomedical scientific knowledge to the end that illness might be prevented or alleviated and human health improved. But the differences have substantial implications for resource allocation and for research administration.

The NIH, both reflecting and shaping the dominant attitude within the biomedical scientific community, has been strongly committed to basic research in the life sciences and relatively cautious about clinical research. During the years of its rapid budget growth, the agency saw its primary mission as laying down the scientific base for modern medicine across the entire range of the biomedical sciences. This meant, among other things, an evenhandedness would apply in allocating resources across all NIH institutes and that no one institute should be permitted to grow more rapidly than any other.

On the other hand, NIH managed to make its peace with the categorical research. Mindful of congressional preferences, NIH has valued the disease emphasis of its individual institutes for its identification with health as distinct from an exclusive research identity. The usefulness of the categorical emphasis in securing appropriations, moreover, was long ago recognized by the observation of one wag that "no one ever died of microbiology." In the main, though, NIH has acted as a trustee of the biomedical scientific community and emphasized the importance of basic, much more than categorical, research.

In the opinion of many medical scientists today, the major obstacle to the prevention and treatment of the chronic and degenerative diseases, especially cancer, lies in our ignorance of the underlying biological processes of health and disease. While dramatic advances in molecular and cellular biology have occurred in the past quarter-century, in this view, much more basic scientific knowledge needs to be developed. The needed scientific advances will come only as qualified scientists, in consultation with their professional colleagues, are free to choose their own research priorities.

Mrs. Lasker and her associates, however, have frequently emphasized clinical research over basic research in a way which has diverged rather sharply from that of the NIH. This clinical research emphasis, typically focused on diseases like cancer, heart disease, stroke, and mental illness, has often been used to justify new medical initiatives. In 1952, Dr. Sidney Farber, advocating an expanded NCI program in cancer chemotherapy, stated, "Today we have the very peculiar and what I regard as unfortunate situation of having laboratory research far in advance of the application of the results . . . to the patient with real cancer. I think . . . it is sad indeed that chemical compounds which might conceivably be of great help to patients now having cancer cannot be administered because there are not enough teams and facilities to do it."[18] In 1959, in urging support for a network of clinical cancer research centers, Farber said this: "Those who have watched . . . progress in cancer research . . . have come to the conclusion that if we are to exploit to the full the opportunities created by [such] progress . . . and if we are to meet the needs of patients who now require facilities so that the results of research may be used for their benefit, an expansion of cancer research in centers in a number of places in the country must take place."[19]

The Laskerites' advocacy of clinical research on categorical diseases has been based upon the belief that the costs of the major killer diseases are so great that the world cannot wait until all the fundamental science results are in. It is imperative in their view that a wide range of clinical research possibilities be pursued and that whatever scientific results there are should be mobilized at that moment into the most effective clinical attack possible. NIH, with the great majority of biomedical scientists, has often regarded Lasker-sponsored initiatives as premature pursuit of certain lines of inquiry lacking an adequate scientific base.

In the mid-1950s, this long-standing conflict between basic and categorical research became further complicated by administrative developments within the National Cancer Institute. Kenneth M. Endicott, M.D., first as director of the cancer chemotherapy program and later as director of NCI, pioneered the use of the contract, rather than the grant, as an instrument for the support of research. This represented an extension of the categorical approach to research support, though it did not initially find favor with Mary Lasker. The contract became the instrument of directed or targeted research programs. These programs—chemotherapy, the search for a viral cause of human leukemia, and carcinogenesis—supported basic as well as clinical research. They were characterized by administrative determination within NCI as to which lines of research cancer should be pursued. This was quite unlike grant-supported research where the determination of what to support was basically left to the scientific community through the NIH peer-review system and where the direction in which research should go was largely a function of the prevailing views within the scientific community. Directed research, in addition to relying upon the contract, also involved research planning—another activity that was anathema to the scientific community.

The NIH leadership was always somewhat uneasy about the use of the contract instrument, especially when contract programs began growing rapidly within NCI. The scientific community, moreover, was and still is highly critical of the use of contracts rather than grants. At root, the conflict stems from a difference of view about the most effective way to see that biomedical research leads to results which can be applied to the conquest of major diseases.

The conflict between the fundamental research strategy and the categorical disease strategy, then, actually masks five closely related issues. What kind of research is to be supported or favored—basic or clinical? What instrument of support is to be used—the grant or contract? Who is to make the authoritative decisions allocating support—the external scientific community, the professional staff of an institute, or the advisory council to an institute? Who is to be supported—university scientists or industrial researchers? What is to be the extent of formal research planning—limited, significant, or very extensive! This potpourri of issues was basically rolled into one in the debate over the National Cancer Act of 1971. The overarching issue concerned the most appropriate strategy of research management for conducting the war against cancer.

The Panel of Consultants also interjected a radically new issue into the debate. Its proposal for a new cancer agency independent of the NIH was, to say the least, extraordinary.

To be sure, the National Institute of Mental Health (NIMH) had been severed from the NIH in 1967 and established as a separate agency in its own right. But NIMH, unlike the other NIH institutes, had substantial programs in training health professionals and in providing community-based services in addition to its research program. NCI had no comparable involvements at the time of the Panel of Consultants.

Mental-health research, moreover, had its scientific roots in both the biological and the behavioral sciences. This bifurcated research tradition reflected a continuing dispute over the issue of whether mental health was biological or behavioral in nature. The dispute was further reinforced by different groups of professionals clustered on either side of the matter. But there was no question in anyone's mind that cancer research was integral to the fabric of biomedical research. To argue otherwise was to deny reality.

It was true that mainstream medical researchers had long condescended to cancer scientists. Much of science, including biomedical science, attacks problems guided by what might be called the principle of the drunkard's search. Like the drunkard searching for his lost keys under the street light "because there is more light here," most scientists pursue those problems that offer some possibility of being solved. The scientific problems associated with cancer seemed unapproachably complex to many scientists. Consequently, the competence of those who worked directly on cancer was often suspect, though not their good intentions.

But such differences within the scientific community as might exist were questions of research management or strategy, not matters affecting the organization of NIH. To think that they accounted for the recommendation for a separate cancer agency is to suggest that pique was a major motivating force behind the Panel of Consultants. No, the panel had larger motives than that of remedying slights of bygone years. But the true background for its organizational recommendation is best discussed in the context of our story.

The existence of the panel, the recommendations of its report, and the legislation that resulted therefrom raise a fourth issue. Were there major scientific or clinical advances in our knowledge of cancer that justified the establishment of a national cancer program? Or were such advances just over the horizon? This issue was the essence of the controversy over the National Cancer Act of 1971. It is a matter on which the participants themselves should be heard. In the final chapter, we shall return to this question more directly.

The reader may wish to ask how well this question was addressed in 1970 and 1971 as both protagonists for and antagonists of an expanded cancer program state their case in the pages that follow. A second question may also help the reader interpret the events of this story: How persuasive was the case that money, management, and organization could influence the pace and direction of scientific and clinical progress related to cancer?

Money can, of course, accelerate the pace of science. Established cancer researchers can be permitted to conduct their research on a larger scale. The entry of newer scientists, both younger ones and those from related fields, can be facilitated for cancer research. But at some point, the law of diminishing returns sets in and the next new scientist is not as capable, not as productive as the last. So, in principle, there are limits to what money can do.

Management can presumably alter the efficiency with which research funds are used. Unfocused and diffuse research efforts can be shifted toward directed or targeted research programs. Human resources, moreover, can be concentrated in cancer research institutions with the beneficial effect of bringing all related talents together. But, when confronted with underlying scientific ignorance of and great uncertainty about the origins and processes of the growth of cancer, there are presumably limits on the ability of management to generate the right answers, let alone to raise the right questions.

Organization is important as it constrains or encourages scientific and clinical advance. An organization unable to control its own destiny without external interference may conclude that bureaucracy is a major obstacle to the realization of its goals. Yet if that organization's goals, and the means to achieve them, are inextricably intertwined with the goals and activities of others, the aspiration for organizational autonomy may be a false hope.

These questions of money, management, and organization and how they relate to scientific and clinical opportunities in the struggle against cancer, are the immediate issues of the cancer legislation debate. The reader may wish to reflect on the adequacy of evidence presented on these issues by either proponents or opponents of an expanded cancer program, and on the degree to which the legislative outcome of 1971 turned on careful assessment of the factual situation and projection of the most probable scientific and clinical developments. Alternatively, the reader may wish to remember the question posed in the Preface of the extent to which the legislative outcome was determined by the prelegislative agenda-setting activities of Mrs. Lasker and her associates. Finally, the influence of contextual factors like the public's deep-seated fear of cancer, the congressional desire in 1969 and 1970 to reorder national priorities, and the impact on domestic policy of the successful U.S. effort to land a man on the moon, also need to be considered as determinants of the National Cancer Act of 1971.

Such deliberation need not be idle speculation. One result of the 1971 law is that the federal government is now financing cancer research on a scale fully four times as great as before the new law. Given the high stakes associated with the cancer crusade, all of us—whether citizens, scientists, administrators, or legislators—have an obligation to exercise our most thoughtful judgment as we collectively confront the continuing problem of cancer.

2

The Benevolent Plotters

But, however partisan I have been, I believe the most rewarding of my political activities were nonpartisan ones performed, as a result of my work in the American Cancer Society, on behalf of American health care generally. My involvement with the Cancer Society brought me into contact with a small group of active people—someone called them "benevolent plotters"—who set out during and after World War II to revolutionize American medical education, research, and health care. They include Albert and Mary Lasker, Dr. Alton Ochsner, Dr. Michael DeBakey, Emerson Foote, Dr. Frank Adair, Anna Rosenberg, my old friend Jim Adams, and a number of other well-known and not so well-known people who were willing to devote time and energy to this noble cause. All of them did not agree all of the time, but all were united in one purpose: to stimulate federal support of medical research and education. To accomplish this required the ear of Congress, and to gain that ear required enormous voluntary expenditures of time and energy, cultivating senators and congressmen, educating them and creating forceful advocates for medical progress.

Elmer Holmes Bobst
*Bobst: The Autobiography of a
Pharmaceutical Pioneer*, 1973

Neither the Senate Panel of Consultants on the Conquest of Cancer nor its report were the product of chance. They owe more to Mrs. Mary Lasker than to any other individual. She was "present at the creation" of the panel, concerned with every step of its work, and active at each subsequent legislative juncture. She was an invited guest when President Nixon signed the National Cancer Act of 1971, and was then appointed a member of the new National Cancer Advisory Board. Her success, however, was attributable in large measure to the skillful enlistment of a number of her influential friends and associates and to her mobilization of a remarkable political coalition variously dubbed "the Lasker-ites," "benevolent plotters," "the health syndicate," "the cancer mafia," and even "Mary's little lambs." The enactment of cancer legislation in 1971 was a reflection of the prodigious investment of time, energy, and political craftsmanship by these "benevolent plotters," and another reminder that certain individuals may be very influential in getting problems to the agenda of government.

LASKER ENTERPRISES, INC.

Mary Woodard was born in Watertown, Wisconsin, in 1900.[1] Her father was a successful, frugal banker, while her mother was an active, civic-minded woman concerned with public parks and urban smoke control. Educated at Donner Seminary, Milwaukee, Mary attended the University of Wisconsin briefly, and was graduated *cum laude* in 1923 from Radcliffe College in art history. She studied at Oxford University, England, and then worked as an art dealer in New York City. Mary Woodard became a successful businesswoman during the depression by introducing Hollywood Patterns, inexpensive dress designs, at a time when women were increasingly making their own dresses.

Mary met Albert D. Lasker in April, 1939, and they were married fifteen months later, she for the second and he for the third time.[2] He was 60 years old at the time and the millionaire president of Lord & Thomas Company, a Chicago advertisiting firm. Albert Lasker decided to close Lord & Thomas in 1942 after forty-four years with the firm and, with Mary, devoted the rest of his life to art, politics, and health. He had an operation in 1950 for intestinal cancer which, it was later discovered, did not remove all malignant tissue. Albert Lasker died of abdominal cancer on May 30, 1952.

Her marriage to Albert, and the resources provided by his estate, permitted Mary Lasker to pursue her life-long interest in health. She had experienced frequent illness as a child and had learned early in life that physicians often knew very little about certain diseases. She also realized that most people could not afford proper medical care, and that a family's entire savings could be wiped out by a single prolonged illness. Albert Lasker's death from cancer reinforced her desire to see medical research provide answers to the major killer diseases.

Mrs. Lasker has pursued her interests in health and medical research in concert with a number of long-standing, like-minded friends.[3] One such friend has been Emerson Foote, one of the three senior executives of Lord & Thomas when Albert Lasker closed the firm, and a senior partner in the successor firm of Foote, Cone, and Belding. Another, Anna Rosenberg, now Anna Rosenberg Hoffman, former assistant secretary of defense for manpower under President Truman, has been an active supporter of Mary's efforts in cancer research since the mid-1940s. The extensive network also includes many prominent medical men, like Michael DeBakey, M.D., the internationally known heart surgeon, and now president of Baylor University College of Medicine, and Howard Rusk, M.D., professor and chairman of the Department of Physical Medicine at the New York University Medical Center. In cancer, the prominent figure has been Dr. Sidney Farber, the scientific director of the Children's Cancer Research Foundation, Boston, until his death in 1973, and the discoverer in the late 1940s of the success of the antifolic acids in combating acute childhood leukemia.

Mrs. Lasker has had an impressive organizational resource in the Albert and Mary Lasker Foundation, of which she is president.[4] Established in 1942, the foundation gives the prestigious Albert Lasker Awards in Medical Research each year to individuals "who have made significant contributions in basic or clinical research in diseases which are the main causes of death and disability." By 1970, no fewer than twenty-three Nobel laureates had received Lasker Awards, all save two prior to their international recognition. In addition, annual awards are given in medical journalism to encourage newspaper, magazine, and television writing and production of outstanding reports on medical research.

Mrs. Lasker has long been active in the American Cancer Society, of which she is an honorary member of the board. It began one day in 1943, when she marched into the office of Clarence C. Little, D.Sc., managing director of the American Society for the Control of Cancer, and demanded to know how much the society was spending for research.[5] "Nothing," she was astounded to learn. The physician-dominated organization was chiefly concerned with the improvement of treatment facilities and the professional education of physicians. Dr. Little, a research scientist himself, had been unable to persuade his board of the merits of supporting research. Mary gave him $5,000 for a pamphlet on the existing state of research, and pledged that she would return.

Fund-raising for the society was then done by the Women's Field Army, an association of national women's organizations. Though the Women's Field Army had raised $171,000 in 1939, $269,000 in 1942, and $356,270 in 1943, it had no representation on the board. Dr. Little thought this arrangement was untenable and pressed his board to add lay representation. They agreed, and Albert Lasker was invited to join them in 1944. He could not do so, but persuaded Emerson Foote, whose parents had died from cancer, to accept.

Foote found the organization unbusinesslike and incapable of conducting a first-rate fund-raising drive. On his advice, the name was changed to the American Cancer Society (ACS), the Women's Field Army became simply the Field Army, and plans for a research program were widely publicized. This new lay influence was apparent when $832,000 was collected in the 1944 fund drive.

The ACS board then recruited Eric Johnston of the motion picture industry to head the 1945 fund drive, and created a lay "governing board," headed by Foote and including Albert Lasker, to help him. Mary Lasker was also active. She arranged for three short *Reader's Digest* articles on cancer to appeal for funds, and these netted more than $100,000 to help finance the drive. Mrs. Lasker also personally retained a professional fund-raising firm to help the campaign. The Laskers' price was that the ACS earmark one-fourth of the funds for cancer research. The drive raised $4,292,000, and as a result, $960,000 was allocated for research.

The aggressive lay "governing board" was not content at being limited to fund-raising. Soon after the drive, Foote and the Laskers suggested to the board of trustees that it be divided equally between medical and nonmedical members, a proposal that most board physicians resisted strenuously. Albert Lasker had recruited Elmer Bobst, who had run the 1945 New Jersey fund-raising effort, to the "governing board,"and Bobst spearheaded the effort to secure greater lay representation. A bitter fight ensued.

More money, a clearer statement of objectives, and reorganization were needed, Bobst thought. He later wrote:

> I decided that the first priority was to move aside the scientists and physicians who were in administrative control of the organization. They were good men, but they were not experienced leaders, and they were not getting results. I wanted majority control to be in the hands of qualified lay leaders. The physician members could form a scientific committee to make recommendations about scientific matters and advise the executive committee.

Such a sweeping revolution, Bobst knew, meant that the society's constitution and bylaws had to be rewritten and that a showdown with the society's well-entrenched and self-satisfied leadership was required. That showdown came at a meeting where Little defended himself and attacked Bobst personally. The acrimonious debate ended when Bobst said, "Now, Dr. Little, I would like to conclude by saying that this society is too small to have both you and me in it. I intend to stay."[6] Little chose to resign. The lay individuals had won a decisive battle.

The board of the American Cancer Society was reorganized to provide equal medical and nonmedical representation, and a strong, primarily lay, executive committee was created. An agreement between the national organization and the local chapters gave substantial autonomy to the latter. James Adams, president of Standard Brands, became chairman of the new executive committee. Other dynamic business and professional people were recruited who, as one writer put it, "brought more kinetic energy to bear on society affairs in a half-hour than the poor little organization was accustomed to experiencing in a year."[7]

For Mary Lasker, this episode was the basis of a continuing relationship with the American Cancer Society. It also drew together a number of dedicated and energetic individuals on whom she would later call for other equally important crusades. And in the episode, both Mary and Albert Lasker and their friends acted upon their fundamental conviction that progress in the war against cancer was too important to be left in the hands of conservative physician-scientists.

Immediately after World War II, the Laskers were initially not persuaded that the National Institute of Health should be the medical research agency for the federal government. But Mary Lasker had been impressed with Leonard Scheele, M.D., a Public Health Service officer she met in 1946, who became director of the National Cancer Institute in 1947. She approached Scheele and pointed out that Congress had shown increased interest in heart research. Why not create a heart institute comparable to the National Cancer Institute?[8]

Scheele liked the idea and had Mary talk with Thomas Parran, M.D., the Surgeon General, who also favored the idea. A bill was drafted based on the 1937 National Cancer Institute Act. Introduced in the Senate, the bill moved quickly through both houses of Congress, and was signed by President Truman on June 16, 1948. The legislation established the National Heart Institute. It also created the National Advisory Heart Council, on which provision was made for lay representation. Scheele, who had been named Surgeon General by Truman in early 1948, appointed Mary Lasker as the first nonmedical member of this advisory council, and she served on it from 1948 through 1952.[9]

Mary Lasker has been attentive to presidents, especially Democrats, and to their wives. She was an early supporter of Senator John Kennedy in his quest for the presidency, and was later named by him to several presidential boards and commissions. Mrs. Lasker was especially close to President and Mrs. Lyndon B. Johnson. Lady Bird Johnson, for instance, presented the Albert Lasker Medical Research Awards in November 1962. Mary Lasker, with her lifetime interest in beautification, was later deeply involved with Mrs. Johnson's national beautification efforts. Furthermore, the Lasker Foundation honored President Johnson in 1966 for his public contributions to health and medical research.[10]

Mary Lasker and her friends also have an unparalleled knowledge of how to publicize and popularize a major initiative. Albert Lasker was an acknowledged advertising genius. Emerson Foote became a leading advertising man in his own right. The Albert Lasker Awards in Medical Journalism actually helped establish that specialized branch of news reporting. The annual Science Writer's Seminar of the American Cancer Society, which brings prominent cancer research scientists together with medical journalists, bears the stamp of the Lasker operation and represents an explicit effort to encourage national press coverage of cancer research progress a few weeks before each annual ACS fund-raising drive. These extensive public-relations resources have often contributed greatly to Mrs. Lasker's success.

By limiting her philanthropy to a few related issues Mary Lasker has attained a great deal of leverage in them. She has also been sophisticated in the support of politicians having similar views on health and medical research. She was identified, for example, in 1971, as having contributed $69,400 to the Democratic party.[11] In short, Mrs. Lasker has a keen sense for the use of money in the pursuit of her objectives.

Mary Lasker is a woman of great charm, talent, and energy, a woman of many interests, strong convictions, and the wealth to pursue them. She and some of her closest associates devote themselves essentially full-time to the causes of health and medical research. Her presence is ubiquitous, her influence great, her contributions substantial. She is never frivolous, always purposeful. She has a keen understanding of the processes of government, a first-name acquaintance with most key officials, and a refined sense of timing. She is, without question, a remarkable political figure, and she and her friends, on more than one occasion, have attempted to redraw the map of medical research in the United States.[12] She has understood as few have that one is always building the agenda for government action.

These qualities, not surprisingly, have made Mary Lasker a very controversial figure in many quarters. She has never seen eye to eye with the leadership of the National Institutes of Health, for instance, and the latter have often viewed many of her initiatives with strong misgivings. Nor has her emphasis upon categorical research ever found wide acceptance among the broad ranks of academic medical scientists. The story of the National Cancer Act of 1971 can be viewed by some as one more controversy generated by Mary Lasker.

WHO SHAKES THE MONEY TREE?

Mrs. Lasker and her friends, in their dealings with the federal government, came to understand very quickly the importance of securing adequate appropriations for medical research. They devoted considerable attention to those congressional leaders responsible for NIH funds, and with notable success. In 1955, a constellation of events occurred that brought together a remarkable combination of people concerned with medical research funding. In January, Representative John Fogarty (D., R.I.), member of the House Appropriations Committee, resumed the chairmanship of the subcommittee that reviewed the budgets of the Department of Labor and the Department of Health, Education, and Welfare. Senator Lister Hill (D., Ala.) became chairman of the comparable Senate subcommittee at the same time. In July 1955, James A. Shannon, M.D., was named director of the National Institutes of Health, on the same day that President Eisenhower appointed Mr. Marion B. Folsom as the new secretary of health, education, and welfare.[13]

Both Fogarty and Hill had been concerned with what they regarded as inadequate funding for medical research. Shannon shared this view and was instrumental in persuading Folsom to seek an increased budget request for medical research from the austerity-minded Eisenhower administration. This developing consensus among legislative and executive branch officials was strongly reinforced by the Lasker associates, and money for medical research began to increase very rapidly.

The NIH budget jumped from $98.5 billion in fiscal 1956 to $213 billion in the next fiscal year, the first year in which these individuals functioned together. Ten years later, by fiscal 1967, the appropriation had grown to $1.4 billion, almost a seven-fold increase over 1957. From fiscal 1959 through 1962, the annual rate of growth exceeded 25 percent.[14]

A regular pattern developed for the annual appropriation process.[15] Shannon would present the overall NIH budget request to Representative Fogarty's subcommittee. Fogarty would inquire about the successive budget cuts imposed by the Surgeon General, the secretary of HEW, and the Bureau of the Budget.

Alarmed by these reductions, the congressman would direct Shannon to have the directors of the individual NIH institutes produce for him "professional judgment" budgets based on their institute's needs. These budgets were consistently higher than the administration's request. "Citizen witnesses" from the medical scientific community would also present recommendations for the budget of each institute that consistently exceeded the "professional judgment" budgets. The House recommendation was usually well above the administration's request and often quite close to the institute directors' recommendations.

In the Senate, Shannon and the institute directors would defend the administration's budget before Senator Hill. Hill, however, paid close attention to the "citizen witnesses," men like Dr. DeBakey, who frequently testified for the National Heart Institute budget, and Dr. Farber, a regular witness on behalf of the National Cancer Institute budget. The "citizen witnesses" were extremely persuasive. One close observer described their talents this way: " 'DeBakey is unique; he has the aura of the surgeon, he's articulate, enthusiastic. Most doctors are not enthusiastic, not used to the verbal give and take. The Rusks, Farbers, DeBakeys have evangelistic pizzazz. Put a tambourine in their hands and they go to work.' "[16] The Senate budget recommendation, therefore, was often close to what the "citizen witnesses" had recommended. As a result, the House-Senate compromise was frequently much greater than the budget request that a more frugal administration had submitted months earlier.

Mrs. Lasker and her colleagues never overlooked the important Republicans on the appropriations committees. Representative Melvin Laird of Wisconsin, long the ranking minority member on Fogarty's subcommittee, played an invaluable role in lining up Republican votes for the NIH appropriation bill each year. Senator Styles Bridges of New Hampshire fulfilled a similar function for Hill with respect to Senate Republican votes. Both men were courted by the Laskerites. In return for their help, both received help for their political campaigns.

The Albert and Mary Lasker Foundation, moreover, created a special Public Service Award as a judicious means for rewarding both legislative- and executive-branch officials who had assisted medical research. Hill and Fogarty each received this award in 1959. Laird and Representative Oren Harris (D., Ark.), chairman of the House Interstate and Foreign Commerce Committee at the time, were recipients in 1963. Representative Claude Pepper (D., Fla.), a supporter of medical research in the mid-1940s when he was in the Senate, received an award in 1967.[17] All of these congressmen had been important to Mrs. Lasker's interests in one way or another, and she made it clear that she did not forget such favors.

This annual process by which funds appropriated to NIH consistently exceeded those requested by the administration was occasionally challenged. In 1958, in connection with the fiscal 1959 budget for NIH, Senator Leverett Saltonstall (R., Mass.), the senior Republican on the full Appropriations Committee, questioned whether the committee was doing the right thing in raising medical research funds so rapidly. He suggested that an outside evaluation be made to determine whether the funds to NIH were being efficiently used. Senator Hill graciously obliged Saltonstall, but thought a better question might be whether funds for medical research were at the proper level.

Hill selected the members of the outside review committee with some assistance from Mary Lasker. The chairman was Boisfeullet Jones, then vice-president for medical affairs at Emory University, and relatively unknown to Hill. The other members, however, included a number of stalwarts among the citizen witnesses for medical research: Dr. DeBakey, Dr. Farber, Dr. Cornelius Traeger of the Multiple Sclerosis Society, General David Sarnoff, chairman of the board of the Radio Corporation of America, and others.[18]

The "Bo" Jones committee reported in May 1960, just as the fiscal 1961 NIH appropriation bill was being considered by the Senate. The committee, after consulting one hundred expert witnesses, had unanimously concluded that funds for medical research had been used "with remarkable efficiency." The NIH had maintained high standards in the research it supported, had earned the confidence of the scientific community, and had protected the independence of institutions and investigators. But though Congress had been generous in the past, the committee's judgment

was that funds "have not kept pace with new opportunities in research. . . . [T]he great advances already forthcoming from this
program justify expectation that, through medical research, the
span of useful enjoyable life can be still further lengthened, and
that the benefit to society of longer, healthier and more productive
lives will be far greater than the cost of the research required to
reach that goal."[19]

Specifically, the committee recommended $264 million *more*
than the $400 million President Eisenhower had requested, and
the Congress did appropriate $547 million for that year. The
promise of "great advances" from medical research argued by
this committee of experts was compelling. The strength of the
medical research coalition was such that challenges to budget
growth could be turned into opportunities to secure more money.
Nor was the utility of the blue-ribbon committee of experts ever
forgotten.

Beginning in fiscal 1968, however, the National Institutes of
Health began to experience a rapidly declining rate of growth in
appropriated funds that affected all institutes, including the National Cancer Institute (NCI). The NCI's fiscal 1968 appropriation
was $183 million, up 4 percent from the prior year; the 1969 ap-
propriation was $185 million, up only 1 percent; but the 1970 ap-
propriation was $181 million, down 2 percent from 1969. Given
inflation, a serious financial squeeze was beginning to make itself
felt.[20]

A number of factors were responsible for this change in the fortunes of medical research. Federal priorities in health were shifting
from research to medical manpower and the delivery of health
services and NIH was encountering more competition for the federal health dollar. Also, health activities in HEW had undergone
major organizational changes in 1967 and 1968, some of which
directly affected the NIH: the PHS Bureau of Health Manpower
was transferred into NIH, as was the National Library of
Medicine; the giant National Institute of Mental Health and the
fledgling Regional Medical Program were transferred out; a new
National Eye Institute was established; and, in 1969, the Division
of Environmental Health Sciences became the National Institute
of Environmental Health Sciences. These changes, coming at a

time of increasing financial constraints, created great difficulty for the NIH, for individual institutes, and for the biomedical scientific community.

Perhaps of greater immediate concern to Mary Lasker, and many others, was the wholesale leadership change among those officials—elected and appointed—who were responsible for NIH. Representative Fogarty died suddenly in January 1967, just as the 89th Congress was convening. One year later, in January 1968, John W. Gardner, the secretary of HEW, who had been sympathetic to the NIH, resigned. On March 31, 1968, Lyndon Johnson dramatically announced that he would not seek renomination as the Democratic party's candidate for president, and thus would not be seeking reelection in November. The end of an era occurred at NIH on August 31, 1968, when Dr. Shannon retired as director, a position he had held for 13 years. To cap things off, Senator Lister Hill retired from office in early January 1969 upon completion of his term and 45 years in Congress. When Richard M. Nixon was sworn in as president on January 20, 1969, he brought a new group of officials with him to the executive branch. Coming in rapid sequence, these events meant a total change in the policy hierarchy and political environment affecting medical research.

Mrs. Lasker was greatly concerned with funding for medical research, and therefore quite concerned with finding a replacement for Senator Hill. Lister Hill had been chairman not only of the HEW appropriations subcommittee but also of the Senate Committee on Labor and Public Welfare and its health subcommittee. He was fourth-ranked in seniority in the Senate at the time of retirement, a skilled parliamentarian, and respected by liberals and conservatives alike. His loss was greater than the loss of Fogarty in Mrs. Lasker's judgment. She had reason to be concerned.

The financial picture for medical research throughout 1969 was discouraging. For one thing, many of the important officials were new to their jobs. Marston had been NIH director only since August 1968. Representative Daniel Flood (D., Pa.), in his second year as Fogarty's successor, was sympathetic to medical research but lacked the strong personal commitment to it which Fogarty had held. Senator Warren G. Magnuson, Hill's successor on the appropriations subcommittee, was preparing to guide the appropriations bill for the Departments of Labor and Health, Education, and Welfare through the Senate for the first time.

It was also President Nixon's first year and he was being watched carefully. One distressing sign was the prolonged vacancy in the HEW position of assistant secretary for health and scientific affairs. The position was not filled until Dr. Roger Egeberg was appointed and confirmed on July 12, 1969.[21] Moreover, Dr. Lee DuBridge, a physicist, the president's special assistant for science and technology, and director of the Office of Science and Technology, had failed to appoint a life scientist as his deputy, as was customary. NIH officials had a sense of a policy vacuum at the highest levels of the new administration.

The research community found little cheer in the budget submitted by President Nixon for fiscal 1970. The request, which was sent to the Congress in early 1969, was $1,452,065,000 for NIH, an increase of only $50 million, 3 percent over the appropriation for the prior fiscal year. The budget request for the National Cancer Institute was $180,725,000, which was actually down 2 percent from the prior year appropriation of $184,952,000.[22]

Farber was active in decrying the implications of this proposed budget cut for NCI. He and James T. Grace, M.D., director of Roswell Park Memorial Institute, warned in late March 1969 that biomedical research in the U.S. was in a state of crisis and the federal support of research was plainly losing ground. Since the previous year had been a "standstill" budget, they argued, the anticipated effect of the proposed reduction, when the heavy toll of inflation was taken into account, was grim.[23] Two months later, Farber and Garb appeared before the House appropriations subcommittee and recommended $65 million more than had been requested by the Nixon administration.[24] The House subcommittee, however, recommended an appropriation equal to the president's request.[25]

The gravity of the situation was not fully apparent until late summer and early autumn. In August, to comply with expenditure-control legislation, NIH announced that all renewal research grants would be reduced 20 percent across-the-board.

The protests were just beginning when higher level HEW officials revised the cuts downward to a level requiring only 5 to 10 percent reductions.[26]

NIH also announced in mid-September that it was closing 19 of 93 small regional clinical research centers.[27] Costs for these centers—four- to ten-bed advanced clinical research facilities— had been climbing steadily for several years at fifteen to twenty percent annually. The announced closing of a pediatric clinical center in California, where several children with leukemia were being treated, generated a good deal of negative publicity and many protest letters for the White House.

All institutes were affected by budget cutbacks. NCI ordered that 380 monkeys, of a colony of 1,400, be put to death.[28] The monkeys, inoculated some five years earlier with human cancer material believed to be caused by viruses, were a year or two away from providing useful research results. The stories multiplied—medical schools were facing "insolvency," the number of NIH grant awards expected in fiscal 1970 would be lower than 1965, sharp reductions were planned in research training grants, a long-time epidemiological study of heart disease was being terminated—and the distress deepened.[29]

Extraordinary efforts were made to bring the plight of medical research to the public and the Congress. Farber, with Dr. Michael E. DeBakey, held a Washington, D.C. news conference on November 11, 1969, to make a last-minute plea that Congress reverse the budget cuts for medical research. Farber described the cumulative effect of past and prospective budget cuts as "a threatened catastrophe to progress in cancer research."[30] These pleas had some effect because Senator Magnuson indicated two days later at the Albert Lasker Medical Research Awards luncheon in New York that the proposed cuts in medical training and research constituted "one goal of the Bureau of the Budget that I hope I can frustrate."[31]

That hope was widely shared. But there was little expectation of any dramatic reversal of budgetary trends. On the contrary, the fiscal gloom was thick. Dr. Carl Baker discussed the prospects for fiscal 1970 with the Association of American Cancer Institutes (AACI), the directors of the major cancer research centers in the country, in early November.[32] He noted an increased interest in health-care delivery in the Congress, and a leadership vacuum with respect to medical research created by the death of Representative Fogarty and the retirement of Senator Hill. Though the House had voted the $181 million for NCI requested by the administration, the amount available for cancer research was likely to be lower still. The Expenditure Control Act of 1968, he reported, required that overall federal government expenditures be reduced by $3.5 billion, and that government employment be reduced to 1966 levels. This would probably reduce the Cancer Institute budget by $8 million or $10 million more. The cuts would be absorbed by reducing the funds promised as moral commitments on continuing research projects, by funding perhaps only 50 percent of competing renewal grants, and by restricting the funding of new competing grants to an all-time low of 15 percent. Baker expected training grants to be cut, and that there would be larger cuts in some contracts than in the grant programs. The expected budget for fiscal 1970, in reality, then, would not be $181 million but only about $173 million.

The Senate subcommittee on appropriations, however, was not prepared to acquiesce. It recommended an appropriation of $200 million for the National Cancer Institute and observed, in its report, that "it is difficult to understand why the budget request . . . was nearly $4.5 million less than the comparable 1969 appropriation."[33] The joint House-Senate conference committee, which reported on December 20, split the difference between the two bodies and recommended $190 million for NCI.[34]

The NIH and NCI appropriations, however, were part of a larger bill appropriating funds for fiscal 1970 for the Departments of Labor and HEW, and for the Office of Economic Opportunity. The larger bill became entangled in a dispute between the president and the Congress over national priorities.[35] The Labor-HEW-OEO appropriation bill added $1.3 billion more than had been requested by President Nixon. As a result, the president warned the Congress on December 18, just as the conference committee was completing its work, that he would veto the bill if it reached his desk as recommended by the joint committee. The "battle against inflation" would not permit him to approve such an increase in federal spending.[36] The Senate chose to lay the bill aside and adjourned for the Christmas recess.[37] It took the bill up again when it returned on January 19, 1970, and passed it unchanged a few days later. The president, as he had warned, vetoed the bill, and the House sustained his action.[38] Both houses went back to work to fashion an acceptable spending bill. On March 5, 1970, fully eight months after the fiscal year had begun, the Labor-HEW-OEO appropriation was signed into law. Included in the act was $190 million for NCI.[39]

President Nixon, in his veto message to the Congress, had made clear that medical research provided no basis for the veto. He said:

> The amounts added by the Congress for health research represent less than one-half of 1% of the total appropriation. Taken separately, I would not have vetoed these increases. On the contrary, when the budget for 1971 is submitted to the Congress it will make a strongly increased commitment for health research, where advances can be made to serve the health needs of the Nation—cancer, heart disease, population research and environmental health.[40]

But this was little consolation in a period of acute budgetary stress, especially since the Office of Management and Budget had placed in reserve the additional $10 million for cancer research provided by Congress.

The budget prospects for NIH in fiscal 1971, moreover, appeared as bleak as the previous year. The formal budget request, submitted in January 1970, estimated an increase of $104 million over the not-yet-passed fiscal 1970 appropriation. But $64 million was earmarked for categorical research, as the president had promised, $29 million involved the overseas use of excess foreign currencies, and $30 million was for health manpower. Most of the NIH research institutes, especially those supporting basic research, faced further budget reductions. The acute distress being experienced by the life-science community was diagnosed in clinical detail in February by three NIH officials. The implied prescription was more money, but prospects for therapy being administered in a timely manner were not good.[41]

The National Cancer Institute fared much better than did the NIH as a whole. The 1971 budget request was for $202 million, up $20 million over the expected appropriation for fiscal 1970.[42] But Mary Lasker was not satisfied. She invited Robert Marston and Robert Berliner to luncheon at Deeda Blair's home in March. Cryptically, as the two NIH officials were leaving, she asked, "You don't mind if we get you some more money for cancer, do you?" Neither man quite knew what she had in mind. What she had in mind became more apparent when the House appropriations subcommittee, in July, recommended $227 million for NCI.[43] The Senate, three months later, recommended $235 million.[44] The fiscal 1971 appropriation bill, signed into law on January 11, 1971, represented a compromise of $230 million.[45]

THE ONLY GAME IN TOWN

Mary Lasker's interests in health have always been broader than medical research. She has been a strong supporter of national health insurance since the mid-1940s, an issue with legislative antecedents in the New Deal. But this has been vigorously opposed by the American Medical Association (AMA) over the years. Not until 1965 did the U.S. government extend health insurance to the aged and the poor through Medicare and Medicaid, and the establishment of a comprehensive national health insurance system remains a major, unresolved health matter on the current national agenda. Mrs. Lasker and her late husband, along with many others, also believed that an apparent shortage of physicians in the U.S. could only be remedied through direct federal government support of medical education. Here again, AMA opposition effectively blocked the passage of significant federal legislation until very recently.[46]

Mary Lasker, though strongly committed to national health insurance and federal support of medical education, was never deeply involved in the legislative debate on either issue. She was close enough, however, to have learned two lessons. The first was that the opposition of organized medicine meant prolonged debate and excruciatingly slow progress for any federal health legislation. Always impatient with delay, Mary Lasker concluded that she should devote her attention to where there was greater hope for accomplishment. Second, Mary realized that the AMA did not oppose federal support of medical research, but, in fact, tacitly supported it. Her energies, and those of her friends, therefore, were forcefully directed to medical research partly because other avenues of health policy were effectively closed.[47]

To be sure, Mary Lasker and her friends believed in research. After all, it was the failure of the American Society for the Control of Cancer to support research which so energized Mary in 1943. The Albert Lasker Medical Research Awards, moreover, are a continuing indication of the Laskers' belief in research. And Mrs. Lasker is more aware than most that cancer, cardiovascular disease, and mental illness will become curable and preventable only through new scientific knowledge. That she sees these diseases as "enemies" in highly personalized terms and is strongly committed to the "conquest" of disease, reflects the mentality of a lay disease fighter, however, and sets her apart from the way in which much of the scientific community thinks about disease. Finally, it is inconceivable that John Fogarty and Lister Hill could have bestowed such public largess on the NIH without the support they received from the "health syndicate."

One Lasker initiative, however, the President's Commission on Heart Disease, Cancer, and Stroke, sought to go beyond research towards the provision of patient care.[48] President Johnson, in his health message to Congress on February 10, 1964, said:

The Public Health Service is now spending well over a quarter of a billion dollars annually finding ways to combat [heart disease, cancer, and stroke]. . . . The flow of new discoveries, new drugs and techniques, is impressive and hopeful. Much remains to be learned. But the American people are not receiving the full benefit of what medical research has already accomplished. . . . I am establishing a Commission on Heart Disease, Cancer, and Stroke to recommend steps to reduce the incidence of these diseases through new knowledge, and more complete utilization of the medical knowledge we already have.[49]

The president later indicated that he had created the commission "at the insistence of the lovely lady, Mrs. Mary Lasker."

The commission was dominated by Mrs. Lasker's associates and friends. Dr. Michael DeBakey of Houston was chairman. Other members included Emerson Foote, Mrs. Florence Mahoney, Dr. Sidney Farber, Dr. R. Lee Clark, Mrs. Harry Truman, whom Mary Lasker had known since the Truman White House days, and Dr. J. Willis Hurst of the Emory University Medical School, the president's personal heart specialist. Mr. Boisfeullet Jones was a key consultant. Farber, Clark, and Dr. Frank Horsfall, of the Sloan-Kettering Institute for Cancer Research, were the principal members of the cancer subcommittee.[50]

The Commission on Heart Disease, Cancer, and Stroke recommended the creation of "a national network for patient care, research, and teaching in heart disease, cancer and stroke." This network would include centers for clinical research, teaching and patient care in universities, hospitals and research institutes. At the apex of the network, 25 heart disease centers, 20 cancer centers, and 15 stroke centers were to be established within five years; at the base, 150 diagnostic and treatment stations were to be established for heart disease, 200 for cancer, and 100 for stroke.[51] The estimated cost to the federal government for all recommendations was $357 million in the first year, $739 in the fifth year, and slightly more than $2.9 billion for the entire five-year period.

Were these recommendations realistic? The commission with obvious enthusiasm believed they were: "Every available fact points to the same conclusion—that the toll of heart disease, cancer, and stroke can be sharply reduced now, in this nation, in this time. . . . [Many] things can be done now, without further scientific advance. Meanwhile, new knowledge of the fundamental processes of life promises great new weapons for the immediate future. . . . The way is there. All that is lacking is the national will to give our people the full measure of protection against their three most deadly enemies."[52] That exhortation to "the national will" would echo again in the words of Senator Yarborough.

Though the commission focused on three diseases, Dr. R. Lee

Clark recalls that the cancer subcommittee had the clearest idea of
what they wanted. Clark, as director and surgeon-in-chief of the
M. D. Anderson Hospital for Cancer Research from 1946 onward, had been the architect of that institution's development. He
considered patient care and research as dual responsibilities of a
cancer research center so that "ongoing daily contact" could be
maintained between "the investigator, as the originator of new information, and the physician, as the converter and purveyor of
that information for the benefit of the patient with cancer."[53]

Farber and Clark had been active in the late 1950s in organizing
an association of cancer institutes. Four major cancer institutes
existed then—Memorial Sloan-Kettering Cancer Institute, M. D.
Anderson Hospital and Tumor Institute, Roswell Park Memorial
Institute, and the National Cancer Institute's intramural laboratories and clinical research facilities. Besides these major centers,
each of which was engaged in fundamental and clinical research,
medical education, and patient care, approximately ten other
smaller centers across the country were engaged in one of these
three activities. Farber's Children's Cancer Research Foundation,
for instance, was one such center.

Farber, Clark, and Horsfall naturally drew upon their experience in writing their subcommittee report. A regional cancer center, they said, would bring together in a single location all the scientific, clinical, and developmental skills and resources bearing
on cancer and would have "the eradication of cancer [as] the accepted institutional goal."[54] Based upon the four major comprehensive cancer research centers and the ten smaller centers, the
subcommittee recommended that twenty comprehensive cancer
research centers be established and geographically distributed so
that each center served an area of approximately 10 million in
population.

The cancer subcommittee report was silent on relations between cancer centers and medical schools, though the heart and
stroke subcommittees laid great stress on integrating heart and
stroke centers with medical schools. This was hardly an oversight. It reflected the historical distance between cancer research
institutes and medical schools, nowhere better illustrated than in
Houston. Though the M. D. Anderson Hospital was part of the

University of Texas system, it was independent of the university's medical school. It presented its budget directly to the trustees of the university and defended that budget before the legislature. "Having its own budget and the right to defend it as one of the free standing units of The University of Texas System has been its greatest prerogative and perhaps the basic reason for its singular growth," Clark later wrote.[55]

Farber and Clark knew what they wanted. They desired a national network of cancer centers each like the existing comprehensive centers, with sufficient autonomy to develop without interference from medical schools and closely tied to the National Cancer Institute. They were thwarted in the first instance by the legislative process and in the second by the administration of the resulting program.

The commission presented its report to President Johnson on December 9, 1964.[56] In his health message to Congress in January 1965, the president announced that legislation would be submitted "for an all out attack on heart disease, cancer, stroke."[57] The legislation, however, modified the commission's recommendations in important ways, moving away from the categorical emphasis to a more general approach linking research and medical care.[58] Only those recommendations relating to the network of categorical medical centers were considered. Language was changed from regional medical "centers" to "complexes," thus deemphasizing the prospects of a system of governmental medical care institutions. Payment for patient care was explicitly precluded unless related to research, training, or demonstration activities, while the commission had emphasized care for heart disease, cancer, and stroke patients.

The legislation was introduced on January 19, 1965, and considered by Senator Hill's subcommittee on health on February 9 and 10.[59] The subcommittee made few changes in the proposed bill and the full committee reported the bill to the Senate on June 24. The bill was adopted by the Senate a few days later.

The American Medical Association, preoccupied with opposing the Medicare legislation in 1965, had given little attention to the work of the president's commission. Dr. Hugh Hussey, the AMA's director of scientific activities, a member of the original commission, had resigned in 1964 because of potential conflict between AMA policy and the commission's recommendations. The AMA president, Dr. F.J.L. Blasingame, wrote Senator Hill in mid-February that his organization had not testified because the short notice given before the hearings had not permitted them to give the bill adequate consideration. In late spring, however, the association issued a criticism of the commission's report which ignored the bill. It then asked that consideration of the legislation be postponed, a suggestion echoed by the American Heart Association, one of the bill's original supporters, but this ploy was unsuccessful.[60]

Support for the legislation had come originally from the American Cancer Society, the American Hospital Association, the Association of American Medical Colleges, and the American Heart Association. As the AMA clarified its position, other opposition emerged. The American Academy of General Practice characterized the commission's recommendations as "a far greater threat to the private practice of medicine than the medicare plan."[61] State health department officials were uneasy about prospective invasion of their domain and sought legislation for comprehensive health planning. The scientific research community indicated some opposition to the commission's report and to the legislation on the grounds that the absence of scientific knowledge, not patterns of organization, was the chief limitation in the war on disease.

By the time the House Committee on Interstate and Foreign Commerce held hearings in July, the AMA had gotten to President Johnson. Sensitive to the liabilities of beating organized medicine twice in one year, and more concerned that Medicare get off to a good start, the president made concessions to the AMA and relayed them to Representative Harley Staggers (D. W. Va.). The legislation was rewritten—on orders of the president—and representatives of HEW provided a different justification for the bill in the House from that provided earlier in the Senate.

The bill which finally passed the Congress and was signed by the president bore little resemblance to the recommendations of the DeBakey Commission.[62] Regional medical "complexes" had become "programs," further easing fears of a network of federal medical care institutions; "coordination" became "cooperation," limiting apprehensions that medical schools would run the effort. Where each "complex" was to have included a medical school, each "program" was now only obligated to include one or more medical centers—any institution providing post-graduate medical training. Most significantly, the categorical emphasis had been blunted. The tiered system of categorical research institutes was eliminated in favor of a decentralized "cooperative arrangement" among local health care organizations engaged in research, training, diagnosis, and treatment related to heart disease, cancer, and stroke and also "at the option of the applicant, related disease or diseases."

The Surgeon General, after some deliberation and debate, assigned responsibility for the Regional Medical Program (RMP) to the National Institutes of Health, partly because of NIH experience with clinical research but also owing to lack of administrative competence elsewhere in the PHS. The first RMP director was Robert Q. Marston, M.D. The advisory council to the program was chaired by Dr. DeBakey, and though Sidney Farber was on the council, there was little cancer representation on the program development committee. As the program evolved, it moved steadily away from the categorical emphasis toward locally determined programs that linked medical schools more closely to the other medical care providers in a region. The categorical emphasis was strongest with respect to heart disease, but cancer received only eight percent of RMP money.[63]

The cancer people were quite distressed with the outcome of the President's Commission. Some remember the effort as one that produced a good report but no action. Others remember their clear ideas about the organization of cancer research being mangled in the legislative process. Still others remember the very limited funds that flowed to cancer, due to the absence of autonomy from other interests. Finally, the experience once again showed the difficulty in establishing a system that included federal provision for patient care, even if the justification was bridging the gap between medical research and clinical practice. By 1969, with respect to cancer, at any rate, Mary Lasker and her allies were back to square one. Medical research was indeed the only game in town.

Several important elements in the Lasker style of operations can be identified from the analysis in this chapter. First, Mrs. Lasker has substantial personal resources—money, time, commitment, the capacity to confer status on others—which she has used in a focused and skilled manner to further both general and specific ends. Her own resources have been augmented through the access to political leadership and the press and broadcast media that she has so assiduously cultivated. Second, Mrs. Lasker has frequently acted through an elite group of long-standing associates whose underlying commitment is to the support of medical research for conquering specific diseases of man. This elite has functioned oftentimes—as in the cases of the "Bo" Jones Committee, the President's Commission on Heart Disease, Cancer, and Stroke, and the Senate Panel of Consultants on the Conquest of Cancer—through a blue-ribbon panel of committed experts. Though always including medical and scientific professionals, prominent lay representation has been a guiding organizational principle for any health or medical-science initiative. Finally, when mobilized, Mrs. Lasker and her friends have displayed great capacity to overwhelm the opposition through a focused, highly publicized, over-simplified, dramatic appeal to the public and through skillful tactical maneuver within the political process. Time, the budget process, and the normal routines of HEW, however, have often seen the grand designs of the Laskerites altered, if not thwarted; but these balancing processes have sometimes merely set the stage for the next conflict.

3

The National Cancer Institute

Whether one thinks a certain organizational arrangement is "good" or
"bad" depends on what one thinks of the kind of policy it facilitates. . . .
Policy-making is essentially a political process, by which the multiplicity
of goals and values in a free society are reconciled and the debate over
means and ends is distilled into a politically viable consensus on a work-
able policy. But if some organizational arrangements facilitate certain
kinds of policy and other arrangements facilitate other kinds, then
organization is also politics in still another guise—which accounts for
the passion that men so often bring to procedural and organizational
matters.

Roger Hilsman
To Move A Nation

The effort to commit the federal government to the active support of cancer research precedes by sixty years the Panel of Consultants. In 1910, President William Howard Taft proposed the creation of a "Bureau of Public Health" which, in addition to exercising federal quarantine powers, would also provide opportunity "for investigation and research by competent experts into questions of health affecting the whole country."[1] Taft's proposal prompted the American Association for Cancer Research to instruct Dr. H. R. Gaylord, director of the New York State Cancer Laboratory in Buffalo, to write the president and request the inclusion of a cancer research division within the proposed bureau.[2] Gaylord, who had been studying cancer of the thyroid in trout, personally delivered his memorandum to the president.[3] He stressed that the cause of cancer was not yet known, that promising lines of research were being pursued on cancer in lower animals, especially fish, and that careful study of cancer in fish would provide information of an invaluable character for humanity. The memorandum concluded with a request for an appropriation to immediately establish and operate a laboratory for the study of fish diseases.

President Taft received supporting recommendations from the Bureau of Fisheries and the Department of Commerce and, on April 9, 1910, sent Congress a request for $50,000 to build one or more laboratories for the investigation of cancer in fish. "Were there a bureau of public health such as I have already recommended," his message read, "the matter could be taken up by that bureau." "[W]e have every reason to believe," he added, "that a close investigation into the subject of cancer in fishes . . . may give us light upon this dreadful human scourge."[4]

The President's interest in cancer research was further heightened when he visited Dr. Gaylord's laboratory in Buffalo on April 30. "I am delighted with what I have seen here," he said after his visit. "There should be more institutions of this kind in this country, and I hope some day we will have one under official auspices."[5] Congress was less enthusiastic. The House Committee on Interstate and Foreign Commerce held five days of hearings on the proposal to create a "Bureau of Public Health" but gave no attention to a laboratory for cancer research.[6] The Senate, likewise, held no hearings on the proposed laboratory. Presidential initiative had been met by congressional indifference.

The Congress did create the U.S. Public Health Service (PHS) in 1912. Ten years later, initial activities in cancer research were begun within the PHS.[7] Dr. Joseph W. Schereschewsky, a public health official located at the Department of Preventive Medicine, Harvard University Medical School, began field investigations of cancer and other malignant diseases in August of that year. Toward the end of 1922, Dr. Carl Voegtlin of the Division of Pharmacology at the Hygienic Laboratory in Washington, D.C., began work on the experimental production of malignant growth in rabbits. Schereschewsky's work led to major statistical analyses of cancer mortality and was expanded to include various laboratory research problems, while Voegtlin and his group worked on the content of glutathione and its reducing power in normal and tumor tissue, the electrical potential in living cells, and the effect of lead and other chemicals upon animals with tumors. These parallel efforts led the Surgeon General to call a conference for April 9, 1928, which recommended systematic investigations into cellular biophysics and biochemistry, the effects of the spectrum of radiant energy on living cells, occupational cancer, and cooperation in statistical research with the Bureau of the Census. These two governmental research groups, consisting of a total of ten scientists, became the basis in 1937 for the staff of the National Cancer Institute.

Congressional interest in cancer research flourished during these years. On February 4, 1927, Senator Matthew Neely (D., W. Va.) proposed that Congress authorize an award of $5,000,000 to the first person finding a practical and successful cure for cancer.[8] The suggestion attracted little congressional support but generated substantial public interest. Neely received almost 2,500 letters during the next year proposing numerous infallible cancer cures—anointed handkerchiefs, mixtures of herbs to be taken internally, various salves to be applied externally—and this response persuaded him that his proposal was "imperfect, if not utterly futile." As a consequence, however, the senator consulted a number of authorities on cancer, among them Dr. Joseph Bloodgood, a famous cancer surgeon from Johns Hopkins University, who introduced Neely to a number of other medical scientists. These consultations led to a bill that Neely introduced on May 18, 1928, which would have authorized the National Academy of Sciences "to investigate the entire cancer subject and report to the Congress in what manner the Federal Government could assist in coordinating all cancer research and in conquering this most mysterious and destructive disease."[9] At the insistence of Senator Joseph Ransdell (D., La.), chairman of the Committee on Public Health and Quarantine, it was amended on the Senate floor to include the Public Health Service as a participant in the study. Without objection, the Senate authorized $50,000 for the study, but the House did not act on the bill. Neely was defeated in his bid for reelection in 1928, and though his bill was introduced in 1929 it went nowhere. Senate initiative was thwarted by House inaction.

The federal government became committed by statute to the support of cancer research in 1937.[10] Dr. Dudley Jackson of San Antonio, Texas, had been attempting to persuade members of Congress and the executive branch since the early 1930s of the desirability of a "Central Government Cancer Committee" that would act as a clearinghouse for information, dispense grant funds, and oversee research within the government. Late in 1936, after his friend, Representative Maury Maverick (D., Tex.), had been reelected to a second term in the House, Jackson persuaded Maverick to introduce legislation to establish a national cancer institute. Maverick first wrote the Surgeon General, however, requesting his comments on the matter. The Surgeon General responded on April 6, and on April 29 Maverick introduced his bill. Meanwhile, Senator Hugh Bone (D., Wash.), had introduced similar legislation in early April, which had all the other 95 members of the Senate as co-sponsors, and had also persuaded his fellow Washingtonian, freshman Representative Warren G. Magnuson, to introduce an identical bill in the House.

Widespread public interest in cancer and cancer research had been developing at that time, which undoubtedly had much to do with the unanimous Senate sponsorship of Senator Bone's legislation. The medical section of the American Association for the Advancement of Science had discussed the subject at its annual meeting in December 1936. *Fortune* magazine carried an article, "Cancer—The Great Darkness," in its March 1937 issue, and the other Luce periodicals, *Time* and *Life*, gave prominent coverage to cancer. Congressional offices were deluged with mail. The Congress moved with record speed. A special joint Senate-House conference on July 8 quickly ironed out differences between the Bone-Magnuson bill and the Maverick bill. A bill was passed by both houses of Congress on July 23, and President Roosevelt signed the National Cancer Institute Act into law on August 5, 1937.

The Act created the National Cancer Institute and established the National Advisory Cancer Council. The Council was authorized to review all research projects for approval before recommending funding by the Surgeon General. The Institute was authorized to conduct and foster research and studies relating to the cause, prevention, and methods of diagnosis and treatment of cancer, promote the coordination of cancer research, provide fellowships at the institute, secure advice from cancer experts in the U.S. and abroad, and cooperate with state health agencies in the prevention, control, and eradication of cancer.

This brief historical excursion points out the fact that a policy issue often has roots deep in history. Cancer has been a matter of public concern for some time, concern that manifested itself in the 1910 initiative of President Taft and the 1927-1928 initiative of Senator Neely. Cancer has been on the formal government agenda since 1937 by legislation, and since 1922 by action of the Public Health Service. A long policy history, it would appear, is a precondition for policy reformulation.

In 1970, when the Panel of Consultants began its work, the National Cancer Institute was an integral part of the National Institutes of Health. To many, it seemed that this had always been so. When the Cancer Institute was established in 1937, however, it was as a division of the Public Health Service, which was independent of the National Institute of Health. The chief of NCI reported directly to the Surgeon General on policy and program matters and took only administrative direction from the director of the National Institute of Health. Not until the Public Health Service Act of 1944 was NCI made a division of NIH.

The National Institute of Health had been established in 1930 as a research laboratory of the Public Health Service. It traced its lineage to the Hygienic Laboratory founded in 1887 on Staten Island, New York. In keeping with the semimilitary personnel rotation system of the PHS, a public health officer could expect a three- or four-year tour at the NIH during his career. Only after World War II did NIH acquire its extramural research program, and then by the transfer of certain medical research contracts from the wartime Office of Scientific Research and Development to the PHS unit.[11]

It was also after the war that the organizational development of the modern NIH took place. In 1948, Congress established the National Heart Institute and the National Institute of Dental Research, and changed the name of the National Institute of Health, singular, to the National Institutes of Health, plural. Other categorical institutes followed, reflecting public and congressional emphasis upon particular disease entities. The National Cancer Institute served as the organizational model for successive categorical institutes.[12]

PROFILE OF AN INSTITUTE

The early organization of the NCI occurred between the passage of the 1937 law and the entry of the United States into World War II.[13] Dr. Carl Voegtlin, who had headed the Washington, D.C.-based Division of Pharmacology of the National Institute of Health, was appointed the first chief of the Institute in January 1938. The research group under Voegtlin had been organizationally merged with the Boston group under Schereschewsky in 1937. The two groups were physically integrated in late 1939 upon completion of a building in Bethesda, Maryland, to house the new institute. The original research staff of the NCI, then, consisted of scientists from these two groups and approximately twenty fellows—young researchers recruited from medical schools and research institutes. Voegtlin was assisted by an active National Advisory Cancer Council (NACC), which met at least every three months from November 1937 to December 1941. The six NACC council members, all medical and scientific authorities, selected one member as a full-time executive director in 1939, an arrangement that lasted until mid-1944.[14]

A Committee on Fundamental Cancer Research, appointed by Surgeon General Thomas Parran, submitted a report to the NACC in mid-1938 that served NCI for several years thereafter as a statement of research priorities.[15] The committee report indicated that three lines of scientific research had contributed to understanding the nature of malignancy: the study of transplantable tumors, which had yielded information on the biology of the malignant cell; the conditions governing the experimental induction of malignant tumors; and the role of genetic factors in the development of cancer. A major problem identified by the committee, though, was that each of these three lines of research had been developed independently and had taken little account of the knowledge gained in other fields. In the committee's view, the cancer problem could be divided into the causal genesis of malignancy and, then, into the formal genesis. The formal genesis of cancer, or the factors responsible for the cancer cell and its tendency for unlimited growth, was the fundamental scientific problem that had been "almost entirely neglected" and deserved substantially more attention. The committee raised a question, which has remained central to cancer research: was understanding the development of cancer more likely to come from the study of normal cellular development or from direct observation of the development of the malignant cell?

The research program of NCI through World War II was conducted mainly in the institute's own intramural laboratories. While the NIH laboratories at this time were staffed and run by officers of the Public Health Service, Voegtlin organized his laboratories in a different fashion. The fellows who came to NCI in 1938 were not commissioned officers but were recruited because of their scientific competence. Laboratory organization was deliberately kept fluid and scientific competence was the basis for establishing laboratory leadership.[16] In this respect, the Cancer Institute set the pattern for the later organization of NIH intramural research as the latter increasingly escaped the paramilitary constraints of the Public Health Service.

The intramural research effort, guided by the report of the Committee on Fundamental Cancer Research, focused on six general areas: the experimental production of neoplasms; carcinogenic agents and factors modifying their action; cellular physiology; the chemical characteristics of neoplastic tissue and of animals with tumors; the experimental treatment of tumors; and epidemiological studies.[17]

The informal pattern of organization yielded after the war to more formality. Concurrently, there was a shift from emphasis on tumors to an emphasis upon scientific disciplines. There were six sections—biology, biochemistry, biophysics, chemotherapy, endocrinology, and pathology.[18] The spirit of the intramural effort was captured in 1957 by G. Burroughs Mider, M.D., then associate director in charge of research:

> It becomes increasingly evident that one is unlikely to control cancer by studying cancers alone. Definition of their attributes in terms of physics and chemistry rests on advances in the natural and physical sciences, the discovery of new facts, new compounds of biological importance, and the ability to measure

decreasing quantities of many things with increasing accuracy.
. . . Progress has been most rapid in scientific research when
imaginative, talented, technically competent, intellectually
curious, and above all, sincerely interested investigators are
encouraged to search for new facts beyond the curtains that
limit our knowledge and to pool their specialized resources and
skills on a basis of mutual interest and respect.[19]

Clearly, the intramural research leadership of NCI was sensitive
to the need for fundamental research in the life sciences, as well
as the need for study of particular cancers.

The NCI sought to initiate clinical research in 1939 when the
Public Health Service added a 100-bed wing to the U.S. Marine
Hospital in Baltimore.[20] However, the war prevented this facility
from becoming a major cancer clinical research center, even
though PHS beneficiaries did receive medical services for cancer
and PHS personnel were trained in the diagnosis and treatment of
cancer. After the war, NCI and the University of California, San
Franciso, jointly established the Laboratory of Experimental Oncology, which functioned until 1953. In that year, the Clinical
Center for the entire National Institutes of Health was opened on
the Bethesda campus and the Cancer Institute received an allot-
ment of clinical research beds within this new facility. Concurrently, the Bureau of the Budget forced NCI to close both the Baltimore facility and the Laboratory for Experimental Oncology.[21]

The extramural research program to support scientific investigators in university and medical school laboratories began
slowly. Dr. James Ewing, of Memorial Hospital, New York, actually denounced the policy of grants-in-aid at the first meeting of
the Advisory Council as "unsound." In 1938, from an appropriation of just over $500,000, only ten extramural grants totaling
slightly more than $90,000 were approved.[22] Expenditures re-
mained very limited throughout the war years. The total extramural expenditures were $45.2 million from 1937 to 1957, of
which $44 million were spent in the second decade.[23] The post-
war growth of the program also saw changes in the distribution of
extramural funds: carcinogenesis accounted for 35 percent of expenditures in 1950, and 30 percent in 1956; tumor development

received 28 percent in 1950, and 30 percent in 1956; diagnosis, 9 and 4 percent respectively; while therapy, the principal growth area, changed from 18 percent to 33 percent, owing primarily to the rapid increase in funds for chemotherapy from 1954 onward.[24]

The National Cancer Institute's research program, then, consisted of an intramural effort in its Bethesda laboratories and extramural grants to university and medical school investigators. Clinical research was conducted both intramurally in the NIH Clinical Center and by grant-supported investigators in medical schools. This became the pattern for all categorical institutes within NIH. The NCI director and all institute directors operated under authority delegated to them by the Surgeon General, in whom statutory authority for medical research in the Public Health Service was vested. Moreover, appropriations were made directly to the Cancer Institute and the other institutes, not to NIH, the parent organization.[25]

Unlike the other NIH institutes, however, three institutes—cancer, heart, and mental health—had statutory obligations for disease control. The 1937 cancer act authorized the Surgeon General to cooperate with state health agencies in the prevention, control, and eradication of cancer. NCI funded state studies of cancer mortality and epidemiology and also provided advisory services to state health departments during World War II.[26] A very modest amount of money was involved. A 1946 report to the NACC laid out the basic features for expanded postwar comprehensive state and local government cancer control programs.[27] Statistical research, public and professional education, and the provision of preventive, diagnostic, and treatment facilities and services through the use of demonstration projects were the main elements of such a program.

Initially through another PHS organization, then directly, NCI began a grant-in-aid program to the states. A total of $23.3 million was spent from fiscal 1947 through 1957; the annual program level ranged from $2.2 million to $3.5 million. The application of existing knowledge about cancer was the emphasis from 1947 to 1952, when the emphasis shifted to the development of new knowledge applicable to cancer in humans. The funds were used for cancer clinics, home nursing care, follow-up services, limited laboratory services for the indigent, statistical studies, and education.[28]

The cancer control program had a number of problems. It was tied by the tradition and working relationships of the PHS to the state and local departments of health, not to universities, their teaching hospitals, or to the private practice of medicine. Consequently, the program could not reach those who needed attention, since cancer patients were treated mainly by private physicians. It was simply the wrong way for the federal government to be in the health-care area, in the judgment of many NCI officials.

James Shannon, who as NIH director transferred the cancer control program out of NCI in the mid-1950s into the PHS Bureau of State Services, had other difficulties.

> Nothing was oriented to what to do after detection. When you have a detection capability, then that ought to be transferred to the medical care system. . . . The programs weren't worth a dam. . . . Control programs are highly attractive to the lay person. They represent the transition of research to practice. The question is not at the conceptual level but at the level of quality. When we closed them out, there was not much that couldn't be done better in the university research center. Scientific advances move quickly unless there are complex technologies associated with them.[29]

There was not strong resistance within NCI to Shannon's actions against the control program, since many NCI staff shared the same reservations.

The development of the National Institutes of Health, with the continuing elaboration of categorical disease institutes, confronted the NIH director with the problem of establishing and maintaining administrative order. A number of NIH-wide integrative mechanisms were developed. One mechanism, not highly visible to the outside world, was developed to facilitate the overall coordination of intramural research across the NIH.[30] Dr. Norman Topping, the associate director of NIH from 1948 to 1952, was principally concerned with the intramural research effort. He initiated and chaired twice-monthly meetings of the scientific directors of each of the institutes, which were continued by his successor Shannon. These meetings became a regular part of the NIH. The scientific directors had no statutory charter as a group, represented their institutes to the Office of the Director, NIH, and were advisory only. Yet, according to one former participant, they ran the intramural program, poked their noses into everybody else's business, and were seen by some as having more loyalty to the NIH front office than to their own institutes. In addition to the intramural research program, the scientific directors exercised control over clinical research in the NIH Clinical Center. Collectively they became one of the most powerful groups within the NIH. Through them, the NIH director exercised policy control over intramural research, developed the capabilities of the NIH laboratories to an unexcelled international position, and maintained these capabilities over the years at a very high level.

Though extramural research projects were funded by appropriations to individual institutes, and not through NIH directly, a dual process of peer review of research proposals was developed to integrate the entire extramural effort into a more comprehensive research program than would be undertaken by an individual institute. Begun in 1947, and extended to the Cancer Institute in 1949, all research proposals submitted to any institute were received by NIH's Division of Research Grants (DRG).[31] The DRG assigned the proposal to a study section for reviewing scientific merit and to an institute advisory council for reviewing program relevance.

Study sections normally consist of about fifteen scientists who are experts in the particular area of research. Most of these scientists are drawn from the university scientific community, but scientists from NIH laboratories and other government agencies are also among the study-section members. A full-time DRG employee is assigned as executive secretary to each study section. Study sections represent areas of scientific inquiry, like physiology, metabolism, pathology, physiological chemistry, pharmacology, and biochemistry, as well as areas of clinical medicine, like endocrinology, hematology, and surgery. NIH philosophy places great emphasis upon the way this peer-review process was designed to support competent investigators, with maximum freedom, within a broad concept of relevance to disease.

The institute advisory councils were statutorily charged with recommending approval or disapproval of research proposals, but administrative restrictions were placed upon their ability to act independently of study sections. If a proposal received a disapproval recommendation by a study section, a council could not approve it unless it was sent back for a second study-section review. Legally, they could recommend approval of a proposal after two negative reviews, but this seldom happened. The normal process of dual review resulted in each advisory council receiving a list of research proposals from the study sections rank-ordered in terms of scientific merit. Councils could alter this ranking to reflect a "priority to pay" based upon the relevance of proposals to the programs of an institute. Changes were seldom made, however, and councils usually accepted study-section rankings as the basis for their recommendations. Thus study-section review came to be a powerful device by which the external scientific community determined which research proposals submitted by it to NIH should be considered acceptable for funding. It further constituted an administrative device for integrating the overall NIH extramural research program beyond that which would be embraced by any single NIH institute.

The NIH director, in addition to seeking administrative order among a large number of categorical disease institutes, also was responsible for achieving scientific balance and coherence within the nation's biomedical research effort. One way he sought this balance was by seeing that the growth in research funding for NIH was more or less evenly distributed across all NIH institutes. If new research initiatives were undertaken, a deliberate effort was made to see that funding grew in reasonable relationship to the full range of biomedical research endeavors.

By the mid-1950s, the National Cancer Institute had developed in a manner that had become the prototype for other NIH institutes. Concurrently, the National Institutes of Health had evolved as the center of a vast and complex network of biomedical research in this country. NCI, no less than any other institute, had become an integral part of NIH through a variety of policy and procedural means.

The integration of the Cancer Institute within the National Institutes of Health took place without unusual difficulty. But the relations between cancer research and biomedical research in general have never been without strain. There has been a history over the years of mutual disdain between cancer researchers and other biomedical researchers. Former NCI director, Kenneth M. Endicott, observed:

> At the basic science level even today, cancer scientists hold themselves apart and basic virologists still look down their nose at cancer virologists. This goes back to the time before NCI and is rooted in the view that the scientific problems were so insuperable, so little understood, that anyone involved in cancer research didn't have his head screwed on right.[32]

One manifestation of this distance between the cancer research community and the larger community of biomedical scientists has been the remote relationship between the cancer problem and the medical school curriculum. The aspiring physician often learned little about cancer in his formal medical education, partly because no set of scientific disciplines was closely related to cancer. Within departments of internal medicine, for example, attention was frequently directed to cardiovascular disease and to the broad but closely-linked set of medical-scientific disciplines—biochemistry, physiology, pathology, pharmacology—that were related to it. Cancer education tended to cluster around treatment (especially surgery) and rehabilitation, not around the underlying scientific problems. Education about cancer, therefore, tended to be separated from the basic medical-sciences curriculum of the medical school.

The NACC and the NCI gave attention to the need for more-comprehensive and better-integrated courses on cancer in the medical curriculum in a 1947 report. The report noted that the faculty most concerned with cancer treatment were the surgeon and the radiologist, not the internist, though neither had a major role in the instruction of the medical student. "The internist," the report said, "is the member of the teaching faculty who has the grasp of the total medical problem and is the key figure in introducing the student to the proper methods of clinical study."[33] It was concluded that the internist must become interested in the cancer problem if medical teaching about cancer was to be improved.

This report led to a program of grants to medical schools to help improve teaching about cancer. Annual grants of $25,000 were made to four-year medical schools and schools of osteopathy, while $5,000 grants were made to two-year medical schools and schools of dentistry. Between 1947 and 1957, $18.6 million was spent for undergraduate medical education in medicine, dentistry, and osteopathy.[34] The Cancer Institute terminated this program in 1957, however, because it had not fulfilled its purpose. The program was replaced by one directed to the level of residency training. The concept of the medical-school cancer coordinator was also developed at this time. One former coordinator said: "We held a series of meetings during those years . . . in conjunction with the American Association for Cancer Research, of which we were all members. We discussed how to broaden the teaching of cancer. We got, for example, the anatomists to use cancerous lymph nodes in explaining the lymphatic system."[35]

Progress has been made in recent years in incorporating the study of the cancer problem into medical education, but the problem has never been fully resolved.[36] In fact, the development of the big-three cancer research institutes—Memorial Sloan-Kettering, M. D. Anderson, and Roswell Park—has been largely independent of established medical schools and offers explicit testimony to the existence of this problem.

Beginning in the mid-1950s, a program in cancer chemotherapy was initiated within the National Cancer Institute. This program was for many years the subject of great controversy within the scientific community, dividing cancer researchers from other biomedical scientists, and among cancer scientists dividing those committed to fundamental research from those oriented toward directed research. This program resulted in the development by NCI of a new approach to research management which varied from the prevailing NIH management pattern—the use of the contract to support directed or targeted research. In a complex way, it set the stage for severe strains between NCI and NIH.

CHEMOTHERAPY

Chemotherapy—the use of drugs and other chemicals in the treatment of disease—received a powerful impulse in the 1930s and 1940s. The sulfonamides were developed and shown valuable in treating bacterial diseases. Penicillin was developed during World War II and was used with brilliant success in the war against infectious diseases. It was made widely available to civilians soon after the war. Another wartime effort was the large-scale, directed, and successful program of antimalarial-drug development, undertaken because the Japanese military successes after Pearl Harbor had cut off America's supply of quinine.[37] These historic clinical advances created widespread interest in chemotherapy throughout medicine.

Until the late 1930s, the treatment of cancer was limited to surgery and radiation. Huggins and Herbst, however, from 1939 to 1941, suggested a rational basis for the use of estrogen therapy in treating disseminated prostate cancer. During the war, it was discovered that nitrogen mustard could produce spectacular, if temporary, remissions in chronic leukemia and Hodgkin's disease patients. This work done under the U.S. Army's Chemical Warfare Service was the basis on which Dr. Cornelius P. Rhoads, director of Memorial Hospital, New York, pioneered the rapid development of the Memorial Sloan-Kettering Cancer Institute to a position of world leadership in cancer chemotherapy. In 1948, Dr. Sidney Farber published his very important findings that dramatic, though also temporary, remissions in acute leukemia in children could be produced by a folic acid antagonist, aminopterin.[38]

These clinical developments aroused great interest and led to suggestions that NCI should provide increased support for chemotherapy. A 1952 American Cancer Society study, among others, showed that confusing and inconclusive results had been obtained from both clinical and experimental studies. The National Advisory Cancer Council, in response to these developments, held a special meeting in late 1952 to consider the scientific promise in the clinical evaluation of chemical agents for cancer therapy.[39]

The serious efforts to promote cancer chemotherapy, however, were directed to the Congress, and especially to the Senate. Rhoads, Farber, and James S. Adams, a member of the NACC, appeared in April 1953 before the Senate appropriations sub-committee as representatives of the American Cancer Society. Farber, in a highly effective slide presentation, talked about acute leukemia. One slide showed "this little fellow who has been on borrowed time now for 45 months, since the onset of treatment, and he is still in excellent condition. He is leading the way, and those who are working in this field believe that if it can be achieved in one it must be achieved in others."[40]

The subcommittee's report indicated that the Senators had been "impressed" by the prospects of treating leukemias with chemical agents. The report encouraged an intensification of this work:

> It appears that the serial examination of chemical agents in the clinic has certain developmental aspects which could suitably be engineered, for the sake of economy of effort and time, in a manner which has facilitated the exploration of other therapeutic agents in the fields of infectious and parasitic diseases. . . . [C]ertain aspects of this problem may have reached a stage where a large-scale integrated effort of a developmental nature may be warranted.

The report referred to the wartime development of penicillin and to the "extensive biochemical, pharmacological, and clinical testing program [which] has produced a medical cure for malaria," in suggesting the approach which might be followed.[41]

Some within NCI and among its grantees thought they detected the influence of Dr. James Shannon in the language of the subcommittee's report. Shannon, then associate director of NIH, had been responsible for the clinical testing of antimalarial drugs in World War II,[42] and had been one of seven members of the wartime Board for the Coordination of Malarial Studies. Shannon's known preference, moreover, was for a small chemotherapy program directed at improving the effectiveness of the agents then known to be moderately effective against leukemia.

A number of individuals, Farber being among the most prominent, thought a program should be broader in scope, larger in scale, and less-directed in management. "There was no enthusiasm," a later account reported, "for an engineered-directed program controlled by such a small committee as guided the wartime malaria program. Neither was there much favor toward limiting the effort to leukemia since a number of other forms of cancer had responded to some of the drugs then available." Consequently, an ad hoc committee, formed to make recommendations on the chemotherapy program, concluded that a research program engineered and directed by the federal government would be inappropriate for technical reasons and for the adverse psychological effect it would have upon the scientific community. The atmosphere of independence was best suited to fundamental research, so NCI's greatest contribution would be to bring together the scattered chemotherapy groups to improve the exchange of information. In December 1953, it was agreed that there would be a program of voluntary cooperation among institutions engaged in chemotherapy research.[43]

Important changes in the composition of NACC were then taking place. Joining the NACC in October 1953 were Farber, Elmer Bobst, and Mary Lasker—who had just completed a term on the National Advisory Heart Council. They lent their efforts to program expansion for chermotherapy. The $21.7 million appropriation to NCI for fiscal 1955 included $3 million for chemotherapy. The Senate subcommittee retreated from its position of the prior year, which favored the engineered program.[44] The proponents of a large-scale, broad-scope, nondirected chemotherapy program had prevailed.

They were also successful in getting more money. From fiscal 1956 through 1962, the successive annual amounts for chemotherapy were: $5 million, $14.9 million, $23.8 million, $31.4 million, $36.5 million, $37.4 million, and $43.7 million.[45] Moreover, these are partial figures, since they do not include research training grants or funds for direct support of NCI laboratories engaged in chemotherapy research.

The chemotherapy program experienced a good deal of administrative floundering. The NACC established the Cancer Chemotherapy Committee in 1953, chaired by Sidney Farber.

This group was expanded in mid-1954 to include representatives of the American Cancer Society, Damon Runyon Fund, National Research Council's Committee on Growth, Department of Defense, and Veterans Administration, aided by one full-time NCI staff member. The expanded committee "managed" the program until NCI, under pressure from the Senate, established the Cancer Chemotherapy National Service Center (CCNSC) in April 1955.[46]

Shannon, who became director of NIH on July 1, 1955, named Kenneth Endicott to be the first chief of the CCNSC, and charged him with bringing administrative leadership to the effort. Endicott recruited eight full-time senior staff for each of the major program areas—organic chemistry, screening, pharmacology-biochemistry, documentation, and biometry. Concurrently Dr. John R. Heller, director of NCI, announced the formation of the Cancer Chemotherapy National Committee to assist the CCNSC. This new committee included representatives from the six "sponsoring" organizations—the National Cancer Institute, Atomic Energy Commission, Food and Drug Administration, Veterans Administration, American Cancer Society, and Damon Runyon Fund. In addition, the committee included the chairman of the NACC's Cancer Chemotherapy Committee, another NACC member, a pharmaceutical industry representative, and a member-at-large. Endicott, as chief of the CCNSC, was executive secretary to the new group, and the program was further assisted by an industry subcommittee and scientific panels on chemistry, drug evaluation, endocrinology, and clinical studies.[47] The program appeared to Endicott to be in danger of strangulation by the plethora of committees, subcommittees, and panels.

Since research grants for chemotherapy were handled through the traditional NIH extramural review process, Endicott turned his attention to that portion of the program over which he could exercise some control—mainly the contract research portion of the program.[48] The contract mechanism was important because the major emphasis of the chemotherapy program was on drug development, including the large-scale screening of chemical compounds. A related need was for substantial animal production facilities. The integration of the various components of such a program was administratively much easier through a centrally managed contract effort than through a program of extramural grants. A large portion of the contract work was performed by private industrial firms, moreover, and administrative policy within HEW had stipulated that the contract was to be the exclusive instrument used by NIH with these organizations.

The search for chemical cures for cancer involved a number of steps. The first was the procurement of large numbers of chemical compounds that might have antitumor characteristics. The second step was the screening, or testing, of these compounds for their antitumor effects. Third, compounds that showed promise had to be evaluated for toxicity in animals, the preclinical pharmacology stage. If a compound passed this far, it entered the stage of human pharmacology where it was further examined for toxicity and appropriate dosage in humans. This was followed by the clinical-trials stage, or the cooperative clinical evaluation of compounds in humans. The final stage was an evaluation of results.[49]

The critical scientific problem in the chemotherapy program was screening.[50] Since numerous chemical compounds were procured for examination, there had to be a system for identifying those that had the promise of antitumor activity. This system involved testing each compound against a specific tumor, or screen, which had been transplanted in a test animal. If the compound showed activity against the animal tumor, it then had to be determined if it also showed activity against human tumors. Only when good correlations could be established between a compound's activity against animal tumors and against any tumors in humans, could it be considered a potential chemotherapeutic drug, a process which required a number of years.

The problem was to select the tumors to function as screens. Prior to 1955, many transplantable tumors were being used throughout the country, but the overall effort was uncoordinated. A report in that year reached the following conclusions. First, "[t]here is no single tumor which could be expected to select all useful agents and therefore a spectrum of tumors provides a greatly improved screening system." Second, no evidence pointed to a nontumor biological system as a possible substitute for a tumor screen. Third, "comparisons of studies of anti-tumor activity in experimental and human neoplasms are hazardous because of the relative paucity of reliable clinical data."[51] The study resulted in selecting three mouse tumors—Sarcoma 180, Carcinoma 755, and Leukemia L1210—for the screening process.

In 1959 and 1960, the screening system was modified once again.[52] Dr. Stuart Sessoms, who succeeded Endicott as CCNSC chief, reported to the NACC in February 1960: "It is fantastic to expect that only three mouse tumors, with only three syndromes of biochemical defects, should be able to predict for dozens or hundreds of human [cancer] diseases."[53] The primary screen, therefore, was broadened to include selected new multiple mouse, rat, and hamster tumors.

The screening problem was critical because rapid expansion of the program meant that a great many compounds were being examined. One long-time veteran of chemotherapy research reflected on the problem this way: "If any fault existed with those early efforts, it was that too many materials were being run through too rapidly. The criteria were not very well established for screening them. We were not taking the best materials and following them up closely. We were generating a lot of data and getting a lot of numbers but that was not telling us very much. It involved a lot of 'first look' at materials and little follow-up. This is what happens when you accelerate things too rapidly."[54]

Rapid growth, however, was a primary characteristic of the chemotherapy program and was driven by powerful political forces. Endicott, in December 1957, presented a conservative administrative view of the program. He wrote: "It is the impression of CCNSC that the chemotherapy program should plateau at approximately the present level for the time being in order to permit the staff, advisory groups, grantees, and contractors to work out the many problems related to achieving a smooth, efficient cooperative program."[55] Within months, however, the Congress had upped the ante for fiscal 1959 by almost 50 percent over the "present level" referred to by the CCNSC director. An agreement was reached between NCI and Shannon about the same time that a ceiling of $25 million would be applied to programmed chemotherapy grants, contracts, and direct CCNSC operations. Five years later, in 1962, these activities had reached a plateau slightly in excess of $30 million.[56]

Congress, encouraged by Sidney Farber, Mary Lasker, and their allies, pushed for program expansion. For example, the Committee of Consultants on Medical Research—"Bo" Jones committee—submitted its recommendations to the Senate appropriations committee in May 1960. These, in turn, were incorporated by the committee into its recommendations for the fiscal 1961 NIH appropriations. Within the substantial budget increase recommended for NCI, the Senate report singled out chemotherapy: "The committee recognizes the urgency for the rapid application of newly discovered anticancer chemicals to patients with advanced cancer, and directs that expansion of clinical investigation of cancer chemotherapy proceed as rapidly as the clinical and scientific resources of the country permit."[57] The proponents of the chemotherapy program were determined to move it forward at a rapid pace.

Not surprisingly, the chemotherapy program drew criticism from the scientific community. Though the particular concerns might change, the basic issue was always the relative emphasis upon fundamental versus applied research. Dr. Alfred Gellhorn, a frequent critic of the program but one who was engaged in chemotherapeutic research, viewed the matter this way: "The situation with respect to chemotherapy is a highly attitudinal one. The conflict is a continuing one of differences of intellectual views. There is no final resolution of this matter. The only fundamental issue is whether the emphasis, both past and current, is prejudicing the adequate support of basic biological processes of malignant diseases." A more applied view was expressed by Dr. Howard E. Skipper, a long-time chemotherapy researcher: "In the biological sciences, particularly in medical research, there is always a bit of bickering between those who do basic research and those who do practical or applied research. It's childish in the extreme. You have to do both. I think the government is putting up money to find a cure for cancer, not to have a thousand scientific papers written on esoteric science."[58]

The chemotherapy program also divided the NCI professional staff. Some favored a highly organized program, while others preferred greater reliance upon individual enterprise to produce the important scientific advances.[59] Intramural scientists did not object to participation in CCNSC activities on a voluntary basis, but resented any requirement that the chemotherapy program be their sole base of NCI support.

The chemotherapy program of the Cancer Institute has had its share of success, especially in developing chemical agents that are effective against the fast-growing, disseminated forms of cancer, like leukemias and lymphomas. It has had far less success in developing drugs that can be used against the slow-growing, solid tumors. The incidence of fast-growing cancers, however, is relatively small compared to the total incidence of cancer. Even so, the hope of discovering a chemical cure for cancer has been so strong that Congress, over the years, put substantial amounts of funds into the quest, always encouraged by Mrs. Lasker, Dr. Farber, and their allies.

One large consequence of the chemotherapy program was the first use of the contract to support large-scale, directed research. This development contributed strongly to the evolution within NCI of a research management style and philosophy that differed from the prevailing NIH philosophy. Research management became a controversial element in its own right.

RESEARCH MANAGEMENT

Dr. Kenneth M. Endicott had joined the Public Health Service in 1940, one year after he had received his M.D. from the University of Colorado. He had come to the NIH in 1942, as a bench scientist in the division of pathology. He later served as the scientific director of the Division of Research Grants of NIH, as the first chief of the Cancer Chemotherapy National Service Center, and from 1958 to 1960 as an associate director of NIH. When Dr. John Heller resigned as NCI director in 1960 to become president of Memorial Sloan-Kettering Cancer Institute, Shannon appointed Endicott to succeed him. In this appointment, Shannon obtained a more decisive administrator than Heller and one with whom he was more compatible. NCI obtained a director who became its contemporary architect, at least for the time prior to the National Cancer Act of 1971.

The use of the contract mechanism by the chemotherapy program made NCI different from the other NIH institutes with their traditional intramural and extramural research programs. Moreover, several years of controversy swirled around NCI's contract research and the issue is still unresolved today.[60] A 1966 report suggested three reasons for the misunderstanding within the scientific community about the contract activities of the National Cancer Institute.[61] One was that the deep involvement of the scientific community in the grant-review and -award process may have led them to see their role on contracts in a similar light. Second, large-scale research and development programs in health had been infrequent in the past and the scientific community may have been unfamiliar with the administrative requirements of such efforts. Finally, other federal agencies supported biological research by contract in much the same way as NIH research grants. Why, then, an observer might ask, should NCI be different?

The biomedical research community's consistent view has been that review of contracts should be performed by some peer-review mechanism. The award process, moreover, should sharply constrain administrative discretion and require that all funded contract research be judged meritorious by the scientific community. Finally, the contract instrument should be used sparingly for special purposes like animal procurement and not for imparting direction to the research enterprise. Kenneth Endicott and his staff, on the other hand, thought that the review process should lead to a judgment about program relevance of research as well as an assessment of the quality of work being proposed. Consequently, the award process should place decision authority in the NCI's professional staff, for legal as well as functional reasons. Finally, the contract should be used when the integration of components of a complex program of directed research is desired. The historical unfolding of these issues, however, was a rather complicated experience.

After the establishment in 1955 of the Cancer Chemotherapy National Service Center, the original cancer chemotherapy national committee had been supplemented by five technical review panels of outside scientists in chemistry, screening, pharmacology, clinical studies, and endocrinology. The rapid expansion of the program made the work load of these panels impossible, so more and more committees and subcommittees were appointed to help. By the late 1950s, there were more than 300 outside scientific advisers to the chemotherapy program! "With this proliferation of panels, committees, and subcommittees," according to one account, "the time required for them to communicate with one another through minutes and reports brought the process of decisionmaking nearly to a halt."[62]

When Endicott became director of NCI, he eliminated all committees and subcommittees of the chemotherapy program and reduced the technical panels from five to four. Contract proposals now were initially reviewed by one of the four panels, and then by a newly created contract review board consisting of three members of the NACC, the chairmen of the four panels, and several at-large members. The panels could use outside advice without restriction, but no standing committees of consultants were to be maintained.[63] Thus, the contract review process was streamlined, the number of participants sharply reduced, but control still remained in the hands of the outside scientific community.

In 1962, President Kennedy issued a directive to all federal agencies on conflict-of-interest of outside advisers. Endicott seized on this and interpreted the existing contract-review procedure as being inconsistent with the president's directive. The contract was a procurement instrument, he reasoned, to be used where the government had a reasonably clear idea of the research product it wished to obtain. Grants, by contrast, were appropriate for basic research where the best guide to research was the scientist's own understanding of what should be studied. Procurement law and regulation, moreover, vested legal authority and responsibility for contract awards in government officials and this could not be delegated to outsiders. Furthermore, since panel and contract review board members were often closely associated with contract proposals they were reviewing, as principal investigators, program managers, or employees of proposing organizations, there existed substantial room for conflict-of-interest in the award process.

The Cancer Institute, therefore, abolished the contract review board and the four panels. External advice to the CCNSC for several years thereafter was limited to an informal steering committee of carefully selected outsiders who relied heavily upon NCI staff, a formal steering committee of full-time intramural staff, and one outside panel. The influence of the scientific community was practically eliminated from the contract-review and-award process, and authority and responsibility for contracts were placed in the hands of the NCI staff.[64] Endicott had established a management pattern consistent with what he saw as the requirements of directed research.

Criticism was not long in coming. The NACC, in 1963, expressed concern about its inability to obtain information about the chemotherapy program.[65] More importantly, Representative Fogarty in the House expressed his view as early as 1963 that NCI contract research proposals should be reviewed by the NACC like grant proposals. In September 1964, Fogarty reiterated this view: "I repeat my belief today that Council review of all individual research contracts is mandated by law."[66] Endicott and Shannon ignored these protestations.

Communication on chemotherapy between NCI and the scientific community was "left in virtual suspense," according to one report, and "doubts, uncertainties, and vague feelings of dissatisfaction with the quality of the program" resulted.[67] These vague feelings found clear expression in the February 1965 report of the Wooldridge Committee. The report, which was generally concerned with whether Americans were getting their money's worth from NIH expenditures for medical research, singled out the chemotherapy program for harsh criticism. "[M]any medical scientists," the report indicated, "question whether the value of the Cancer Chemotherapy Program . . . is commensurate with its cost." The report noted that the review group for the chemotherapy program had found that "a substantial fraction of the contract work on the program was of lower scientific quality and showed evidence of inadequate central supervision." Furthermore, the report continued, "NIH has rendered itself vulnerable to criticism by a policy of gradual divestment of non-government advisors from the Chemotherapy Program." The report recommended that an ad hoc committee of scientific and management experts make a thorough review of the program and make recommendations for its future development.[68]

The overall Wooldridge Committee report had reflected the strong preference of academic scientists for basic research rather than categorical research. Even so, Mary Lasker and Sidney Farber seized on its criticism of the chemotherapy program to press their case for council review of contracts. The council met on August 13 and 14, 1965, and minutes of that meeting suggest a compromise was reached: "The Council should not participate in the difficult steps involved in the procurement and deliverance of contracts . . . [since] to do so would be at variance with the trend in Council activities which is directed less and less towards the minute details and more and more towards the scope, character, and direction of programs and the broad strategy related thereto." The Council also adopted unanimously a resolution proposed by Dr. Farber that indicated general information to be presented to the council, a resolution some believed signaled the establishment of "an agreeable modus operandi" between NCI and the Council.[69]

Mrs. Lasker had other ideas. She went to Senator Hill and persuaded him to add very explicit language to an appropriations subcommittee report on a supplemental appropriation bill for fiscal 1966. The report, issued on October 19, 1965, included this statement:

> The committee . . . restates its intent that review and approval of research contracts is to be handled in the same manner as that for research grants—through the appropriate study sections and advisory councils . . . [and] directs the Director of the National Cancer Institute to prepare a report on the review and approval of these research contracts subsequent to each of the meetings of the National Advisory Cancer Council during fiscal 1966, and . . . make these reports available to the Committees of Appropriations of the House and the Senate.[70]

This language sought to draw the noose tight around the neck of the NCI.

Endicott went to Shannon for help. Shannon went to Fogarty who said he would need a letter from Secretary of Health, Education, and Welfare John Gardner if he were to intervene. Shannon went to Gardner and, on October 20, Gardner wrote to Fogarty. In his letter, Gardner said, "Review and approval of each individual contract by the Advisory Council to the National Cancer Institute introduces such a fundamental change in policy that I would be extremely reluctant to see it introduced without a thorough study. I assure you that I will initiate such a study immediately and will report back to the congressional appropriations committees early next year."[71] The restrictive language was stricken from the Senate bill by the joint House-Senate conference committee, who indicated in a report of October 21 that they had agreed to hold contract review in abeyance on the understanding that the secretary would submit his report not later than the end of February 1966.[72]

Gardner appointed a committee chaired by Dr. Jack P. Ruina, the president of the Institute for Defense Analysis, and a person very familiar with both grant and contract support of research and development in the Department of Defense. The principal findings of the Ruina Committee report, released in March 1966, were: advisory councils were not required by law to approve individual contracts and should not be required to do so; the grant mechanism was inappropriate for directed research or development programs, and the contract should be used for such programs; programs for directed research—including objectives, justification, expected funding levels, management plans, and types of contractors—should be submitted to the appropriate advisory council for review and approval prior to initiation, termination, or substantial change in scale or direction of effort; once initiated, execution of such a program should be the full responsibility of a program manager; the NIH should take significant steps to make career opportunities and status for program managers more attractive; and "a strong management structure" for directed research should be established independent of the intramural or extramural research efforts.[73]

Endicott presented the results of the Ruina report to the NACC at its spring 1966 meeting. The report upheld the position of Endicott and Shannon on the role of the advisory councils in contract proposal review, and effectively thwarted Mary Lasker and Sidney Farber on this issue. Shannon was present at this meeting and Endicott recalls "a really heated session, a Council confrontation, with Shannon backing me completely."[74] Mrs. Lasker was upset. Ironically, in Kenneth Endicott, she was dealing with someone more strongly committed to directed research for categorical purposes than most NIH officials. But though Endicott "was dedicated to cancer, and he was application-oriented," a former NIH official said, he was also "very single-minded and temperamentally unable to work with the Advisory Council."

Endicott extended the use of the contract mechanism to develop two other directed-research programs, one in chemical carcinogenesis in 1962 and the other in virology in 1964. The chemical carcinogenesis program was an effort to evaluate suspected chemical compounds for their cancer-causing properties. There were several million existing chemical compounds "on the shelf," and new ones were being introduced into the environment at a rapid rate. Many of these compounds had been identified as carcinogenic in animals. "The problem," Endicott said, "was how to anticipate the effects of those compounds before we introduced them and had a disaster."[75] Two approaches were taken. One analyzed those occupational settings in which humans were exposed to certain chemicals in measurable amounts. The other studied the major differences in the forms and incidence of cancer by country and culture. The generic problem was the same as that for chemotherapy, how to develop biological test models or screens which would predict carciogenic effects in human beings. "My feeling was why wait," said Endicott. "Let's get on with it. I could not see any way to do this other than by large-scale support. These were not the problems for the laboratory scientists."[76]

The viral oncology program was concerned with the role of viruses in the cause of cancer in man and animals.[77] It began in 1964 with an appropriation of $10 million for the special virus leukemia program. Prior to this, scientists had established that viruses were the causative agents for certain leukemias in chickens, mice, and rats. The program was undertaken to determine whether viruses were also cancer-causing in humans and to find preventive means for man. In 1967, demonstration of the presence of viruses in induced solid tumors in animals led to the initiation of the solid-tumor virus program. In 1968, the leukemia and the solid-tumor efforts were merged into a single special virus cancer program since the two programs had substantial scientific overlap and the top management was the same. The virology effort was also a large-scale, directed-research program.

Overall, contract research grew to be a substantial portion of the NCI budget. By fiscal 1971, contracts for chemotherapy amounted to $23.8 million, those for viral oncology were $31.2 million, and those for chemical carcinogenesis were $9.7 million, while the overall level totaled $78 million.[78]

Endicott introduced other changes into NCI. Where the other institutes had a single scientific director for intramural research, Endicott appointed three—one each for chemotherapy, etiology (viral oncology and carcinogenesis), and general laboratories and clinics. The chemotherapy and etiology scientific directors, moreover, were given the dual responsibility of managing a large-contract, directed-research effort and the related intramural research activity. Though the Ruina report had suggested that contract management capability be separate from the intramural research program, NCI had decided much earlier on the basis of the CCNSC experience to consolidate staff responsibility for these activities. They did so because of the difficulty of recruiting outside scientific talent to manage programs, and to insure up-to-date scientific competence in their program managers.[79]

Endicott's actions on research management had effects on others. Carl G. Baker, M.D., had become interested in research management in the late 1950s. He organized evening discussion groups which met over a two-year period. The first year dealt with history and philosophy of science, and the second placed the emphasis on management. A former associate recalled him as "imbued with books on administration and management." In 1961, Endicott appointed Baker associate director for program, the official responsible for planning and analysis within the Cancer Institute. Baker recalled that period: "I was associate director but felt great dissatisfaction with [NCI] policies. I never felt we were getting to the heart of the matter. We didn't have our objectives clearly in mind."[80] Baker recruited a young, management-oriented industrial sociologist named Louis Carrese, and the two developed the "convergence technique."

The "convergence technique" was essentially an adaptation to research management of the networking and planning techniques of systems development that had been designed for the scheduling, control, and costing of production problems.[81] It involved the detailed specification of program objectives, consideration of alternative program approaches and selection of priority approaches, identification of major program elements and their relationships in an integrated program, and indication of specific research projects of high program priority. This planning information was arrayed in chart form in order to draw together the performance of multidisciplinary research and the needed resources of personnel, material, equipment, facilities, and funds. Unlike production-planning models, however, the technique could not utilize tight time-to-completion estimates because of the greater uncertainty associated with research. After years of analysis of the cancer problem, Baker's main motivation in the development of the "convergence technique" was to tackle high-priority research problems using improved management methods.

At first, the technique was applied to the special virus leukemia program. Baker, Carrese, and Frank J. Rauscher met for three weeks and worked out a detailed plan. This plan was approved internally by NCI on October 6, 1964, and was presented to the NACC on October 14. The council unanimously recommended that "the NACC and the NCI Board of Scientific Counselors go on record as enthusiastically endorsing the scientific plan of attack and the scientific management program as outlined to us, and therefore the making of speedy progress in this problem of the causation of acute leukemia and the means of eradicating acute leukemia." The existence of a plan, however, hardly insulated the program from scrutiny. The fiscal 1969 report for the special virus cancer program indicated, somewhat defensively, that it had been reviewed no fewer than 18 times, that the "overall consensus . . . has been that substantial progress had indeed been made toward specific program objectives," and that the effort should continue "as planned and as new leads allow."[82]

The convergence technique was next applied to the chemotherapy program. Baker, Carrese, C. Gordon Zubrod, M.D., scientific director for chemotherapy, and Saul A. Schepartz, Ph.D., Zubrod's associate director, worked for six weeks to develop a comprehensive plan. This was presented to the Advisory Council at their August 13 and 14, 1965, meeting. It was later published as part of a detailed review of the entire chemotherapy program.[83] The convergence technique, then, was being incorporated into the chemotherapy plan just as the contract authority dispute was being settled in favor of an NCI staff-directed research effort. Following this, the convergence technique was then applied to the carcino-genesis program by Baker and Dr. Umberto Saffiotti.

Endicott, during his tenure as director of NCI, forged a new pattern of research management within the NIH. It is true that the Cancer Institute continued to support a large amount of undi-rected, grant-supported research, for which the dual review of NIH study sections and the NACC continued to hold. Neverthe-less, large-scale, directed-research programs were established in chemotherapy, viral oncology, and carcinogenesis. The NACC was limited to review of major policy directions for such pro-grams, but was barred from review of individual contract propos-als. The external scientific community was essentially out in the cold. Contract management authority was vested in and exercised by the NCI staff. The "convergence technique" was being applied to the management of the directed research programs. NCI staff recruited to manage the programs were also given responsibility for the related aspects of the intramural research program. The thrust toward central control over research management was firmly established and a cadre of subordinate officials, brought along by Endicott, were personally committed to the new order. But the development of this directed-research effort set the teeth of the scientific community on edge.

"IMPEDIMENTA"

Kenneth Endicott left as director of the Cancer Institute on November 9, 1969, after nine years in the position. Carl Baker was named acting director on November 10. Eight months later, after NIH had conducted an extensive search for a permanent suc-cessor, Baker was named director of the institute. The contrasts between the two men were revealing.

Endicott's background and experience were broader than

cancer, and when he left NCI he became director of the NIH Bureau of Health Professions Education and Manpower Training. Baker's entire career was occupied with cancer and NCI. He had served in the biochemistry laboratory when he first came to NCI in 1949, then successively in the [extramural] research grants branch, the office of the NCI director, and as assitant to the associate director of NIH for intramural research. Baker was named assistant director of NCI in 1958, then became acting scientific director for intramural research, and was named associate director for program in 1961. He was scientific director for etiology from 1967 until his appointment as acting NCI director in 1969.

Though a strong administrator, Endicott had a pragmatic view of the limits of research management. Baker had a stronger belief in management per se, and was more rigid in his approach than his predecessor. While Endicott had a strong concern with the application of research knowledge, Baker's commitment was to solving the problem of cancer. At home with the scientific and administrative aspects of cancer, Endicott was equally professional in dealing with the Congress and the external political world of medical research. Baker knew more about cancer, but was untried in dealing with the Congress, especially with other than the appropriations committees, and with the high politics of the cancer enterprise. Endicott's style was a judicious blend of aggressiveness and cunning, while Baker tended to be blunt, argumentative, and even undiplomatic. In sum, Endicott responded both with vigor and adaptability to the world as he found it, while Baker's response was more frequently based upon prior analysis of the cancer problem and great confidence in knowing the right thing to do.

Relations between the National Cancer Institute and the NIH underwent a subtle shift in 1968 and 1969. Shannon, who left NIH in 1968, had had a long, working professional relationship with Endicott. While the two men did not see eye to eye on everything, Shannon gave Endicott considerable latitude and the latter, in turn, readily acknowledged Shannon as the NIH "boss."

Robert Marston was firmly installed as NIH director when Carl Baker became acting director of NCI. Marston had appointed Robert Berliner as deputy director of NIH for science. Berliner, a renal physiologist, had done his residency at New York University's Goldwater Memorial Hospital during World War II, where Shannon was director of research, and the two coauthored a number of scientific papers during this time. Prior to coming to the "front office" of NIH in 1968, Berliner had served fourteen years as the director of intramural research at the National Heart Institute. He was the living embodiment of the traditional NIH commitment to fundamental research. He had also, over the years, developed a good deal of skepticism about the NCI's efforts to push toward directed-research programs run by a small number of agency staff.

Baker, by contrast to Berliner, was perhaps more strongly committed to directed research than anyone else at the top level of NCI. He chafed under the administrative constraints on NCI that emanated from the NIH front office and a number of strains between the two organizations came to the fore. As Berliner saw it, "there were no major issues between NIH and NCI, just continuing skirmishes." But from Baker's viewpoint, the skirmishes were quite bothersome.

The most persistent issue of conflict had to do with the promotion and tenure of intramural research staff.[84] The scientific directors of all the NIH institutes, in their collective capacity, reviewed all decisions on the tenure or promotion of intramural research staff, regardless of the particular institute in which the research scientist was located. The criteria for promotion stressed scientific accomplishment and productivity, so there was a good deal of emphasis placed upon published scientific articles. The National Cancer Institute had given their intramural research staff administrative responsibilities for the directed research programs. But research administration reduced the time available for the actual conduct of research, so NCI staff were not always as productive in scientific papers as scientists from other laboratories.[85] Since the scientific directors were strongly committed to research, both fundamental and clinical, they tended to look down upon contract management by NCI staff. Baker took the view that scientific directors from other institutes were not the appropriate ones to judge the value of NCI scientists to the cancer program. But no changes in NIH practice occurred, and the issue remained outstanding.

There were other issues between NCI and NIH. Contract authority, which was vital to the Cancer Institute, had been held by
the NIH front office. Approval was needed, therefore, from several officials in the office of the director of NIH before NCI could
sign contracts. Delegation of this authority from NIH to NCI was
in process, however, in 1970. Another issue, especially vexing to
Carl Baker, was that the hiring of NCI staff at higher grade levels
was controlled by the U.S. Civil Service Commission, involved
the personnel apparatus of HEW, and took such a long time that
qualified individuals seldom were prepared to wait while the
paper shuffling took place.

In general, the National Cancer Institute, and Carl Baker as director, chafed under the administrative constraints imposed by
NIH, HEW, and others. Baker felt there was too much emphasis
upon detail at too high a level in the organization, that there were
too many layers above him that were too slow in response. He
desired administrative simplification. But when the Panel of Consultants heard the NCI litany of complaints, they amplified these
difficulties into an argument for complete administrative autonomy from the NIH. This came as a mild surprise to NIH. One
official commented, "I don't think we were ever informed regarding the arguments raised about administration impedimenta." But
where NIH may have seen the problems as "impedimenta,"
others looked upon the difficulties between NCI and the "front
office" as far more serious.

From President Taft's concern in 1910 for a laboratory to study
cancer in fish, to the 1922 initiation of cancer research within the
Public Health Service, to Senator Neely's proposed study in
1928, to the establishment of the National Cancer Institute in
1937, cancer research has always had a prominent position in the
history of medical research supported by the United States government. Though an integral part of the National Institutes of
Health from the late 1940s onward, NCI preceeded the modern
NIH by almost a decade and in its statute and pattern of organization became the model on which subsequent institutes were based.
Beginning in the mid-1950s and extending for ten years, developments in research management and planning took place within
the Cancer Institute to quicken the advance of research results into

medical practice. These developments in the history of NCI have been supported by the Congress—indeed, mandated by it on occasion—and elaborated by the professional staff of NCI, not always with the enthusiastic sanctioning of the biomedical scientific community. The net effect of this history in 1970, when the Panel of Consultants began its work, was to amplify certain inherent tensions between NCI and NIH into a rationale for the recommendations of the panel.

A "Moon Shot" for Cancer?

[T]here is a widespread and understandable public demand that this Nation, with its great technological competence, should turn its attention to the finding of the cause and cure of cancer, with the same vigor and ingenuity that were used to win the war. We may expect that this demand will increase; for with our aging population, more and more persons who once died from the diseases of childhood and early life will live into middle and old age to die of cancer. More and more families will see first hand the anguish of a loved one dying with cancer and will know the hopelessness of trying to save him. It will be impossible for them to understand why a Nation which has so perfected the art and science of destruction of life in war does not bend its full effort to improving its knowledge and applying it widely and promptly to the saving of life from so dreadful a cause.

> Thomas Parran, M.D., Surgeon General
> U.S. Public Health Service
> December 5, 1946.

The time has come in America when the same kind of concentrated effort that split the atom and took man to the moon should be turned toward conquering this dread disease. Let us make a total national commitment to achieve this goal.

> President Richard M. Nixon,
> "The State of the Union"
> January 22, 1971.

If I had a choice between a moon walk and the life of a child with leukemia, I would never glance upward.

> James T. Grace, M.D., Director,
> Roswell Park Memorial Institute
> January 8, 1970.

Dissatisfaction with the Regional Medical Program, distress in the decline of funds for cancer research, and the need to find a successor to Senator Lister Hill—these factors led Mary Lasker to search for a new cancer initiative. A book by Solomon Garb, M.D., *Cure for Cancer: A National Goal*, published in mid-1968, helped crystallize her thinking. The time had come, Garb rather stridently asserted, to reexamine the nation's cancer research effort, to move beyond the development of new scientific leads and to exploit existing leads. The latter could be done, he wrote, with characteristic hyperbole, "through a national commitment to make the cure . . . of cancer a national goal, in the same way that putting a man into orbit around the earth was made a national goal, and then achieved."[1]

One thing to be learned from the space program, Garb believed, was that money could buy scientific talent. Another lesson was that an independent agency that reported directly to the president had great advantages over those agencies buried in the bureaucratic hierarchy of a department. Further, he wrote, the space program benefited enormously from having a clearly defined goal. So, with simple if not simplistic logic, Garb urged that the cure for cancer be established as a major national goal. The main task for health professionals then became "the crystallization of the public attitude toward the need for a more effective [cancer] research program."[2] Garb's propagandistic talents were soon to be employed to precisely that end.

Mary Lasker read Garb's book, distributed copies to many of her friends, and invited Garb to visit her in New York. Garb, who was scientific director of the American Medical Center in Denver, did so and recalled being in close agreement with Mrs. Lasker "on practically everything." Through Luke Quinn, she arranged for Garb to testify with Sidney Farber before the House and Senate appropriations committees in 1969 on behalf of the National Cancer Institute budget.

Col. Luke Cornelius Quinn, USAF (ret.), had served as a liaison between the air force and the Congress from the end of World War II until his retirement from the military in 1952. At that time, Representative John Fogarty persuaded Mary Lasker to hire him as a lobbyist for health and medical research causes and draw upon his extensive experience and wide personal friendships on Capitol Hill. She did so and also arranged for Quinn to become the Washington representative of the American Cancer Society.

In time, his clients included the M. D. Anderson Hospital and Tumor Institute, as well as a number of other medical scientific organizations.[3]

Mary Lasker had been accustomed to exercising influence in Washington through "insiders"—powerful figures like Fogarty, Hill, and President Johnson. With an uncertain congressional situation and with Nixon in the White House, an "outside" or "grass roots" strategy began to develop. Conversations between Mrs. Lasker, Luke Quinn, Sidney Farber, Emerson Foote, and Solomon Garb led to the formation in 1969 of the Citizens Committee for the Conquest of Cancer. On Tuesday, December 9, 1969, a full-page advertisement sponsored by the committe appeared in the *New York Times*. Bold print declared: "MR. NIXON: YOU CAN CURE CANCER." The ad, which noted that 318,000 Americans had died from cancer the previous year, read:

> This year, Mr. President, you have it in your power to begin to end this curse.
>
> As you agonize over the Budget, we beg you to remember the agony of those 318,000 Americans. And their families. . . .
>
> We ask a better . . . way to allocate our money to save hundreds of thousands of lives each year.
>
> America can do this. There is not a doubt in the minds of our top cancer researchers that the final answer to cancer can be found. . . .
>
> Dr. Sidney Farber, Past President of the American Cancer Society, believes: "We are so close to a cure for cancer. We lack only the will and the kind of money and comprehensive planning that went into putting a man on the moon."
>
> Why don't we try to conquer cancer by America's 200th birthday? . . .
>
> Our nation has the money on one hand and the skills on the other. We must, under your leadership put our hands together and get this thing done.[4]

The advertisement, written by Quinn, listed Solomon Garb and Emerson Foote as co-chairmen of the committee, and gave the committee's address as 866 United Nations Plaza—the same as that of the Albert and Mary Lasker Foundation. The public drum-beating had begun.

The route to an effective "inside" strategy was also being pursued. In mid-1968, it was widely assumed that Senator Wayne Morse (D., Ore.) would succeed Senator Hill as chairman of the Committee on Labor and Public Welfare and probably as chairman of the health subcommittee as well.[5] Unexpectedly, however, Morse was defeated in his reelection bid in November. Senator Ralph Yarborough was the committee's next-most-senior Democrat after Morse so he became chairman of the full committee.

The health subcommittee chairmanship was for a time an open question. Mrs. Lasker, through Mike Gorman, her other Washington representative,[6] lobbied to secure the position for Senator Edward Kennedy, though without the senator's overt support. But Yarborough, who had devoted most of his legislative career to education, now decided that health was the area that needed emphasis and consequently took the chairmanship of the subcommittee as well. The Lasker-Gorman effort on behalf of Senator Kennedy, however, aroused the Texan's suspicious nature. Since Mrs. Lasker's objectives for cancer required close working relations with the subcommittee chairman, a certain amount of time was needed in early 1969 to permit ruffled feathers to be smoothed.[7]

Toward the middle of 1969, Mrs. Lasker suggested to Yarborough that an outside citizen's committee, advisory to the Labor and Public Welfare Committee, be established to study the needs in cancer research and to recommend steps for mobilizing a major national cancer program. Yarborough was receptive. He recalled testimony years earlier in which cancer-research experts had assured him that one billion dollars a year for cancer research would bring understanding of the cause and cure for cancer in a decade. He wanted to find a way to put a billion dollars a year into cancer research. A citizen's committee might be just what was required.

Several problems had to be overcome. For one thing, Yarborough was not a calm, deliberate individual. Mrs. Lasker described him as "very, very disorganized, very harassed." His staff found that he seldom gave his sustained attention to any matter for more than a few minutes. Though regarded as a man of great integrity and great compassion, his personal style made him difficult to work with.

Moreover, the chairmanship of a full committee was a major job for any senator, and Yarborough found a number of matters requiring his constant attention. Cancer, though important, simply did not have the urgency of the next day's agenda for the senator or the committee staff. Mary Lasker, usually accompanied by Mike Gorman, would frequently visit Yarborough and his staff to discuss the cancer initiative. Yarborough was always interested, but after her departure other legislative priorities would crowd cancer from his immediate agenda.

The biggest problem was how to set up a citizen's committee. Mary Lasker wanted Yarborough to simply appoint a group of consultants. The senator and his staff, however, insisted that it could not be done so easily. "Senator Hill did it," she would respond, recalling the "Bo" Jones Committee, "why can't you?" A good deal of time and a certain amount of acrimony was spent on this matter. Finally, the suggestion was made to the general counsel of the committee that a resolution be introduced to do the job. Mrs. Lasker was initially apprehensive about a resolution's passing the Senate. But this approach was agreed on, and Senator Yarborough, on March 25, 1970, introduced Senate Resolution 376 to establish a Panel of Consultants on the Conquest of Cancer.

Yarborough told the Senate that the panel would be asked to determine the adequacy and effectiveness of the present level of support for cancer research. Though the NCI budget had been adequate from the late 1950s to the early 1960s, he contended, it had been "barely sufficient to maintain the status quo" in the past three years.[8] The panel would also be asked to recommend the necessary action to achieve cures for the major forms of cancer by 1976—the bicentennial of the Republic.

Yarborough, prior to introducing S. Res. 376, secured forty-six of his Senate colleagues as co-sponsors to insure parliamentary success. The resolution was referred to his committee, reported out on April 15, cleared for Senate action on April 23, and adopted unanimously by a voice vote on April 27. The co-sponsors had increased to 51 senators by the time of the vote.[9]

While most of the activity was taking place in the Senate, the House was not idle. Representative John J. Rooney (D., N.Y.), a senior member of the Committee on Appropriations, an old ally of Representative Fogarty, was one of Luke Quinn's close friends. At Quinn's request, he introduced House Concurrent Resolution 526 on March 4, 1970. The resolution declared:

> That it is the sense of the Congress that the conquest of cancer is a national crusade to be accomplished by 1976 as an appropriate commemoration of the two hundredth anniversary of the independence of our country; and
>
> That the Congress appropriate the funds necessary for a massive program of cancer research and for the buildings and equipment with which to conduct the research and for whatever other purposes are necessary to the crusade so that the citizens of this land and of all other lands may be delivered from the greatest scourge in history.[10]

Slightly amended, this was unanimously adopted as H. Con. Res. 675 by the House on July 15 and the Senate on August 28.

The significance of this resolution lay not only in its unanimous adoption, but also in its sponsorship. Rooney was chairman of the appropriations subcommittee that approved budgets for the Departments of State, Commerce, and Justice, and had firmly established himself as one of the Capitol's most feared budget cutters. It was a measure of the importance the Congress attached to doing something for cancer that a fiscal conservative would sponsor a resolution calling upon Congress to appropriate the "funds necessary" for an expanded program of cancer research.[11]

In August 1970, when the Senate adopted the Rooney resolution, Ralph Yarborough rose on the Senate floor to say that he felt the resolution complemented the earlier action by the Senate in establishing the panel of consultants. The Senator found it appropriate that the House and Senate had combined "their common sentiment" and had "dedicated the entire Congress to the necessary legislation and appropriations" to seek the conquest of cancer.[12] The sense of the Congress was clear, even though it had received ample encouragement from Mary Lasker and her colleagues.

House Concurrent Resolution 675 reflected not simply the sense of Congress, however, but the sometimes uncritical tendency of the legislative branch to act as if words had no meaning or no consequence. The country could hardly be freed from the threat of cancer by 1976 through legislative edict. Yet no sober voice, in House or Senate, National Cancer Institute, American Cancer Society, or newly formed Panel of Consultants, was raised to say "preposterous" to the inflated language. All could benefit from Rooney's action but none need accept responsibility for it. The scientific community, it might be noted, was also silent.

THE PANEL OF CONSULTANTS

In the past, it had been enough for Mary Lasker to work with Representative Fogarty and Senator Hill to put more money into cancer research on the basis of existing legislation. But she and her allies had concluded that new legislation was needed if the conquest of cancer was to become a highly visible national goal. It was expected that the panel's recommendations would provide the basis for new legislation. This prospect fit closely with Yarborough's personal ambition to become the principal author of major new health legislation as successor to Lister Hill.

The decision to seek new legislation meant that the members of the Panel of Consultants had to be selected in a way that increased the likelihood of their recommendations' being incorporated into law. Senate Democrats would not be a problem since they controlled the body by a substantial majority, liberal Democrats were numerous among the members on the committee, and the chairman was personally committed to the task. The interest of Senator Javits insured the support of many liberal Senate Republicans. Conservative Republican senators, however, would be highly responsive to the Nixon administration on any legislation, so a way had to be found to neutralize possible opposition from that quarter.

It was decided to divide the panel equally between medical-scientific members and lay members, a characteristic of similar

Lasker efforts. The medical-scientific members would represent the full range of cancer research and treatment specialties, and the major cancer research institutes. Enough like-minded veterans of earlier campaigns would be included to guarantee the outcome, with the professional status of the doctors and scientists and their potential impact with the Congress taken into account in the selection process.

Among the medical-scientific people, Sidney Farber was chosen first and was the logical choice to be co-chairman. Yarborough called upon his fellow Texan, Lee Clark, and he too agreed to serve. The research director of Memorial Sloan-Kettering Cancer Institute, Dr. Frank Horsfall, was in poor health so Mary invited Dr. Joseph Burchenal, a vice-president, to be a member. Burchenal suggested Dr. James Holland of Roswell Park Memorial Institute as another member.[13] Solomon Garb was selected because of his book and his publicist activities at the "grass roots" level.

Dr. Jonathan E. Rhoads, a surgeon at the University of Pennsylvania, was selected because he was the new president of the American Cancer Society. Also chosen was Dr. William B. Hutchinson, a Seattle surgeon, brother of the late Fred Hutchinson of Cincinnati baseball fame, and a close friend of Senator Warren G. Magnuson (D., Wash.), Lister Hill's successor on the Senate Committee on Appropriations. Dr. Mathilde Krim, a research associate at Sloan-Kettering, was a basic scientist whose husband, Arthur Krim, had formerly been treasurer of the Democratic National Committee. Dr. Joshua Lederberg, Stanford geneticist and Nobel laureate in medicine, was the most prominent medical-scientific member chosen.[14] Dr. Paul B. Cornely, a Black physician, was president of the American Public Health Association. Dr. Henry Kaplan, a radiologist at Stanford University, was suggested by Burchenal. Another radiologist chosen was Dr. Wendell Scott of Washington University, St. Louis, who had long been active in the ACS. Dr. Harold P. Rusch, professor of cancer research at the McCardle Laboratory, University of Wisconsin, was a basic scientist very familiar with NCI.

The key question, though, was how to pick the lay members of the panel. Here it was also decided to pick some veterans of previous Lasker campaigns. Beyond this, however, the attention focused on the selection of a number of prominent Republican businessmen who would neutralize any opposition from the administration to the legislative implications of the panel's recommendations.

Among the lay members, Mary Lasker chose her old friends Emerson Foote and Mrs. Anna Rosenberg Hoffman. William McC. Blair, general director of the John F. Kennedy Center for the Performing Arts, who with his wife, Deda, was the frequent host for Mrs. Lasker on many of her Washington trips, was selected. Lewis Wasserman, president of the Music Corporation of America, a large contributor to the Democratic Party, and a friend of Mrs. Lasker's, was picked. Jubel R. Parten, a wealthy Houston philanthropist, was picked by Yarborough, who described him as one of the few wealthy men who had consistently supported progressive causes. Two representatives of organized labor were included, Emil Mazey, secretary-treasurer of the United Automobile Workers, and I. W. Abel, president of the United Steel Workers.[15]

No Lasker initiative is complete without several individuals who understand the world of modern advertising and the press. In addition to Foote, Mrs. Lasker persuaded Mary Wells Lawrence, chairman of the board of Wells, Rich, and Greene, Inc., to join the panel.[16] Another panelist was Michael J. O'Neill, managing editor of the *New York Daily News*.[17]

Mrs. Lasker turned to Laurance Rockefeller, chairman of the board of Rockefeller Brothers, Inc., for assistance in selecting the Republicans for the panel. Laurance had long been active in behalf of cancer research as chairman of the board of Memorial Sloan-Kettering Cancer Center and a member of the ACS board. His conviction that more needed to be done for cancer research made him quite willing to serve on the panel. But he also worked with Mrs. Lasker to select several other key panel members. One of these was G. Keith Funston, former president of the New York Stock Exchange, chairman of the board of the Olin Mathieson Chemical Corporation, a director of a number of other major corporations, and also on the board of directors of the American Cancer Society.

Mrs. Lasker and Mr. Rockefeller concluded that Elmer Bobst, honorary chairman of the Warner-Lambert Pharmaceutical Co., should be on the panel. Bobst had been deeply involved in the takeover of the American Cancer Society, was a member of the ACS board, had played an active role in the Society's antismoking campaign, and had known Mary Lasker during this entire time. In addition, Bobst was one of President Nixon's closest personal friends.[18]

Rockefeller was instrumental in choosing Benno C. Schmidt for the panel. Around 1960, Schmidt had joined the board of trustees of Memorial Sloan-Kettering Cancer Center at the request of Rockefeller, but was relatively inactive until 1965.[19] He then became actively involved with Memorial Hospital for Cancer and Allied Diseases, first as a member of the board of trustees and chairman of the executive committee, and by 1970 as chairman of the board. He also became a member of the board of the Sloan-Kettering Cancer Research Institute.[20] In these several capacities, Schmidt had impressed many people and Rockefeller had no hesitation in suggesting him to Mrs. Lasker. She passed his name on to Senator Yarborough.

"Out of the blue," Schmidt recalled, "I received a letter from Senator Yarborough asking me to serve as a member of a panel of citizens—scientists, doctors, and laymen—to assess where we stood and what were the areas of promise in cancer research." He checked with Yarborough to determine if the panel was going to be a serious group and found them, in his judgment, to be extremely well selected. So he accepted the invitation.

One important task was to select a chairman for the panel. Rockefeller suggested to Mrs. Lasker that Schmidt would be an excellent choice. She checked with Dr. Burchenal, who warmly endorsed the idea, and then conveyed the suggestion to Senator Yarborough.

Schmidt, a lawyer, had been a partner since 1946 of J. H. Whitney & Co., a New York investment firm, and managing partner since 1959. Much to the surprise of his staff, Ralph Yarborough knew this New York lawyer-businessman. Schmidt was born in Abilene, Texas, in 1909, had attended the University of Texas, and had received both the A.B. and LL.B. degrees in

1936. Yarborough, who had a law practice in Austin at the time, had taught in the Texas Law School and Schmidt had been in his class. Schmidt recalls having done well in this course, as he had done in law school generally. For three years, from 1936 through 1939, Schmidt also taught on the faculty of the law school. Yarborough, looking back, remembered Schmidt "as a brilliant professor."

But Schmidt was a Republican and his ties to Texas were to Yarborough's political opponents. Schmidt was a friend of John Connally, former governor of Texas, whom he had taught in law school. The conservative Connally wing of the Democratic party in Texas constituted the sworn enemy of the liberal Yarborough wing. Lloyd Bentsen, who defeated Yarborough for the Senate Democratic nomination in 1970, was a Connally Democrat. Schmidt was also close to George Bush, who, as the Republican candidate, ran unsuccessfully against Bentsen for the Senate in 1970. Had Yarborough won the primary that year, Bush would have been his opponent. Schmidt, though a Texan, had no ties to the Texas liberal Democrats and had numerous ties to both Republicans and conservative Democrats in that state.

There were a number of discussions between Yarborough and his staff about the possibility of Schmidt as chairman; some, a staff aide recalled, were rather violent. The staff was unhappy about Schmidt, but Yarborough was interested in cancer and prepared to put aside political considerations to demonstrate his interest. Besides, the senator regarded Schmidt as "a driver. I knew if he agreed, he would do it. I was agreeable to him doing the job."

Soon after he had agreed to be a panel member, therefore, Schmidt was asked by Yarborough to serve as chairman. Laurance Rockefeller, Mary Lasker, and others urged him to accept. He was reluctant. Schmidt said, "I had learned long ago that if you take on the job of chairman, you take on the job." He did not want it unless people were serious about doing something. "I didn't want to be occupied in the last half of 1970," he said, "with a job that came to nothing." Schmidt talked to a number of the senators on the Labor and Public Welfare Committee, and several others as well. If they were serious about getting a report, passing legislation, and providing appropriations, Schmidt said, "then it seemed a challenge which could not be refused." He did discover a serious desire in Congress to do something about cancer and, as a result, agreed to become chairman.

THE MANAGING PARTNER

J. H. Whitney & Co. is a private firm engaged in venture capital financing. It was established in 1946 by John Hay Whitney, one of America's wealthiest men, with an initial capital of $10 million. T. A. Wise estimated that the original capital had multiplied ten or twelve times by 1964.[21] Conservatively, the value of the firm would have increased several times again by 1970.

Benno Schmidt joined J. H. Whitney & Co. in 1946 as one of six partners, and has been managing partner since 1959. In this capacity, he has overseen the investments of a firm worth between one-quarter and one-half billion dollars. He has been director of a number of corporations in which J. H. Whitney & Co. has invested. These have included the Esperance Land and Development Co., an agricultural project of 1,400,000 acres in western Australia being jointly developed by J. H. Whitney & Co., Chase International, and American Factors.[22] Schmidt was no stranger to large, bold, new undertakings.

Schmidt was not well known in Washington, D. C. when he became panel chairman, but the city was familiar to him. He had worked for the general counsel of the War Production Board in 1941-1942. Then, after three years as a colonel in the U.S. Army, he returned to Washington as general counsel to the foreign liquidation commission of the economic division of the State Department. There he worked on lend-lease settlement, the disposal of foreign surplus property, and related matters. In this position, Schmidt recalled, "it seemed to me that I spent the bulk of my time on the Hill explaining to congressional committees what we were trying to do and why we were trying to do it."[23]

Schmidt has been active in New York City civic endeavors. He has been chairman of the board of the Bedford-Stuyvesant Development and Services Corporation, established by the late Senator Robert F. Kennedy to assist in the rehabilitation of one of the major poverty areas of Brooklyn.[24] He was appointed chairman of the Fund for the City of New York, an organization established in 1968 by the Ford Foundation to provide assistance to the City of New York in improving the effectiveness of municipal services.[25] In addition, in 1968 and 1969 he chaired Mayor John Lindsay's Welfare Island Planning and Development Committee, which proposed general guidelines for the development of the 147-acre island in New York's East River.[26]

The appointment of Schmidt as chairman of the Panel of Consultants on the Conquest of Cancer, therefore, brought to the cancer crusade a man with extensive experience in both private and civic activities. Here was an individual of recognized leadership, capable of dealing with large, somewhat unwieldy committees of prominent, accomplished citizens. Solomon Garb would later say:

> There were a number of highly prominent individuals on the panel, and I wondered who would be able to lead such a group. Schmidt was a most remarkable man. He kept things going. He kept focused on the issues. There were arguments and differences of opinion expressed within the panel, but these were ironed out before the meetings closed.[27]

In short, the chairman came to the job with the leadership skills necessary to produce a hard-hitting report with a good chance of being accepted.

The staff for the cancer effort had been selected before Schmidt became chairman. Robert F. Sweek, program manager for the Atomic Energy Commission's (AEC) liquid-metal fast-breeder-reactor program, heard Garb on a radio broadcast in mid-December 1969 while driving home from a graduate seminar at American University. Garb was talking about the need for a national cancer program comparable to the space program. Since Sweek had just come from a discussion of the transfer of modern management techniques from large-scale technological programs to domestic programs, he wrote Garb suggesting that such skills might be applied to cancer control. An increasingly warm exchange of correspondence led Sweek to visit Garb in Denver on April 3, 1970, and the two men hit it off very well.[28]

Though Sweek was looking for a Ph.D. thesis topic, other more important possibilities came to Garb's mind. He got in touch with Luke Quinn who wrote Sweek on April 8 and invited him to his office. That meeting took place on April 21, and Quinn and Sweek also got on well. As a result, Sweek visited Mary Lasker in New York on May 2 to talk about becoming staff director for the Senate's committee's effort. He detailed his interest in the position in a letter the following day. Mrs. Lasker, in turn, passed Sweek's name on to Yarborough, who was receptive to the suggestion. On June 1, it was formally agreed that Sweek would be director of the special staff of the Senate Committee on Labor and Public Welfare[29] to work with the Panel of Consultants. It was also agreed that the staff should be housed at the National Cancer Institute. Carl Fixman was chosen as Sweek's deputy.

The autonomy of the special staff created some problems with some panel members, as well as with Yarborough's regular staff. Dr. Sidney Farber, who had been involved with the panel from the beginning as co-chairman, wanted to get the medical members together and work out a common position before meeting with the nonmedical members. Sweek sought to head off such a move by a series of visits to the medical members of the panel. He visited Dr. Burchenal at Sloan-Kettering, Dr. Holland at Roswell Park, Dr. Rusch at the University of Wisconsin's McCardle Laboratory, Dr. Farber at Children's Hospital, Boston, Dr. Rhoads at the University of Pennsylvania Hospital, and Dr. Clark at the M. D. Anderson Hospital and Tumor Institute. The staff director was interested in learning about cancer research, and in exploring with them the need for a "national plan" for cancer.

The independence of the staff that irritated Farber vanished when Schmidt became panel chairman. Schmidt met with Sweek on June 19, a few days before the panel's first meeting. He indicated his preference for a brief, nontechnical report, with technical materials as appendices, and illustrated this with the Welfare Island report he had just submitted to Mayor Lindsay. Sweek came away from this initial meeting much impressed with Benno Schmidt, and quite aware of his take-charge nature.

Schmidt knew many of the panel members personally.[30] Laurance Rockefeller, a long-time friend, had been instrumental in

getting him to become a trustee of Memorial Hospital. He also
knew Keith Funston and most of the other nonmedical members.
Of the medical members, he had come to know Dr. Burchenal and
Dr. Krim through his work as a Memorial Sloan-Kettering trustee. Schmidt had met Dr. Lee Clark in 1966 when he visited M.
D. Anderson to discuss radiotherapy at Sloan-Kettering. Though
they were not close during this time, Schmidt and Clark did see
each other occasionally from then onward, and their common link
to the University of Texas reinforced the friendship. These personal relationships helped the chairman immensely.

The first meeting of the panel was held on June 29 in Washington.[31] Senator Yarborough opened the meeting and suggested that
the conquest of cancer by 1976 would be a fitting tribute for the
country's bicentennial. He asked for a completed report by the
end of October. Schmidt then outlined his plans for producing the
report and, over luncheon, he and Farber discussed subpanel as-
signments with the members present.[32] It was agreed that the subpanels would begin immediately to generate working papers that
would be distributed to the other panelists. Three of the subpanels
held their initial meeting that afternoon.

Initially, the work of the panel was done through the subpanels.
Since the questions were mainly technical, most of the work was
done by the medical members who drafted working papers for
their groups and the other panelists. The subpanels reported on
their activities at the second panel meeting on July 27 and at the
third meeting on August 24 and 25.[33]

Sidney Farber suffered a mild coronary attack on his way to the
Washington meeting of the panel on August 24. He was forced to
withdraw from active participation in the panel's work, though he
rejoined it by the time it made its presentation to the Senate committee in December. Schmidt, with the panel's concurrence, appointed Dr. Lee Clark to act as co-chairman of the full panel.[34]

Schmidt had also concluded by this point that a full-time coordinator was needed to pull together the technical analyses being
prepared by the subpanels. So, exercising his influence as a trustee of Memorial Sloan-Kettering, he announced on the morning of
August 25 that Dr. Joseph Burchenal would devote all of his time
to the activities of the panel. By late August, then, a number of

technical papers were in various stages of preparation. The chairman had also appointed a full-time coordinator of scientific activities who was no more than a fifteen-minute taxi ride from his Rockefeller Center office.

The nontechnical work of the panel was as important to Benno Schmidt as the technical. Early in the panel's life, he stressed that it was not just the Yarborough panel but a panel of the Senate, not only a panel of the Senate but of the Congress, and not simply a panel of the Congress but of the whole government. He worked to make this a reality, drawing upon his own resources, and those of others.

In particular, he met in mid-August with Sweek, Luke Quinn, and Bob Harris, staff director for Yarborough's full committee. Schmidt wanted to talk about his plans for discussing the work of the panel with senior people in both the legislative and executive branches of the federal government who might be able to influence the reception of the report. He felt strongly that a personal effort by him was needed to insure that the report would be supported by both parties in both houses of Congress and by the administration. It was essential to him that everyone realize the cancer effort was apolitical and that no one was trying to secure any partisan advantage from it. Quinn provided background on some of the major health legislation of the past fifteen years. Harris drew on his experience as staff director to the Senate committee. It was agreed that this group would meet again soon.

Schmidt did meet with Burchenal, Quinn, Sweek, and Fixman on September 16, but for another purpose. He indicated, however, that he and Dr. Burchenal would be having lunch the next day with several members of the House. This luncheon had been set up by the then House Minority Leader, Gerald Ford, at Schmidt's request.[35] Those present in addition to Ford included Rep. John McCormack (D., Mass.), the Speaker of the House; Rep. Carl Albert (D., Okla.), the Majority Leader; Rep. George Mahon (D., Tex.), chairman of the Appropriations Committee; Rep. Robert Michel (R., Ill.), the ranking minority member of the appropriations subcommittee on HEW; Rep. John J. Rooney of the appropriations committee; and Rep. William Springer (R., Ill.), ranking minority member on the House Interstate and

Foreign Commerce Committee. Schmidt found them deeply interested in the work of the panel and willing to support an increased emphasis on cancer.

Schmidt also visited George Shultz, then director of the Office of Management and Budget and one of President Nixon's key advisers on domestic policy, to explain the work of the panel. Moreover, though little publicity was given to the panel, the *Wall Street Journal*, a newspaper with a wide readership in a Republican administration, carried stories on it in late August and again in late November.[36] The existence of the panel and the nature of its work, therefore, was being made known to several important audiences, and the panel chairman was most influential in this.

Schmidt was absolutely clear in his own mind that the crucial issues were organizational and not technical. He was also quite committed to playing the central role in the way those issues were treated in the panel's report. The issues came into focus in July and August, but it was September and October during which the critical choices were made. In those months, Schmidt relied heavily on a number of small working groups of panel members, as well as on the full panel, in order to forge consensus and establish agreement on the major recommendations.

"NO CONFIDENCE" IN
THE NATIONAL INSTITUTES OF HEALTH

The most controversial recommendation in the report of the Panel of Consultants was that there be established an independent agency[37]—the National Cancer Authority, which would absorb all the functions of the National Cancer Institute and be completely severed from the National Institutes of Health.[38] The NIH had, since the end of World War II, acquired an international reputation as the world's leading biomedical research organization. Therefore, the recommendation for a separate cancer agency had to be seen as a vote of "No Confidence" in the NIH. How did the panel arrive at this judgment?

The ambience in which the panel worked was highly favorable to the idea of a separate agency. Mary Lasker, disenchanted with the NIH leadership and frustrated at the outcome of the Heart Disease, Cancer, and Stroke initiative, had concluded by 1969 that the only way to get a breakthrough in health was to have a "space program" approach and to concentrate resources on a single problem—cancer. Solomon Garb, in his book, had deplored the fact that NCI was buried under an extensive "bureaucratic over-lay," and had singled out the space agency "as an example of a particularly successful organization" that might be emulated for cancer.[39] The long concern of Dr. Farber and Dr. Clark for the autonomy of cancer research predisposed them toward a different pattern of organization. Senator Yarborough, moreover, liked the idea of an independent agency. "I remembered World War II," he said later, "how we got nuclear fission ahead of the central pow-ers. I also recalled how we got ahead of the Russians by NASA in the space program."[40] And, when he introduced S. Res. 376 on March 25, 1970, he said that the panel of consultants "should be free to make recommendations in the fields of research, training, financing and administration, *with particular attention directed toward the creation of a new administrative agency*"[41] (emphasis added).

There were intimations of a separate agency in the discussion of subpanel 4 at its first meeting.[42] The National Cancer Institute, it was agreed, was the logical organization to administer the cancer program, but it might have to be strengthened and upgraded. In particular, the group felt that the constraints of the hierarchy of HEW, PHS, and NIH had to be removed. While existing legisla-tive authority permitted administrative streamlining, the judgment of the subpanel was that not much would happen without some "clarifying" action by Congress. But while there was ample evi-dence of support for a separate agency, explicit panel considera-tion of this issue was approached cautiously.

The staff prepared the groundwork for this recommendation. Sweek had worked for Admiral Hyman Rickover in the Navy, and then for the Atomic Energy Commission. Fixman had worked for the Maritime Commission, had been a vice-president of New York Shipbuilding during the construction of the nuclear mer-chant ship, the Savannah, and had also worked for the Atomic Energy Commission. Their experience strongly reinforced the view that an autonomous agency for cancer was necessary. Sweek wrote Schmidt in late July that the staff had concluded that NIH was too cumbersome and inefficient to manage mission-oriented research and development.[43]

The staff also sent panel members a good deal of material critical of the organizational constraints placed upon the Cancer Institute. In mid-August, Sweek sent out a memorandum on the authorities and organization of the Public Health Service, the National Institutes of Health, and the National Cancer Institute, which noted that the stated mission of these organizations did not include a mandate to eliminate cancer at the earliest possible time—a serious omission. In early September, the panelists received an analysis from Sweek on the growth in staff and authority of the office of the assistant secretary for health of HEW. Special attention was given to the problems this office created for the NCI and the negative implications it suggested for an expanded cancer effort. The following day, a copy of a foreword by James Webb, former administrator of the space agency, to "An Administrative History of N.A.S.A., 1958-1963," was sent to panel members, which further reinforced the concern for management and the advantages of an independent agency. Then, on September 15, Sweek mailed out excerpts from the Wooldridge Committee's report, especially those portions critical of NIH's management ability. The cover memorandum alleged that no action had been taken to correct the deficiencies identified in 1965, and declared that this should not be permitted to happen again.

Schmidt himself moved rather discreetly on the matter of a separate agency. In the New York meeting in mid-August with Harris, Quinn, and Sweek, he questioned whether NCI could handle an expanded cancer effort, whether in NIH or not, and, if not, what organization might be established. At the third meeting of the full panel, on August 24, Schmidt raised for the panel's consideration the question of what organizational changes were necessary, or whether the job might require a new organization. That same day the panel heard the NCI director, Carl Baker, give his reasons why an all-out effort in cancer could not be successful in the existing organizational environment. The organizational status of an expanded cancer program had not been directly confronted, but the issue was being put with increasing frequency and focus.[44]

By mid-September Schmidt had made up his own mind. In a meeting in New York, involving Quinn, Sweek, Fixman, and Dr. Burchenal, the panel's recommendations were discussed. It was agreed that legislation would be recommended to establish an independent agency for cancer. The proposed agency would report to the president, have the mission of conquering cancer as a major disease at the earliest possible time, and absorb and build upon the NCI.

The proposal to create an independent cancer agency, separate from NIH, was formally on the panel's agenda at its New York meeting on September 18. The discussion was lengthy and very intense on the pros and cons of taking the National Cancer Institute out of the National Institutes of Health. The basic-research-oriented scientists did not favor dismembering the NIH, an organization they knew well and which had served them well over the years. Mrs. Anna Rosenberg Hoffman, though undecided on the proposition, took a more pragmatic view and questioned whether the idea would be politically acceptable. Support for the proposition was canvassed—some remember this as an informal vote—and only five people favored taking the effort out of NIH.[45] Schmidt did not commit himself.

Quinn, who had spoken to the panel-meeting on the politics of medical research, had told the panel earlier that they would get nowhere if they were not prepared to take on the NIH. After the "vote," he told the staff not to worry. Some of those "votes" would change, he predicted. Quinn was right.

The process of persuasion began in earnest. The scientists constituted the key group, and Benno Schmidt turned his substantial persuasive skills on them. Some of the scientists remained in New York on September 19 to work on the technical portion of the report. This gave opportunity for a more relaxed probing of the issue and for serious back-and-forth talk between the chairman and the scientists. Four of the scientists—Dr. Burchenal, Kaplan, Krim, and Holland—met in New York again on October 5 to continue work on the technical portion of the report. Schmidt joined them the next day and discussed a draft he had just completed of the "summary and recommendations" portion of the report. The draft recommended the creation of the "National Cancer Author- ity," the name selected for the proposed new agency. The scien- tists, reluctantly but surely, began to come around to Schmidt's point of view.

Mary Lasker went to work to insure that her close friends on the panel understood the merits of a separate agency. Sidney Farber, who returned to the work of the panel at this time, was willing to lend his support. Emerson Foote, Anna Rosenberg Hoffman, and William Blair all adopted the idea of the separate agency.

The panel had decided at its September meeting that voting was not an appropriate way to proceed, and had agreed to reach unanimity on all recommendations. When the full panel met again in Washington on October 7, the die had been cast. Schmidt's draft report was accepted in all its essentials, though it was ac- knowledged that more work remained to be done on the details. Mrs. Lasker hosted a party that evening at the Blair's home and the panel, for the moment, relaxed from its labors.

Closely related to the issue of a separate agency was the ques- tion of the philosophy of research management that was appropri- ate to an expanded national cancer program. On this issue the panel was more deeply divided.

THE CATEGORICAL IMPERATIVE

CANCER PANEL ACTION REVIVES DEBATE ON TARGETED V. CRE- ATIVE RESEARCH; NOT ACADEMIC THIS TIME; DEBATE TO BE "JUDGED" BY CAPITOL HILL MONEY MEN[46]

This lead was how *The Blue Sheet* sought to capture the underly- ing conflict inherent in the panel's consideration of research man- agement. The panel of consultants worked in a highly charged atmosphere. A crash program mentality permeated everything. Yarborough repeatedly spoke of finding a cure for cancer by 1976. The Rooney resolution reiterated the same theme. The pressures toward an NASA-type agency also fed this crash program thinking and conjured up visions of finding a cure for cancer by the massive application of money and manpower.

A number of events during the life of the panel reinforced the categorical approach to research management, to the distress of the scientists on the panel. For instance, subpanel 4, with Lee Clark and Solomon Garb, concluded at its first meeting that the primary mechanism for implementing new work in an expanded cancer program should be the contract. This instrument was flexible, it permitted technical effort to be focused in accordance with approved plans, and it increased the accountability of the research community to the funding source.

Elmer Bobst, at the second meeting of the full panel, warned against continued funding of the kind of research supported in the past. He emphatically urged that the most promising scientific leads be identified, and that resources be concentrated on exploiting these leads. An objective-oriented research program, with continuous evaluation of results, was necessary, Bobst believed. If this was not done, he said, the results would not be commensurate with the dollars spent.

Carl Baker was promoted from acting director to director of NCI on July 13, 1970, just as the panel got underway. While he had some contact with the panel—a presentation at the third meeting, several meetings with Benno Schmidt, and conversations with other panelists—he had extensive contact with Sweek and Fixman, since their offices were adjacent to his.[47] Baker pulled a number of NCI planning documents together for Sweek, and Sweek responded by giving him a copy of the 10-volume plan for the AEC's fast-breeder liquid-metal reactor. Baker studied the plan carefully, and he and his staff had ample opportunity to explore with Sweek and Fixman the full range of questions about research management.

Sweek and Fixman brought with them engineer's preferences for a centralized, project-management approach to the cancer program. They saw themselves as experts at the management of large-scale, technological programs, including programs having a high degree of scientific uncertainty, and they found NCI's research-planning and management strategy underdeveloped at best.

The Blue Sheet, one of the few news sources that covered the panel during the summer of 1970, picked up the theme of the targeted, directed research versus basic research very quickly and linked it with the preferences at the staff. As early as July 15, it ran a story about a "targeted, directed, therapy-oriented cancer research program—modeled after the applied and developmental systems approach used by AEC and NASA" that was "being pushed by an articulate group" within the panel of consultants. A month later, the weekly was discussing how industrial participation fit with what the panel had come to see as its main task— "organizing a coordinated, NASA-type approach to tapping all the major resources available in the nation to attack cancer."[48]

In the September 16 issue, the lead to a two-page story highlighted an underlying issue within the panel:

BATTLELINE BETWEEN "CREATIVE RESEARCH" AND "STRUCTURED RESEARCH" BEING ESTABLISHED INSIDE SESSIONS OF SEN. YARBOROUGH'S CANCER COMMISSION

The story alleged that "development of a structured research program, utilizing a systems approach, to speed exploitation or discard of *any* new knowledge in the cancer field—be it a significant advance or a mere tidbit—is believed to be the objective of the 'new thinkers' and 'shake-up-the-establishment' activists on Sen. Yarborough's 'Committee of Consultants for the Conquest of Cancer.' " The article pointed out that "creative research" had failed to produce a "cure" for cancer and this had left it vulnerable to demands for a new approach. "The public," it was emphasized, "wants results for its money!" The primary implication, if the structured research approach won out, was that money for government contracts would be made available at the expense of funds now used for research grants. And all this was going on behind closed doors![49]

In this *Blue Sheet* story, enough references were made to the "systems engineer," and the "full-time staff . . . strongly oriented toward planned targeted research," to indicate that the staff had been talking to a correspondent. The story was sufficiently embarrassing that Sweek complained to Sen. Yarborough that it bordered on irresponsible journalism, and assured the senator that the final report would put everything in perspective.

Before the final report came out, though, *Science* magazine picked up the story. In its October 16 issue it reported that the panel was likely to recommend that "planning and management techniques" developed by the AEC and NASA be applied to cancer. The article focused on Sweek, "an aggressive systems management expert," whom it alleged would have a major influence on the panel's recommendations:

> Sweek sees medical research as lacking the organization and discipline necessary to achieve spectacular results in a reasonable time. He wants the commission to recommend that a national plan to combat cancer be drawn up by a committee of experts over the next year. Without such a plan, he claims, it is difficult to assess progress regularly. Sweek also wants more accountability from the scientists who receive government money. He ridicules projects that can only be justified in terms of the researcher's interests. Most of the funding under the master plan would be done through contracts, but he insists that a large grants program remain intact.[50]

The staff director was acquiring a public profile that could only alarm the scientific community.

This story brought things to a head within the panel one week later. At a meeting in Schmidt's New York office, the four principal scientists—Burchenal, Holland, Kaplan, and Krim—indicated to Sweek and Fixman their great displeasure with the *Science* article, especially with the portions about Sweek. It was decided that it would be unwise to write a letter of clarification, since the report would be out soon and would speak for itself. But that fact hardly assuaged those present. Sweek offered Schmidt his rewrite of Schmidt's earlier draft of the summary and recommendations, but the chairman brusquely said he would work from his own draft. The line-by-line review which then followed was conciliatory toward the scientists who were present. Words like "centralized control" were deleted; "administration" was used in place of "management"; and the panel's emphasis on the importance of freedom in basic research was explicitly spelled out. The scientists, who had been on the defensive for weeks, had recovered some in the home stretch.

Schmidt had remained aloof from the planning and management recommendation until the separate agency question was resolved. In fact, the staff had been frustrated for many months at their inability to get him to focus on the issue. Only after an early October meeting with Carl Baker, at which the NCI director stressed the need for a recommendation favoring a national plan, did Schmidt address this matter. By mid-October, he had settled on draft language, and this had been the basis for the October 23 discussion with the scientists. The full panel met on October 30 for its last meeting prior to presenting its report and agreed upon the final draft of the summary and recommendations. Or so it seemed.

Mary Lasker had not been involved in the drafting of the report, partly because Schmidt did not feel compelled to clear everything with her. With Farber inactive because of his illness, Emerson Foote became her principal representative on the panel. But the report began to take final shape only after the October 7 meeting, and Foote was not a member of the several small working groups on which Schmidt relied during this time. Luke Quinn was at an October 13 meeting with Schmidt in New York, but though Schmidt had once listened to Quinn's appraisal of the Congress, the chairman was now operating on his own.

In mid-October, Sweek heard from Quinn that Mrs. Lasker wanted the "National Advisory Cancer Board" to have the final approval for all contracts.[51] Sweek also heard from Foote that Mrs. Lasker wanted a copy of the report within several days no matter what shape it was in. Schmidt had not released his draft to anyone at that point, but he, along with Laurance Rockefeller, met with Mrs. Lasker prior to the October 30 meeting of the panel.

Not until after the October 30 meeting did it become clear that Mrs. Lasker did not agree with the recommendations of the report, but that she had not raised her objections when she met with Schmidt and Rockefeller. Instead, she had instructed Luke Quinn to prepare alternate language for the prospective legislation that would incorporate the report's recommendations. The sticking point was her insistence that the proposed board have final approval over all grants and contracts. The problem was brought to Benno Schmidt's attention and he immediately met with Mrs.

Lasker to work out an accommodation. It was agreed that the board would have prior approval for a coming year's program-plan and budget, but that the administrator of the "National Cancer Authority" would have full authority for administering the program, including contract approval. With this hurdle cleared, the finishing touches were put on the second recommendation.

The budgetary picture was clear to the Panel of Consultants when it began its work. Equally clear was the expectation of the members that the final report would request more money for cancer research. Mary Lasker, directly and through Luke Quinn, made it clear that she wanted a recommendation for a substantial and rapid increase in funds. On this issue Mrs. Lasker clearly prevailed. The third recommendation did not require extensive discussion. It was, as budget questions often are, not a matter of analysis but of choosing the appropriate set of round numbers. The key procedural problem was to escape the budget trimming and balancing tendencies of both NIH and HEW, and the recommendation for a separate agency accomplished that.

THE PANEL'S REPORT TO THE SENATE

It was Friday morning, December 4, 1970. Senator Ralph W. Yarborough made his way from his office in the Old Senate Office Building to the hearing room in the New Senate Office Building. He had called a meeting of the Committee on Labor and Public Welfare to receive the report of the Panel of Consultants on the Conquest of Cancer.

Yarborough had originally asked the panel to submit its report by October 31, hoping its report might aid him in a November reelection bid. Now, however, Yarborough was a lame-duck senator, having lost the primary election in Texas that spring to Lloyd Bentsen of the John Connally wing of the party. In the waning days of the 91st Congress and his own Senate career, he would perform several eleventh-hour acts on behalf of cancer research. The Panel of Consultants, unlike the senator, was hardly a lame-duck group. Though it had completed its report, its chief instigator, its chairman, and its members fully intended to see their recommendations through to new cancer legislation in 1971.

Senator Yarborough introduced the panel members and praised them for their work. Among those present were Benno Schmidt, the panel chairman, Dr. Sidney Farber, senior spokesman for cancer research in the U.S., and Dr. R. Lee Clark, president of the M. D. Anderson Hospital and Tumor Institute. Also present was Mrs. Mary Lasker, the driving force behind the panel.

The country, Yarborough noted, had not yet displayed the will to make the fight against cancer. But the panel's report led him to be optimistic: "I believe we now have that will—the people have the will and the government must lead and direct its people in this fight, not retreat."[52] His aim had been "to move this government toward the cure of cancer—to make finding its causes and cures a great national goal," Yarborough said. He turned to Senator Jacob Javits (R., N. Y.), the ranking minority party member of the Committee and his "partner" in this effort, and urged Javits to continue the work on behalf of cancer research when he was gone. Javits immediately pledged himself to follow through on the panel's recommendations and expressed the hope that by adopting them "we can do for cancer what the Salk vaccine did for polio."[53]

Senator Edward M. Kennedy (D., Mass.), one of the Committee members present that morning, suggested that the presentation of the report should be seen in relation to the Senate's vote of 52 to 41 on the previous day to discontinue funding for the U.S. development of a supersonic transport.[54] The SST, which had been strongly supported by the Nixon administration, had been defeated for a combination of economic and ecological reasons. A major theme in the congressional debate had been the need to forego expensive technological projects of dubious value and to reallocate funds to pressing domestic needs. Cancer research was an obvious alternative to the SST in the reordering of national priorities.

Benno Schmidt presented the concisely stated findings and recommendations of the report.[55] Cancer was the number one health concern of the American people. Although it was the second-leading cause of death in the U.S., cancer was the most-feared disease. But the funds spent on cancer research were grossly inadequate to the task of finding cures for cancer. Moreover, the incidence of cancer was increasing, owing in part to the growing proportion of the aged in the population, and also to the effects of smoking upon lung cancer. Though the nature of cancer was not fully understood, the cure rate was gradually improving. Equally important, fundamental advances in scientific knowledge in the past decade had opened up far more promising areas for intensive investigation than had ever before existed. As a result, the report stated, "a national program for the conquest of cancer is now essential if we are to exploit effectively the great opportunities which are presented as a result of recent advances in our knowledge." The report indicated that such a program would require three essential ingredients that were not present at that time: "effective administration with clearly defined authority and responsibility"; "a comprehensive national plan for a coherent and systematic attack on the vastly complex problems of cancer"; and "the necessary financial resources."[56]

The panel's assessment led to three far-reaching recommendations. First, it recommended the establishment of a new government agency—the National Cancer Authority—whose mission would be defined by statute as "the conquest of cancer at the earliest possible time."[57] The Authority would absorb all functions of the existing National Cancer Institute and would be completely independent of the National Institutes of Health.

What persuasive arguments led the panel to its radical proposal? *Efficiency* was one reason given by the panel. The dry-as-dust language of the report barely reflected the panel's impatience and frustration with the bureaucratic red tape in which NCI was enmeshed. But the message was plain. "An effective major assault on cancer" the report argued, "requires an administrative setup which can efficiently administer the coherent program that is required in this formidable and complex scientific field." It would be difficult to realize this setup within the Federal government and, at a minimum, would require "a simplification of organizational arrangements and a drastic reduction in the number of people involved in administrative decisions." This "straightline organizational efficiency" did not exist within NCI, NIH, or HEW, and there was "real doubt" whether it could be achieved within those organizations.[58]

Efficiency, however, was less important than *priority* for cancer research, which the panel's report alleged could not be realized within HEW: "Apart from the question of whether it can be done, there is also the question of whether it would be wise to require the Secretary of Health, Education, and Welfare to attempt to give cancer the priority necessary to carry out the congressional mandate in a department charged with the multiple health and other responsibilities of that Department."[59] Cancer was of such overriding importance, the panel was saying, that it deserved to be considered independently of the nation's other health needs, and certainly independently of the many nonhealth priorities of HEW.

If concern for priority led to a critique of existing arrangements, it also led to a justification for the new organization. The report suggested that

> in the past when the Federal Government has desired to give top priority to a major scientific project of the magnitude of that involved in the conquest of cancer, it has on occasion, with considerable success, given the responsibility for the project to an independent agency. Such an agency provides a degree of independence in management, planning, budget presentation, and assessment of progress which is difficult if not impossible to achieve in a large government department.[60]

The analogy to the Manhattan Engineer District, which had developed the atomic bomb, and to NASA was clear to all.

But the real barrier to priority for cancer research, the panel felt, was not so much the secretary of HEW as it was the leadership of NIH. James Shannon had been the undisputed leader of the NIH during his thirteen-year tenure as director. Though he had a cadre of aides—an associate director for extramural research and training, a director of laboratories and clinics, and an assistant director for collaborative studies—by preference, capability, and long experience he had dealt first-hand with the institute directors on both scientific and administrative matters. Robert Marston established a more collegial leadership—a troika—at the top of NIH. In December 1968, five months after becoming director, he appointed Robert W. Berliner as his deputy director for science and John F. Sherman as deputy director for administration.[61]

Berliner had been appointed director of laboratories and clinics of NIH in June 1968, just before Shannon's retirement. The new position of deputy director for science substantially upgraded those responsibilities. He now became responsible for extramural research grants and contracts as well as intramural research. Directors of the individual institutes no longer took research matters straight to the NIH director, but found themselves dealing with Berliner or with Marston and Berliner.

Berliner's new responsibilities were substantial. The position of director of NIH laboratories and clinics was changed to that of associate director for direct research but never filled. Berliner handled these responsibilities in his new and expanded responsibility for both intramural and extramural research. The NIH associate director for extramural research and training reported directly to the deputy director for science, as did the assistant director for collaborative research.

This meant for NCI that its research programs, whether extramural, intramural, or collaborative, were all reviewed by Berliner. The vexing fact of life to a number of members of the Panel of Consultants was that NIH director Marston and Berliner, his deputy for science, would make decisions about NIH research priorities and the allocation of resources in light of what they thought was best for the full range of biomedical science, not simply what was best for NCI and cancer research.[62] Nor was the situation helped for Mary Lasker and her friends by Berliner's reputation for being a "scientist's scientist," and not at all favorable to targeted research.

While the panel's report acknowledged that "from many standpoints it can be argued that any cancer program should be in the Department of Health, Education, and Welfare and indeed that it should be in the National Institutes of Health,"[63] its basic assertion was that priorities of both organizations would never give cancer research adequate attention. The panel concluded that what they saw as NIH domination of cancer research was no longer acceptable. Remarkably enough, the panel reached its sweeping conclusion *without once* consulting the leadership of either the NIH or HEW!

The second recommendation was that "a comprehensive national plan for the conquest of cancer" be developed. Such a plan would continue the research currently being supported by the NCI, provide for initial program expansion to be built upon existing facilities and manpower, strengthen existing cancer centers and create new ones, consider the manpower requirements of an expanded program, and make ample provision for fundamental research.

The report straddled the issue of "targeted versus creative" research. It first leaned toward the comprehensive cancer center as "the best organizational structure for the expanded attack on cancer." Such centers had the advantage of "a critical mass of scientists and physicians committed to the cooperative solution of the cancer problem, of research facilities, of patients, and of financial and other resources." "[T]he solution of the cancer problem," the report continued, "lends itself to a multidisciplinary effort where teams of highly qualified specialists are available to interact on problems of research, both clinical and nonclinical, teaching, diagnosis, preventive programs, and the development of improved methods in the delivery of patient care, including rehabilitation." Such centers, the report indicated, "should also serve as administrative coordinators of those programs which require regional coordination."

The report next leaned in the direction of the "creative" research scientists. It concluded that a national plan "should provide for the generous use of grants . . . in order to stimulate continued independent exploration, particularly in those areas where knowledge is not sufficiently mature for coordinated programs aimed at reaching defined objectives."

But what about the planning and management of research? A coordinated national program, the report suggested, should "be generated by the voluntary productive interaction and joint planning of the scientists who will be responsible for doing the work," a nod to the scientific community. But, the report cautioned, "the effective use of collective planning does not mean that centralized

administration or management of resources should be sacrificed."
"Optimum communication and centralized banks of information," it continued, were needed to "rationalize the decision-
making and to make information available when and where
needed." This moderate language was a far cry from what *The
Blue Sheet* and *Science* had anticipated, and what the staff had desired. Something substantially less than a crash program was
being proposed.

The panel, quite obviously, skated around one of the toughest
issues it confronted. The reason was clear according to one panel
member: "The issues simply were not joined. The staff was hot
for a Manhattan Project. The basic scientists were on the defensive. They were for the individual grant system. We fudged the
issue as a consequence."[64]

Fudging the management issue made good sense. The panel's
bold recommendation for enough money was to satisfy proponents of both "creative" and "targeted" research. "[A] coordinated national program . . . would require an appropriation in
fiscal 1972 of $400 million. Thereafter, the cost of the program
would increase at the rate of approximately $100 to $150 million
per year, reaching a level of $800 million to $1 billion in 1976."[65]
This recommendation would double the funds for cancer research
in a single year, and would redouble them in another four years!

The report argued that "these sums are not large in terms of our
national resources or of the human suffering and economic loss
attributable to cancer."[66] The panel felt that the funds were of
such importance to the American people that they should be provided even if that meant "the raising of additional revenues." Of
"utmost importance," the report argued, was that increased funds
for cancer research not result in reduction of funds for other medical research and health programs.

Ten days later the House and Senate accepted the conference
committee recommendation for the fiscal 1971 appropriation bill.
Congress was well on the way to providing more funds for medical research across the board than the Nixon administration had
requested. The OMB had made it plain to NIH officials, however,
that any increases above the original request would be placed in

reserve. Then, gratuitously, OMB added that if NIH thought the budget for fiscal 1971 was tough, they should wait until January when they would see the proposed budget for fiscal 1972!

Officials of NIH noted the panel's recommendation with more than passing interest. They were quite interested in any prospects for recapturing momentum in medical research. But neither the NIH leadership nor that of NCI felt that large new sums would be forthcoming for research from the Nixon administration. Fiscal gloom was pervasive.

Benno Schmidt read the forceful summary of the final paragraph of the report:

> Cancer is an implacable foe and the difficulty of eliminating it as a major disease must not be underestimated. A top priority commitment by the Congress, the President, and the American people is required if we are to mount and sustain an assault on cancer of the magnitude envisaged. . . . The Panel is unanimously of the view that an effective national program for the conquest of cancer should be promptly initiated and relentlessly pursued.[67]

Senator Kennedy praised the work of the panel as an example of the cooperation that was possible between public and private sectors in health. The senator asked Schmidt if he, as a businessman, thought the country could afford such a program. All panel members, Schmidt responded, especially the businessmen, were sensitive to any additional taxes, but "my strong personal view is that not only can we afford this effort, we cannot afford not to do it." "Your answer," Kennedy replied, "reflects that the businessmen on the panel have come to the uniform opinion that this is a sound economic principle and should be given top priority. Such an answer . . . is extremely compelling."[68] The senator, in the press conference which followed the hearing, was asked whether he supported the recommendation for an independent cancer authority. His answer was an unequivocal yes—the panel recommendations and their persuasive report fully justified such a step.

The scientific portion of the report was completed in November, nearly two months after the principal recommendations of the report had been agreed upon. In fact, the scientific report was relatively modest compared to the boldness of the panel's recommendations. As some critical observers pointed out, the panel's conclusions did not follow from the scientific analysis. But then the purpose of the scientific report had always been understood as the legitimation of the larger efforts of the Panel of Consultants. On July 2, 1970, Dr. Mathilde Krim wrote a number of her scientific colleagues asking for their assistance on identifying research areas of greatest promise. She indicated that the purpose of the Panel of Consultants was "to determine what legislation, new Federal programs, and authorizations, would be required to move us closer to the ultimate conquest of cancer." The panel's final recommendations were to be in terms that could result in "immediate appropriations and implementation." Though the time available for such a vast undertaking was short, the Sloan-Kettering scientist wrote, "it is essential that solid legislative and budgetary proposals be ready for the next Congress."[69]

Krim's letter was one of several approaches the scientists on the panel pursued in writing the scientific portion of the panel's report. Individual panel members also prepared written contributions, and from late August onward, editorial supervision over the report was exercised by Dr. Joseph Burchenal. Since the scientific competence of the panel members was not all-encompassing, many contributions were sent to nonpanel experts for review. Responsibility for drafting the final report mainly fell to Dr. Krim. "We mashed it together, we got it together in a hurry," Burchenal said later. "It is a good solid report, not a brilliant report," was his appraisal.[70]

The report identified four avenues of advance for a national cancer program—prevention, detection, treatment, and research. Prevention could be accomplished through the cessation of cigarette smoking and the reduction or elimination of exposure to the carcinogens of particular occupations, polluted urban atmosphere, and the chemicals of modern life. Early detection of cancer, however, coupled with prompt and appropriate treatment, was the most effective method of reducing deaths from cancer. Early detection meant that a tumor could be treated while still localized, rather than at the advanced stage when it had metastasized, or spread throughout the body.[71]

Surgery, still the most widely used form of cancer therapy, had achieved substantial cure rates in those cases where there was early diagnosis. But, the report indicated, "recent attempts to establish even more extensive surgical procedures have not resulted in increased rates of cure, and there now appears to be relatively little prospect of major advances from a more extensive removal of tissue."[72] Radiation therapy, second to surgery as a means of treatment, had made "remarkable strides" in the past twenty years. Delivery of greater X-ray dosages, more-precise beam definition, and reduced side-scatter had resulted in much increased 5-year survival rates for several types of cancer and much less damage to adjacent healthy tissue. But there remained a number of unresolved scientific and clinical problems that limited its use.[73] Chemotherapy had definitely been established as a curative therapy and could be used against disseminated as well as localized tumors. Its basic premise—that "cancer cells differ in some way from normal cells and so can be destroyed by chemicals that will not produce equivalent injury to normal cells," had been validated.[74] Even so, the mechanisms of cancer chemotherapeutic drugs were complex and their understanding depended upon further fundamental research in the biochemistry of normal and cancer cell function.

The bulk of the scientific report was devoted to cancer research, and specifically to a discussion of cancer biology, cancer etiology, molecular biology, and the immunology of cancer.[75] A major goal of cancer biology research, the report noted, was the identification of "specific and constant chemical differences between normal and cancer cells." "Momentous advances" in physics, chemistry, and molecular biology had made possible the detailed study of such differences at the subcellular level. The hope was expressed that "the cancer cell seems finally close to yielding the kind of information biochemists have been seeking for so long," and optimistically proclaimed that cancer chemotherapy stood now where chemotherapy of infectious diseases had stood in 1937.[76]

The etiology section (cancer etiology deals with causes of

cancer) discussed epidemiological studies of cancer, and chemical and radiation carcinogenesis. The most extensive discussion, however, was of viruses as potential cancer-causing agents in humans.[77] Though viral causes of cancer in animals had long been established, the search for a viral cause in humans had yielded only hypotheses and not confirmed evidence. The suspected viruses and problems of detecting them in human cancers were elaborated. The portion about unconfirmed hypotheses of causal mechanisms made clear the large debt that cancer research owed to the more general elucidation of genetics and molecular biology. Prospects for prevention and treatment of virus-caused cancers were considered, as were the existing scientific limits on pursuing these matters. Finally, it was suggested that the NCI's Special Virus Cancer Program needed to be broadened from a concern with the leukemias and sarcomas to include consideration of carcinomas—those tumors that accounted for almost 90 percent of human cancers.

Molecular biology received its share of attention. The major progress in understanding cellular growth and differentiation of normal cells had opened up broad new vistas for understanding these same processes in cancerous cells. But next to virology, it was immunology, with its promise of immunotherapy, which received the greatest amount of consideration.[78] Yet underlying the discussion of immunology was the recognition of the limits of existing scientific knowledge. There was also the unresolved problem that if an immune response to cancer was stimulated, the cell-mediated response could lead to a rejection of the cancerous cell while the humoral immune response might actually inhibit such rejection.

The scientific report, authored by numerous individuals and put together as a single document in a short time, was not the most lucid prose ever written. The intended audience was not the scientific nor medical communities, or it would have been more technical than it was. If written for a lay audience, however, it appeared more calculated to impress than to inform. It identified 22 areas of promise in cancer biology, 7 in epidemiology, 14 in chemical carcinogenesis, 16 in viral oncology, 11 in etiologic interactions, 23 in immunology, and 10 in clinical investigation.[79]

A close reading of the scientific report, however, revealed numerous references to identified knowledge gaps, limited methodological tools, and indications of need for further fundamental research. Past progress appeared to be attributed more to a series of small steps than to major scientific breakthroughs, though the report conveyed this impression implicitly rather than explicitly. Prior scientific advance, moreover, was not parceled out between cancer researchers and non-cancer researchers, nor were any distinctions made between that research which had been supported by NCI and that funded by the other NIH institutes.

Many medical scientists critical of the cancer initiative were of the opinion that there were no major scientific or clinical advances imminent in cancer which warranted an expanded cancer program. The scientific report did little to refute this view. One can only conclude that the medical scientists on the panel were unable to do so, given their commitment to such a program. Strangely enough, the scientific report was never seriously critiqued by the scientific community, nor was it ever closely examined by the Congress. But then it was not meant to be. The objectives of the Panel of Consultants, Mary Lasker, and the chairman of the Senate Labor and Public Welfare Committee, were directed to securing organizational autonomy and more funding for cancer research.

The last act that Senator Yarborough performed for cancer research was to introduce legislation which embodied the recommendations of the panel. "I realize," he said, "that in all probability I shall not see [this bill] completed before the end of my service in the Senate. However, within this body there are men of courage and conviction of both parties who I know will step forward to take up my work and see it through to its completion."[80] He then hurried over to the Senate chamber to drop S.4564—the Conquest of Cancer Act—into the legislative hopper of the 91st Congress.

The presentation of the panel's report attracted little publicity. The Saturday edition of the *Washington Post* had a front page story but the *New York Times* carried the news on page 22, and the event went largely unnoticed.[81] This was not surprising. The 91st Congress was rushing to complete its work and would not do so until early January, just a few days before the opening of the 92nd Congress. The funding bill for the SST, which had been defeated by the Senate, was now before the House, so that issue was not yet resolved. Both bodies were struggling to write a second appropriations bill for HEW, since the first had been vetoed by the president. It was a busy time in which to recommend a major program initiative in cancer research.

Yet one year later, on December 23, 1971, President Nixon signed the National Cancer Act of 1971, the legislative result of the panel's report. The president expressed great pride in the new law and hoped "that in the years ahead that we look back on this day and this action is shown as being the most significant action taken during this administration."[82] Between the hastily called Senate hearing in December 1970 and that well-attended signing ceremony in December 1971, cancer research legislation had occupied a prominent place in the priorities of the president, the Senate, the House, and the interested public. The outcome of that year of legislative struggle and controversy was an expanded national program in cancer research, a program that held the promise of touching the lives of millions throughout the world in the years to come.

5

Cancer Research and
Presidential Politics

"You know how the State Department has an Israel desk and a Laos desk?" one Kennedy staff man explained. "Well, somewhere down there in the White House there's a Kennedy desk, keeping track of everything we do. At first we thought we were imagining it, but there's no doubt any more. But their responses are nearly all cosmetic, a sort of counter-public relations policy."

Warren Weaver, Jr.
"By His Deeds,"
Esquire, February 1972

Robert Marston received a telephone call from a *Washington Post* reporter immediately after the panel presented its report to the Senate committee. What did the NIH director think of the proposal to separate cancer research from the National Institutes of Health, he wanted to know. Marston's reply was straightforward and to the point. "I'm against it," he said. "Cancer Institute obligational authority is going up this year, and cancer research profits from interplay with other research in the diseases of man."[1]

On Tuesday, December 8, 1970, Marston forwarded a memorandum to Roger Egeberg, the HEW assistant secretary for health and scientific affairs, in which he indicated that NIH and the panel were "in agreement on two important points."[2] First, he wrote, "I concur in their recommendation of enhanced support of biomedical research and in their assessment that the combination of scientific opportunity and urgency of the cancer problem makes it imperative to proceed with additional investigations in the areas they have listed." Secondly, observing that action by the administration and the Congress on the fiscal 1971 budget was "most likely to reverse the downward trend in cancer research support," Marston noted that "in keeping with the report, this proposed enhanced financing of the cancer research program in 1971 is not at the expense of other research programs."

Marston indicated, however, that he was in "unequivocal disagreement" with "the proposed locus of cancer research activity in a separate independent agency apart from the rest of biomedical research." He did not agree that NCI could not administer expanded activities within the framework of NIH. Moreover, the NIH director argued, the NCI already had most of the authorities that the panel had suggested were necessary for the conduct of a comprehensive program. The heart of Marston's argument was the following:

> The separation of the substantive scientific programs of cancer research from other biomedical research endeavors would be counterproductive to progress in cancer and other research. At the present time, various Institutes and Divisions of the NIH are engaged in and support research work that is highly relevant to the areas of special promise listed by the Panel. The virologists, immunologists, cell biologists, epidemiologists, pharmacologists and others involved in such studies are focused on fundamental life processes, which are undoubtedly vital to the understanding of cancer even though they may be addressing their research problems through different motivations.

This reasoning was to become one of the principal recurring arguments made by those opposed to a separate cancer agency.

Two additional matters in the report required "considerably more discussion" in the NIH director's opinion. For one thing, the panel had not adequately justified "the precipitous rate of [financial] growth" being proposed. While the financial situation since 1967 "has left some facilities and research expertise unemployed," Marston said, doubling expenditures in a single year would probably overload both facilities and expert manpower and yield less than the desired return on the investment. Expansion of research support, he continued, should be based upon available manpower and "on probable competition with other desirable endeavors, especially, for our purposes, other areas of biomedical research." In taking this view, Marston fulfilled the panel's expectation that the NIH director would balk at giving increased priority to cancer research to the detriment of other areas of biomedical research.

Finally, the NIH would "explore responsively," Marston indicated, the several recommendations of the panel that dealt with coordination, integration, and additional authorities needed to pursue research objectives in cancer. "This will be done, however," he wrote, "with the realization that the basic lack in the cancer problem is fundamental knowledge of life processes, not developmental capability or central coordination." Rather than directed research, Marston felt the panel had been wise "to recommend that there should be generous use of the grant mechanism to support individual investigator-generated research ideas." Returning to his initial point, he concluded with the observation that "eventually this investigator-generated research, with or without the cancer label, will generate the basic knowledge needed for a coordinated and comprehensive national program on cancer prevention and control."

This NIH viewpoint was, for five months at least, the basis for the official HEW and White House policy on cancer. But while NIH officials reasoned from the circumstances of biomedical research as they saw them, the White House responded to its assessment of changing political realities. One of these "realities" concerned the new chairman of the Senate subcommittee on health.

A MATTER OF SUCCESSION

Edward Kennedy had been elected to the U.S. Senate from Massachusetts in 1962 at age 30 to fill the unexpired two years of the term to which John Kennedy had been elected in 1958. He was reelected to a full six-year term in 1964, though flat on his back in a Boston hospital recovering from a near-fatal airplane crash earlier that year.

Edward Kennedy quickly distinguished himself as a diligent, hardworking legislator who learned the folkways of the Senate. His committee responsibilities included the Judiciary Committee, and the Labor and Public Welfare Committee, on which his brother John had served and his brother Robert would later serve.

When Edward Kennedy entered the Senate, John Kennedy was president of the United States and Robert Kennedy was attorney general. The assassination of the president in November 1963, left Robert Kennedy as a logical contender for the presidency in the judgment of many. The assassination of Robert, then junior senator from New York, in June of 1968, while campaigning for the Democratic nomination for president, left Edward Kennedy in the position of heir apparent. He had resisted approaches in August 1968 to let his name be entered as a presidential nominee at the Democratic convention. During 1969 and 1970, he repeatedly declared that he was not a candidate for his party's nomination in 1972. Even so, neither professional politicians nor public-opinion polls would indulge him his avowed desire of noncandidacy. All of Edward Kennedy's actions were interpreted as a strategy for his seeking the presidency in 1972.

One significant measure of this Kennedy's stature within the Senate, and the political respect accorded him by his Senate colleagues as the surviving member of the family, was his election in January 1969 to the position of Democratic Majority Whip. The whip's position, second only to the Majority Leader, had been held by Senator Russell B. Long, of Louisiana, since 1965. Kennedy's bid for the position had represented a bold challenge, and his election was a major personal victory for a man not yet forty in a body whose average age was nearly sixty. The young senator from Massachusetts suddenly became his party's most prominent national spokesman.

All this changed, however, in July 1969. Just before midnight on the eighteenth, the car which the senator was driving plunged off a narrow bridge on Chappaquiddick Island, causing the death of his only passenger, Mary Jo Kopoechne, one of the young women who had worked in Robert Kennedy's campaign for the Democratic presidential nomination. Senator Kennedy escaped with his life, but the political losses he sustained were very great. His prospects as a presidential candidate for 1972 were considerably dimmed.[3]

The senator also confronted the fact that his reelection in Massachusetts in 1970, previously seen as a sure thing, had become a public test of his viability as a political candidate. Consequently, he spent much time campaigning in the state during the rest of 1969 and throughout 1970. He consistently declined invitations to speak out-of-state in behalf of other Democratic candidates for public office, a service he had previously performed with great frequency and effectiveness. The voters of Massachusetts returned him to the Senate for his second full term in November 1970.

The time spent in Massachusetts, however, was time away from the business of the Senate. There was considerable cloakroom grumbling among his Democratic colleagues that he had slighted his responsibilities as Majority Whip. So there was considerable speculation in early January 1971, when the 92nd Congress was being organized, that there would be a challenge to Kennedy, perhaps from another liberal, perhaps from a conservative. The challenge came from the right, and it was successful. Senator Robert C. Byrd (D., W.Va.), in a surprise upset that shook the Senate leadership, ousted Kennedy as Majority Whip by a vote of 31 to 24 in the Senate Democratic caucus.[4] Byrd's last-minute decision to make the challenge came only after he learned that ailing Georgia Senator Richard B. Russell was still alive at the Walter Reed Army Medical Center. Armed with Russell's proxy, Byrd had calculated that he had 28 votes to 27 for Kennedy.[5] Senator Russell died four hours after the vote.[6] Had he died before the caucus met, Byrd would not have made his move. For Kennedy, it was a stunning defeat.

Kennedy partially recovered from this loss when the Committee on Labor and Public Welfare was reorganized. Senator Yarborough's November defeat meant that two vacancies needed to be filled in the 92nd Congress. The full committee chairmanship could have gone to Senator Jennings Randolph (D., W. Va.), the ranking Democrat, but he would have had to relinquish his chairmanship of the Committee on Public Works. Randolph was unprepared to forego his control over the "pork barrel" of dams, river and harbor improvements, federal court houses and post offices, and similar forms of federal largesse. Next in line for the chairmanship was Senator Harrison Williams (D., N.J.). Reelected in November 1970, after surviving a serious challenge from the New Jersey Republican State Committee chairman, Nelson Gross, Williams readily assumed the position.

Senator Hill and Senator Yarborough had each been chairman

of the full committee and also of the health subcommittee. Williams apparently considered this possibility, but it would have involved giving up the chairmanship of the subcommittee on labor. Such an action would have strained relations with organized labor, which had provided strong support to Williams throughout his Senate career, so he decided to continue as head of the labor subcommittee. This left the way open for Senator Kennedy to become the chairman of the health subcommittee, a position that he had indicated was of great interest to him. Given his recent loss of the whip's position, the chairmanship offered him an opportunity to partially recoup the costs of the Chappaquiddick episode and to address a broad range of important national issues.

Kennedy's chairmanship of the health subcommittee was very good news to Mary Lasker and the Panel of Consultants. She had wanted him to have the position two years earlier. Moreover, she had always been a strong supporter of the Kennedy family.[7] Now he could be counted on to strongly assist the new cancer initiative.

Sidney Farber, the patriarch of Children's Cancer Research Foundation in Boston—the "Jimmy Fund"—also knew the Kennedy family very well. He had known John Kennedy when the latter was senator from Massachusetts. This relationship had helped him persuade the new president, in 1961, to initiate the study of heart disease and cancer that preceded the President's Commission on Heart Disease, Cancer, and Stroke. He was also close to Edward Kennedy.

Significantly, the first major political address made by Edward Kennedy after the accident at Chappaquiddick was at a testimonial dinner in Boston for Dr. Farber. "I have known Dr. Farber for many years," the Senator said:

> Among members of the United States Senate there is no more respected spokesman for medical research. President Kennedy counted him as an adviser and a personal friend. . . . No one will ever estimate the years of life he has saved for countless human beings, now and in the future. We can only be grateful that he has devoted his genius to this work.
>
> I think we must marvel at the strength of Dr. Farber. Day after day, he moves into the midst of the most tragic of human

situations—the tragedy of young children with leukemia. Each day, he gives these victims cheer amidst their doom and gives their parents understanding and faith amidst their tears. Dr. Farber must live each day with death in its most agonizing form. Yet he himself is full of hope and optimism. He has added precious years to children's lives and hope to all of us.

When hope is matched with genius, the result is progress. The greatest tribute to Dr. Farber will be on that day, which is sure to come, when cancer cells can be found and disarmed before they begin their deadly march through the body. The finest tribute we can pay, in the meantime, is to pledge our own abilities to make that day come as soon as possible.[8]

Within a year and a half, the senator would have a great opportunity to make that pledge himself.

Kennedy's succession to the health subcommittee chairmanship was widely anticipated when the Panel of Consultants presented its report on December 4, 1970. Power on health matters was then ebbing rapidly from Senator Ralph Yarborough and flowing to Senator Kennedy. This much was clear that Friday morning. In politics, attention is directed to those with power or those soon to acquire it. Kennedy questioned members of the panel extensively. In turn, they directed many of their responses to him rather than to Yarborough. After Yarborough announced the introduction of his bill, with the knowledge it would die in that session of Congress, Kennedy said, "I'd consider it an honor to reintroduce such legislation next year."[9] And, on January 25, 1971, Senators Kennedy and Javits jointly introduced S. 34—the Conquest of Cancer Act—into the 92nd Congress.[10]

A BUDGET IS REOPENED

Benno Schmidt and the other prominent Republican panel members began directing their attention to President Nixon and his principal advisers soon after the report was presented to the Senate. The groundwork for this effort had been carefully laid. Schmidt had discussed his intentions to seek support from administration officials at the second meeting of the Panel of Consultants in late July. Elmer Bobst, at that meeting, reported that he had recently spoken to President Nixon and had urged him to interest himself in cancer. Bobst also agreed to a request from Schmidt to speak to the president in support of the panel's recommendations at the appropriate time. The panel chairman had personally visited George Shultz, director of the Office of Management and Budget, during the summer and informed him of the work of the panel. In addition, he had lunched in mid-September with the House leadership, including a number of key Republican lawmakers.[11]

In November and December, these contacts were pursued with greater purposefulness. Schmidt gave a copy of the "summary and recommendations" to George Bush, then U.S. ambassador to the United Nations, just before Thanksgiving. Bush was asked to pass this portion of the report on to the president so he would have it prior to the panel's presentation. The ambassador did convey the "summary and recommendations" to John Ehrlichman, President Nixon's principal adviser on domestic policy, with the request that Ehrlichman transmit it to the president.[12]

In mid-December, Schmidt paid another visit to George Shultz. He delivered a copy of the panel's report to the OMB director, discussed its main provisions, and expressed the hope that the recommendations—developed after extensive study and serious consideration—would be accepted and supported by the administration.[13]

Schmidt also wrote President Nixon directly soon after the presentation of the report. He reviewed for the president the work of the panel, enumerated the members of the House and the officials in the president's administration he had seen. The panel chairman stressed that the controversial recommendation for a separate agency was necessary if cancer were to be declared a national goal, and expressed the hope that the president would find the panel's recommendation worthy of support. Laurance Rockefeller and Keith Funston wrote similar letters.

Late in December, Benno Schmidt and Mary Lasker went to Elmer Bobst with the suggestion that "it would be 'a wonderful idea' " if the President would mention the cancer initiative in his State of the Union message in January. Bobst was very receptive to the idea and broached it to the president. He and Schmidt followed up with a joint memorandum asking that the president indicate his general support for a national cancer program.[14] The "managing partner" and his Republican colleagues were exploring every avenue by which they could get their message directly to the president.

Inside the government, other forces were at work. December of each year is the month in which the president's budget request for the coming fiscal year is determined, both at the overall level and for each department and agency. Officials in OMB were putting the finishing touches on the budget request for fiscal 1972, which the president would send to Congress in January. The assistant director of OMB responsible for the National Institutes of Health was Richard Nathan. Nathan saw the budget choice confronting the administration in health as one between health-care delivery or medical research. He tended to favor increased funding for health-care delivery, while he strongly opposed increases for medical research. Officials in the HEW secretary's office shared this view.[15] But it was the OMB that exercised control over the purse strings, and Nathan had been the one who had conveyed the rather bleak picture of continued "austerity" to the NIH leadership in mid-December. Thus, while Schmidt and his fellow panelists were pressing their case for an expanded national cancer program, the largely invisible but powerful budget officials in the administration were seeking to hold down medical research across the board.

George Shultz had come to Washington, D. C., in early 1969 as secretary of Labor in the first Nixon cabinet. He had made a strong, favorable impression upon the new president. In mid-1970, when the old Bureau of the Budget was abolished and the Office of Management and Budget created in its stead, the former dean of the Graduate School of Business of the University of Chicago was named director of the new agency. Shultz had an office in the Executive Office Building across the street from the White House, like his Budget Bureau predecessors. But unlike them, he also had an office in the west wing of the White House, very near the president's office. Also unlike the men who had preceded him, Shultz involved himself very little in the day-to-day budget review process, preferring to leave those responsibilities to his deputy director for budget, Caspar Weinberger. He was thus somewhat isolated from the professional OMB staff, but very close to the president.

In mid-December, with the printer's deadline for the budget drawing near, Shultz was reviewing the detailed budget recommendations presented to him by Weinberger, prior to making his own recommendations to the president. One concern he had was the adequacy of federal government support for research and development. Within the administration, discussions were underway about the loss of the U.S.'s international trade advantage in high technology exports, the relationship between declining federal R&D and declining industrial productivity, the increasing unemployment of scientists and engineers as the space program was cut back in the wake of the Apollo success, and the ways in which the nation's technological resources could be directed to pressing domestic problems.[16]

Consequently, Shultz, who often sought outside advice, asked Dr. Edward E. David, Jr., the president's science adviser, whether the fiscal 1972 research and development budget was adequate.[17] David had no immediate response, but initiated his own inquiry. From several different sources came back the word that medical research, as reflected in the budget for the National Institutes of Health, could probably use $50 to $100 million more money than had been provided in the recommendations prepared by Nathan and transmitted by Weinberger.

Marston and his staff at NIH were informed that additional funds for medical research for fiscal 1972 were under consideration.[18] What would they do if they had an extra $50 to $100 million? Tentative draft language was prepared by NIH for both the State of the Union message and the special message on health. Marston recalls that he and his staff had just about concluded that $20 million might go to cancer research, $20 million to heart research, and $60 million to the rest of medical research. But, as is usual in this kind of exercise, the NIH officials did not know all the factors that had prompted the inquiry to them in the first place.

Traditionally, the White House staff judge issues in terms of their benefit or harm to the president. Shultz and John Ehrlichman were in a position to put a number of pieces together.[19] They knew that Senator Kennedy would be introducing legislation calling for an expanded national cancer effort as soon as the 92nd Congress convened in January. Whatever the Chappaquiddick accident and the loss of the Democratic Majority Whip's position had done to impair Kennedy's prospects for being the Democratic presidential nominee in 1972, no one at the White House took him for granted or wished to enhance his political reputation in any way. The prospect of the senator being out in front of the president on an issue so politically potent as cancer was unacceptable. Not with the 1972 presidential elections drawing ever closer.

Moreover, Shultz, Ehrlichman, and the president himself had heard from important Republican businessmen, a key constituency, on the significance of endorsing and supporting an expanded national cancer program. Finally, from within the bureaucracy, though not through the normal channels, there came the message that $100 million for medical research was a very reasonable addition to the fiscal 1972 budget. The White House staff seized the opportunity and recommended that the president add an additional $100 million to the budget for cancer research and announce this step in his State of the Union message.

On January 22, 1971, before a joint session of Congress, President Nixon delivered the annual State of the Union message. He announced six great goals for the "New American Revolution"—welfare reform, full prosperity in peacetime, restoration and enhancement of our natural environment, improvement of America's health care, strengthening and renewal of state and local governments, and complete reform of the federal bureaucracy. The president promised a set of proposals "for improving America's health care and making it available more fairly to more people." The proposals would eliminate the inability to pay as a barrier to obtaining basic medical care, augment the number of physicians and health professionals through increased, redirected aid to medical schools, improve the adequacy of health services and control their costs, and encourage preventive medicine. Then, with respect to cancer, the president dramatically announced:

> I will also ask for an appropriation of an extra $100 million to launch an intensive campaign to find a cure for cancer, and I will ask later for whatever additional funds can effectively be

used. The time has come in America when the same kind of concentrated effort that split the atom and took man to the moon should be turned toward conquering this dread disease. Let us make a total national commitment to achieve this goal.

If there was to be a political controversy over cancer research, the president had for the moment assumed the high ground.[20]

THE PRESIDENT PREEMPTS THE SENATOR

Exactly one week later, President Nixon sent *The Budget of the United States Government for Fiscal 1972* to the Congress. It included a request for $232 million for the National Cancer Institute and a special request of $100 million for "cancer research initiatives" that would use "all pertinent institutes and agencies" of the federal government.[21] The total amount of NIH funds requested for cancer research, therefore, was one-third of a billion dollars. The Panel of Consultants had already had a substantial impact on the budget for cancer research.

The president elaborated on the new cancer effort when he spoke on February 4 to the twentieth annual scientific session of the American College of Cardiology. But he also paid tribute to those who did research in cardiovascular diseases and sought to allay any anxieties which might have been raised by the cancer initiative. "[T]he fact that we are putting emphasis in that area [cancer]," the president said, "does not mean reducing emphasis in the area in which you are so vitally interested." "[I]f you look at the . . . numbers . . ." he continued, "you will find that the amount . . . we have asked the Congress to appropriate in . . . research for cardiovascular ailments has increased by 16 percent over 1970."[22]

On three occasions in late January and early February, then, the president defined his position on the cancer effort. That position included a request for more money and a pledge to ask for whatever additional funds might be necessary. It included a commitment to avoid expanding cancer research at the expense of heart disease research. In addition, the president expressed a preference for using the entire resources of NIH and was conspicuous by his silence on the matter of a separate agency for cancer. He did, however, call for "a total national commitment" to finding a cure for cancer and suggested that the skills which "split the atom and took man to the moon" be employed for cancer. On a political level, this was a very reasonable position for the president to take given the contending forces acting upon him. On a public level, the cancer initiative brought him front-page coverage across the country, as well as substantial editorial support. The president had moved swiftly to a position of leadership on an important political issue.

In the absence of a detailed statement by the president, it fell to his advisers to fill in the details on a given policy issue. In this case, Dr. Edward David assumed this responsibility. Slightly alarmed by the apparent enthusiasm with which some White House staff had embraced a "moon shot for cancer," David met the issue head-on in a Chicago speech to the assembly of the Association of American Medical Colleges (AAMC) on February 13. The science adviser had come, he told the AAMC, "to enlist your aid and advice in a totally revolutionary undertaking which President Nixon has proposed," the campaign to find a cure for cancer. It was the president's hope, David said, that this campaign "will invigorate the best of biomedical science to provide a major improvement in the health of our society."[23]

But David had come not only to seek advice, but also to signal the directions in which the administration intended to move. "Cancer," he said, "is not a simple disease; it is probably many. There is likely not to be a single cure, but a series of steps"[24] leading to effective treatment and hoped-for prevention. Therefore, cancer research could not be separated from the rest of biomedical research:

> The problem of cancer straddles virtually all the life sciences—molecular biology, biochemistry, virology, pharmacology, toxicology, genetics—any one of these, or all of them, will contribute to the final solution. No one is wise enough to pick and choose just those components of the total biomedical spectrum that will be vital. Who knows what new discovery will become vital even next year?[25]

One implication of the national cancer effort, David suggested, was organizational and managerial. The "cancer crusade, as it has been called," he said, "will call on very different talents from scientists, researchers and managers than Apollo and the Manhattan Project did." In organizing the cancer research effort "we must take account of the differences between this effort and the Apollo and Manhattan projects." He later added, "Indeed, we do not believe in an AEC or NASA for cancer."[26] A separate agency had been ruled out.

On the other hand, it appeared to David that cancer research had advanced to the point where a major initiative was warranted. One year earlier, the science adviser noted, in *The Budget of the United States Government: Fiscal Year 1971*, the president had proposed a major effort "to investigate viruses as a cause of cancer . . . with the goal of eliminating viral cancer." The response to the proposal by the Congress, the medical community, but most of all by the biomedical scientists, had confirmed for the president the validity of his judgment. Moreover, David noted that in his discussions with biomedical scientists he had found a quiet sense of confidence and sober recognition of the opportunity that stimulated the president to propose his campaign against cancer. "Molecular biologists, cell biologists, and biochemists of world reknown tell me that of all the medical problems facing us," he said, "they would choose cancer as the one in which to attempt a concerted effort of research." Undue optimism was unwarranted, since the results might not come for 10, 20, or even 50 years. "But," Dr. David said, "the judgment seems widespread that the time has come to try."[27]

The appropriate vehicle for the cancer research effort, the president's science adviser indicated, was the NIH. For more than two decades, the NIH had developed the resource base of biomedical research institutions. The impressive results had more than overcome initial apprehensions. Biomedical research in other countries had received critical assistance from the NIH. Two Nobel laureates were among the intramural research staff at Bethesda. Academic science in American medical schools and universities had flourished. Not only had new scientific knowledge been developed, but the effort had significantly affected disease patterns in the United States. David enumerated several successes: "Vaccines for poliomyelitis, German measles, mumps, and other infections are available to us all. Victims of Parkinsonism have been raised from the status of cripples to productive members of society. Children who would have been dead from leukemia are alive today and are regarded as cured in a surprising number of instances."[28] The historical record was there for all to see.

Dr. David, though, was not merely reaffirming the existing order of things. Cancer research could not be separated from the rest of the life sciences, nor would organizational responsibility be taken out of the NIH. On the other hand, both the biomedical research community and the National Institutes of Health would have to adapt to this new cancer initiative. "The President has issued his call for action by setting a clear goal for the biomedical research community," said the science adviser. This goal was "a logical, yet imaginative, extension of today's biomedical research enterprise." And the challenge was a "unique opportunity . . . to show that we can concentrate our fire effectively when a possibility for progress emerges from research."[29]

"What is the most rational and effective way to undertake this effort," David asked, "considering that Apollo and the Manhattan efforts don't provide as any infallible blueprint?" It was the president's belief, he said, that having "honed and sharpened" the NIH, that organization should now be used "as we embark on this new adventure." Though cancer research must be closely related to all the other life sciences, the effort should be a partnership between the scientists and the public. Laymen could not program scientific efforts, but they were entitled "to know the strategy, to know the short-term objectives, and to receive progress reports." Specifically, on organizational matters, David indicated that there must be one director responsible for the cancer research effort: "He must plan the effort, direct it, and report to the American people." This individual should be part of the total life sciences effort, and thus part of NIH. But the cancer research director must not be "confined to a single institute," Dr. David said, and "he must have the administrative freedom to approach the problem with daring and imagination, and to catalyze the transfer of results from scientific discipline to scientific discipline." Then, having led his audience this far with intimations of changes to come, the science adviser concluded by saying, "The President is in the process of considering the best design for the administration of this effort and will report his decision to the American people in the near future."[30] The science adviser, while concurring with the biomedical community on several basic issues, was simultaneously signaling the need for some new responses to cancer.

The president's "National Health Strategy" message to the Congress clarified things a bit more. "[We] must reaffirm—and expand—the Federal commitment to biomedical research," the president said, but two critical areas—sickle cell anemia and cancer—deserved special attention.[31] Cancer, of course, was the major problem. The statistics were frightening. In 1971, 650,000 new cases of cancer would be diagnosed in the United States, and unless the present incidence was reduced, one of four Americans now alive would develop cancer.

Then the president evoked the image of the space program once again.

> In the last seven years we spent more than 30 billion dollars on space research and technology and about one twenty-fifth of that amount to find a cure for cancer. The time has now come to put more of our resources into cancer research and—learning an important lesson from our space program—to organize those resources as effectively as possible.

But recognizing that such analogy always provoked apprehensions among medical scientists, he added a cautionary note:

> When we began our space program we were fairly confident that our goals could be reached if only we made a great enough effort. The challenge was technological; it did not require new theoretical breakthroughs. Unfortunately, this is not the case in most biomedical research at the present time; scientific breakthroughs are still required and they often cannot be forced—no matter how much money and energy is expended.

This balanced statement from the president seemed very reassuring to the scientific community.

The president, in defining the broad research goals for the scientific community, went on to say that "of all our research endeavors, cancer research may now be in the best position to benefit from a great infusion of resources. For there are moments in biomedical research," he observed, "when problems begin to break open and results begin to pour in, opening many new lines of inquiry and many new opportunities for breakthrough." It was for this reason, the president said, that he had asked for the additional $100 million for cancer research.

Finally, in his clearest exposition of the organizational issue, the president indicated that he was directing the secretary of HEW "to establish a new Cancer Conquest Program in the Office of the Director of the National Institutes of Health." This program would have its own director, appointed by the secretary, and supported by a new management group. Furthermore, a new Advisory Committee on the Conquest of Cancer was being created for the purpose of advising the program "in establishing priorities and allocating funds," as well as for advising other officials including the president.

The Nixon administration's position appeared to be emerging with some clarity. The administration was providing more money for cancer research. It was equivocating on research management, as had the Panel of Consultants. And while it was suggesting internal organizational changes within the National Institutes of Health, the administration was consistently and firmly ignoring the panel's recommendation that an independent agency be established for cancer. The conflict was coming into focus.

The proponents of an expanded national cancer program had realized a substantial measure of success by the end of January 1971. They had secured President Nixon's general support for such an initiative in both his State of the Union message and in the special request for an additional $100 million for cancer research included in the fiscal 1972 budget. Beyond this, legislation had been introduced into the Senate that incorporated the specific recommendations of the Panel of Consultants. It was understood that hearings would soon be held on this legislation and it was expected that the administration would be highly attentive to its progress, if only because the chief sponsor was the most threatening potential opponent to President Nixon in the 1972 election. A reformulation of cancer policy was clearly on the formal government agenda for 1971.

The elite group responsible for the Panel of Consultants and its report had substantial resources in their own right—wealth, status, power. But their success in advancing cancer to the formal agenda of government was a product of access to President Nixon and to Senators Kennedy and Javits and of the incentives that all parties had to act. An "inside" strategy of agenda-setting had already paid significant dividends. Not until the end of 1971, however, would it also be clear that the Panel of Consultants, by their report, had successfully shaped the legislative debate.

6

Two Days in March

The creation of an independent Cancer Authority, removing the NCI from the ambit of the NIH, would, in my opinion, not accomplish anything that could not be done within present NIH processes, or trivial and easily realized modifications thereof. On the other hand, it would unleash forces of a divisive character which would quickly destroy the integrity of the NIH. I predict that in a very short time, orderly governance would be replaced by anarchy, and that instead of a judiciously balanced program of biomedical research, program emphasis would be entirely determined by uncritical zealots, by experts in advertising and public relations, and by rapacious 'empire builders.' These latter forces are not to be disdained and they have played an invaluable role in the past quarter century in making the lay public aware that, through research, there was a real possibility of realizing inchoate public hopes and aspirations to control disease. As forces modulating the scientific judgment process, their contributions have been positive and important. As determinants, however, I would expect them to create chaos.

James A. Shannon, M.D.
Letter to L. Hollingsworth Smith, Jr., M.D.
February 24, 1971

If it is not necessary to create another agency, and none of us wishes to create additional agencies if they are not required, then that case will have to be clearly and incontrovertibly made by those who hold that view.

Senator Edward M. Kennedy
Conquest of Cancer Hearings
March 10, 1971

President Nixon had moved rather swiftly and very effectively. While the Panel of Consultants up to this time had functioned largely behind closed doors, the president was acting in full view of the American people. By the end of February, he had preempted much of the national publicity on cancer research and he had denied Senator Kennedy an unchallenged leadership role on the issues. He had seized the cancer initiative and he had left his options open.

But the proponents of the panel's recommendations were hardly inactive. Luke Quinn was busy trying to line up more co-sponsors for S. 34—the Conquest of Cancer Act. In addition to Kennedy and Javits, the principal sponsors of the legislation, twenty-five other senators had become co-sponsors of the bill. All were Democrats, and most were liberal Democrats. Quinn, on February 15, in one of the many memoranda he sent out to a mailing list of several hundred important individuals concerned with cancer, described these twenty-seven senators as the "hard core of support." The addition of other co-sponsors had slowed down and perhaps stopped, he noted. "We need a total of at least fifty-one," he wrote, "so that we can show at the outset that enough senators are in favor of the bill to insure its passage after hearings. We must now bend our efforts to getting the additional support."[1]

Senator Kennedy, meanwhile, had asked Benno Schmidt and the other members of the Panel of Consultants to continue in an advisory capacity to the Senate health subcommittee. They agreed to do so and turned their attention to building support within the Senate.

Senator Kennedy was also turning his attention to the legislative business at hand. He had scheduled hearings for early March. The Conquest of Cancer Act, which Senators Kennedy and Javits introduced, attached the "highest priority" to improved means of prevention, diagnosis, and cure of cancer and declared "that a great opportunity is offered as a result of recent advances in the knowledge of this dread disease to conduct energetically a national program for the conquest of cancer." This "great opportunity" was to be pursued through the National Cancer Authority, an independent agency to be headed by an administrator and deputy administrator, both appointed by the president and confirmed by the Senate. All functions of the existing National Cancer Institute would be transferred to the proposed agency, and all authorities of the secretary of Health, Education, and Welfare that pertained to cancer would similarly be transferred to the new administrator.[2]

A National Cancer Advisory Board was also proposed, which would absorb all functions of the existing National Advisory

Cancer Council. The 18-member board would be equally divided between physicians, scientists, and lay individuals. The primary responsibility proposed for the board was to advise and assist the National Cancer Authority in the development and execution of its program, and to report annually to the president and the Congress on progress toward its objectives.[3]

The bill also required the administrator to prepare within one year "a comprehensive plan" for a national cancer program "with appropriate measures to be taken, time schedules for the completion of such measures, and cost estimates for the major portions of such plan." Finally, S. 34 authorized "such sums as may be necessary" for each fiscal year to carry out the purposes of the legislation.[4]

The proposal to create an independent agency for cancer began to draw fire immediately. The February speech by Dr. David was the most forthright attack on the idea. Dr. James A. Shannon, the former NIH director, concurred with the president's science adviser. One week before the hearings, he was quoted as saying that taking cancer research out of NIH "would strip a broad and complex area of science away from contiguous areas [which] would be bad for cancer research, and . . . bad for science."[5] Still others voiced concern about the dangers of comparing the effort to conquer cancer with that of going to the moon.

Senator Kennedy, as he opened two days of hearings on March 9, 1970, conceded that there had been criticism about the need for establishing a National Cancer Authority independent of the National Institutes of Health. A separate agency, he said, would "maximize the chances to make real progress against cancer." The rationale was administrative, organizational, and budgetary, according to the senator, not essentially scientific. The Panel of Consultants had unanimously supported the recommendation, and now there were 52 senators, co-sponsors of the legislation, who wished to write the provision into law. The burden of proof, he made very clear, rested with those who opposed a separate agency. It was they who would have to "clearly and incontrovertibly" make the case against such a step.[6] His own position was not in doubt.

Nor was the position of Senator Jacob Javits in any doubt.

Though not the ranking Republican on the health subcommittee, he was the ranking member of the full committee, and he spoke with the authority of one of the senior liberal Republicans in the Senate. He addressed himself to the administration representatives sitting at the table before him:

> I believe that the American people feel about cancer the way they felt about the moon shot and the way they felt about splitting the atom, that these are exceptional and unusual circumstances. I deeply feel, and I think Senator Kennedy has the same feeling, that our people suffering with this particular dread disease . . . have a general feeling that this is unnecessary, that we can break through, not in the years it will take the bureaucracy, but with the concentrated, high level project of the character proposed in S. 34. Therefore I believe that the likelihood is, unless you gentlemen have some very good alternative, that the Congress will make the decision. It is not necessary to have the doctors make it, or even the Commission [sic] make it. The Congress will make it. This is our top priority in health, and we think we can break through here, toward the conquest of cancer and that we have been presented by the Panel with the methodology for doing it.[7]

He hoped that the administration witnesses would come to grips with these questions, instead of getting off on routine explanations about the relationships between HEW, NIH, and cancer research.

Senator Peter H. Dominick (R., Colo.), was the ranking minority member of the subcommittee and the spokesman for the administration on this legislation. He was the only senator to criticize S. 34. Dominick recognized that there was a general agreement that it was time for "a concerted effort . . . to bring [cancer] under control." The disagreement was over the best means to this end. The president, he pointed out, had proposed a new cancer conquest program as an integral part of the NIH.

> I am supporting the President's proposal rather than S. 34 because, although I support the objectives of both, I believe the President's approach to be more effective. S. 34 seems to be based on the assumption that cancer research is adaptable to the

same programmatic approach which has been used to solve defined technological problems on a crash basis, such as putting a man on the moon, or developing the atomic bomb. The validity of that assumption is questionable. . . . The necessary scientific breakthroughs have not yet occurred . . . and such breakthroughs cannot be forced.

Under the circumstances, Dominick was convinced that retaining the cancer effort within the National Institutes of Health was essential.[8]

THE ADMINISTRATION'S POSITION: MARCH

Elliot Richardson, secretary of Health, Education, and Welfare since mid-1970, sent his regrets at being unable to testify. The administration was represented by Dr. Roger O. Egeberg, HEW assistant secretary for health and scientific affairs, Dr. Jesse L. Steinfeld, deputy assistant secretary and Surgeon General of the Public Health Service, Dr. Robert Q. Marston, director of the NIH, and Dr. Carl G. Baker, director of the NCI.

Steinfeld presented the testimony. He was the logical choice to do so. As Surgeon General, he had come to that position from the National Cancer Institute where he had been assistant director for program. His association with cancer research was a long one, going back to the immediate post-World War II years and the Laboratory of Experimental Oncology in San Francisco. Steinfeld, in the testimony, held out an olive branch to the Senate subcommittee on health. "A substantial consensus had been reached across the Nation," he noted, on the need and timeliness for a campaign to conquer cancer, a consensus shared by Congress and the executive branch. The president had asked for an additional $100 million for cancer research, and had also added $20 million to the fiscal 1971 budget request for NCI over the previous year. The health subcommittee, through the Panel of Consultants, he said, deserved special commendation for focusing attention "on the existing possibilities in cancer research that have now become evident." Acknowledging differences between the administration and the subcommittee, Steinfeld volunteered that the explanation of the administration's position "on this complex mix of scientific, organizational and management issues," would reveal "the great areas of agreement we have in seeking solutions to the cancer problem, and how small—in all truth—are our differences." There was agreement, first, that though cancer was the second leading cause of death, it was the disease most feared by Americans, and second, that in the last decade "major advances in our fundamental knowledge about cancer have opened new and promising vistas for research into the treatment and prevention of cancer."[9]

The key difference, Steinfeld testified, was that "from both a scientific and a managerial standpoint, we believe that the formation of a separate authority outside the NIH and HEW would be a serious mistake." Four scientific considerations supported this position. First, the Surgeon General argued, biomedicine was "a vastly complex and interdependent" universe, and "the separation of any segment from its innumerable major and minor connections in the network of health science would not be in the best interests of either cancer research or biomedical research in general."[10] Further, the existing excellent relationships between the scientific community and the NIH would be disrupted by the proposed arrangement. Third, Steinfeld said, emphasis on cancer would affect the distribution of scientific talent and might lead to a serious loss of scientific momentum in fields like heart and lung disease, environmental health, and so on. Finally, if cancer research were separated from the NIH, the latter organization's contribution to biomedical knowledge might well be limited by being cut off from expanding cancer research.

On the management side, Steinfeld noted that the Panel of Consultants had cited a need for more autonomy for cancer within the federal bureaucracy. But NCI already had extensive management autonomy, he said, and was completely responsible for day-to-day operations within broad NIH policy. Moreover, substantial costs would be required to duplicate management resources for a separate cancer authority, while the NIH framework provided great flexibility for the development of institute programs at the forefront of biomedical science.

The more serious management consideration, Steinfeld indicated, was whether cancer should be given the overriding national priority urged by the Panel of Consultants.

> The report is . . . asserting that a priority should be determined for cancer wholly outside the context of requirements for other bio-medical research activities and for health as a whole. . . . This ignores not only the question of what constitutes a balanced and . . . coordinated approach to the problems of health, but also the question of competing needs for resources, including scientific and supporting manpower.

It made little sense to HEW, in short, to treat cancer independent of health and medical research in general. To the extent that funding reflected priority, the Surgeon General pointed out, that depended "less on organizational status or location than on the degree of national commitment to the objective sought through the program." Beyond the additional $100 million request, the president had pledged himself to seek "whatever additional resources . . . can be effectively used," Steinfeld reminded the Senators.[11]

Steinfeld did concede that some new management techniques were needed for research coordination, data-base improvement, and scientific communication. There was little detail on what this implied, however, beyond the president's "National Health Strategy" message of February 18. Implementation of the president's directive to the secretary was now well under way, the Surgeon General said. The NIH, the office of the secretary of HEW, and nongovernmental scientific and management consultants were "developing detailed management and administrative mechanisms" to accomplish the goal of the conquest of cancer. These mechanisms would be discussed with the Congress as they were developed. Steinfeld concluded the presentation of the administration's position on a note of caution:

> We must prepare ourselves and the American people for a waiting period of unknown and possibly anguishing duration. . . . [T]he battle against cancer must be fought on many fronts. We must continue basic research while concurrently advancing available "practical solutions"—even though we have not yet discovered the basic causes of the diseases.[12]

Senator Kennedy wasted no time on Egeberg, Steinfeld, or Marston, but turned immediately to Carl Baker.[13] Did "a comprehensive overall program for the conquest of cancer" exist? he asked. There were "managed programs," Baker responded, "where we can define our objectives with some clarity." But, in areas of cancer research that were not well-developed scientifically, he continued, "we have insisted on maintaining what I call a more loosely structured approach." The NCI director thought it "healthy" to have both approaches and said that NCI had tried to develop this kind of mix.

The senator was dissatisfied, however, and persisted in his questioning:

> Perhaps I ought to clarify . . . the question that I was driving at. . . . [T]o what extent can you say that you have a comprehensive plan . . . to meet the crisis of cancer? . . . I would be interested to know if you have in any one place a comprehensive plan to meet the crisis of cancer, and how you move away from that in terms of achieving that objective.

Was the work plan or organization at NCI any different from the other NIH institutes? the senator wanted to know. A larger proportion of money went to "structured programmatic work," Baker said. Apart from funding, Kennedy asked, "I am interested in your approach—is it any different? Or is it that you are just doing the best you can out there? . . . I am trying to clarify in my own mind exactly whether you have a comprehensive program to meet this need."

Baker, when able to respond, noted that the grant-supported research and the intramural laboratory activity were quite similar to other NIH institutes. But NCI had "gone further down the road of multidisciplinary integrated programming" in etiology and in chemotherapy, he said. An overall framework was being developed, the NCI director stated, to link the more loosely structured and the more tightly structured research into a single plan. But Senator Kennedy had obviously found what he considered a weakness of NCI in the absence of a comprehensive plan.

Other senators were very interested in the matter of the separate agency. Senator Thomas Eagleton (D., Mo.), observing that Dr.

David had criticized the NASA approach to cancer because fundamental scientific knowledge was unavailable, put this question to Dr. Egeberg: "Now if there were what appeared to be a comprehensive explanation of the complex phenomenon that we call cancer, what then would you think of a NASA-type approach? If we were, either on the threshold of that type of comprehensive explanation, or indeed, if we had it, how would you then view a NASA type approach?"[14] Egeberg said that under such circumstances he might view it more favorably than he did now, but even then he did not understand how the close association between cancer research and all other scientific fields could be avoided.

But, Eagleton pointed out, the Panel of Consultants recorded no dissents on the NASA-type approach. Were there any? Marston interjected that the panel's report clearly had not proposed a NASA-type approach if that meant a single approach to a cure for cancer. Furthermore, he added, the panel members had not reached agreement on the recommendation for a separate agency without difficulty:

> [M]embers of the Panel did meet with Secretary Richardson, and they emphasized the fact that . . . the single most difficult question that they struggled with . . . was the question of a separate authority for cancer. And there was considerable discussion by the chairman of the consultants and of members of the committee with the Secretary, and I was present for that.[15]

In other words, the NIH director was suggesting, the senator should not invest too heavily in seeing the panel's separate agency recommendation as easily arrived at or unequivocally held.

Senator Claiborne Pell (D., R. I.) was the next to ask about the space-program analogy: "Now . . . if the same effort had been put into the elimination of cancer that we did put into getting a man on the moon in . . . 10 years, and the same budget, in your view as intelligent medical men, do you believe we would have conquered cancer?" Steinfeld responded that "we are at a place now which we could not have anticipated 10 years ago," especially in virology and viral oncology. Marston was more explicit: "I want to be a little blunt and I want to say absolutely no, we would have put the money in the wrong areas on the basis of the knowledge that we have today. We would have chosen the wrong places to put the money 10 years ago."[16]

Toward the end of the morning, Kennedy returned to the matter of the separate organization. The "casual or intentional association" that had been suggested between an independent agency and a single approach to the cancer problem was simply not valid. "I think it is important," he said, "that with the establishment of a separate agency you are still going to have a multifaceted approach and a variety of different kinds of opportunities to meet the problems that are suggested by the disease." At the heart of the organizational question was an issue of judgment:

> I think finally that how we establish the most effective unit to house the administrative, scientific, and managerial skills which are going to be necessary to do the job is a matter of judgment. I know from talking to the members of the Panel that these things aren't black or white; there is so much that involves individuals who are going to take positions of leadership and responsibility. We can stress the obvious or apparent advantages of one and you can stress the apparent advantages of another, but we are going to have to make a judgment somewhere along the line.[17]

Those at the hearing were left with the strong impression that Kennedy had already made up his mind.

Senator Dominick was critical of S. 34. He indicated that he was receiving messages on both sides of the issue from his constituents. Through questioning the administration witnesses, he drew out the point that many discoveries of assistance in understanding cancer were likely to come from work totally unconnected with cancer research. He suggested that separating cancer from NIH might lead to pressures from other institutes for similar status, and Dr. Marston assured him that this would occur. For the record, he introduced letters he had received from several medical and scientific associations—"all of these opposing the separate agency and supporting the utilization of NIH or the Administration proposals."[18]

When the subcommittee recessed for lunch, it was clear that the differences between the administration and the subcommittee were not small, as Dr. Steinfeld had tried to suggest. The proposed legislation sought an independent cancer agency, while the administration favored keeping cancer research within NIH. Senators Kennedy and Javits had made it unmistakably clear that they favored the separate agency and would not be easily moved from that position. Among the other subcommittee members, only Senator Dominick supported the administration. The rest indicated no real differences with S. 34—which they had cosponsored.

ADVERSARIES

Frequently, congressional hearings are organized to allow the proponents of legislation to make the initial presentation. In the case of S. 34, however, the health subcommittee decided to draw out the opposition to the bill first, and then present the case in its favor. In addition to the administration's views, therefore, the entire first day was given over to testimony by the adversaries of the legislation.

The president of the Association of American Medical Colleges (AAMC), John A. D. Cooper, M.D., accompanied by Joseph Murtaugh, director of program and planning, testified immediately after the administration spokesmen. The AAMC has two main constituencies—medical school deans and medical school faculties (its testimony reflected the views of the latter)—the biomedical scientific community. The academic medical centers, Cooper noted, had been "in the thick of the battle" in cancer research, they were "the single most important source of research activity and research information in medicine and the life sciences," and they included the "scientists and institutions which have had much to do with the great progress in medicine which we have witnessed in our lifetime."[19]

The association had become engaged with the cancer legislation for a variety of reasons. For one thing, the extensive cosponsorship of S. 34 led AAMC officials to conclude that the Senate hearings would tend to be pro forma, rather than an open debate on the critical issues. They disagreed with the implication of the panel's report that administrative problems, not scientific ignorance, constituted the main barrier to the solution of the cancer problem. They also felt that the organizational implications of the bill had not been sufficiently aired.

Beyond this, the AAMC was trying to counteract the impression that there was extensive support for a separate agency for cancer within the scientific community. Representative Cornelius Gallagher had made it appear that a long list of scientists supported his resolution calling for a separate agency, and the AAMC had been alerted to the misuse of these names.[20]

Finally, the AAMC acted because it was receiving requests to do so from various sources within the administration. Edward David, after all, had essentially appealed to them at their February meeting for opposition to the Senate bill. Furthermore, conversations with NIH officials reflected a growing concern in Bethesda with the prospects for the legislative success of S. 34. And connections between the two organizations were very close, Joseph Murtaugh having been the strong right arm of James Shannon for many years before going to the association.

Cooper came armed with a resolution adopted by the AAMC assembly at its February meeting. The association "wholeheartedly" endorsed the "broad-based and intensive attack on the cancer problem" called for by President Nixon at the magnitude envisaged by the Panel of Consultants. It was important that such a program "be undertaken as an integral part of the existing national framework for advancement of biomedical knowledge for the nation's health."[21] This resolution had been reaffirmed by the Board of Regents of the American College of Physicians one week earlier, and by the American Association of University Professors of Pathology on March 8, the previous day.

The AAMC president was concerned with the relation of cancer research to medical research. Current scientific knowledge of cancer, he said, was "a consequence of broad advances across the full scope of the biomedical sciences," and further advances would depend on research "in the broad unknowns of biological processes and the nature of life." Pointedly, Cooper remarked, "we do not [now] possess the kind of scientific understanding and formulations that underlay the great development of our space efforts and the Manhattan Project."[22]

Concern for cancer research led immediately to a concern for
the institutional setting of medical research. The NIH, Cooper observed, had become the world's greatest institution for the support
of biomedical research. Within the NIH, each institute contributed to understanding the fundamental problems of disease
through a complicated pattern of interaction with the other institutes. Two vital conditions, Cooper argued, were essential to a
scientific victory over cancer:

> The direction and conduct of our new cancer effort must be in a
> framework that assures close integration of this work with the
> full scope of scientific endeavor in biology and medicine, [and]
> . . . national policies must provide for a full and vigorous culti-
> vation of the broad base of biomedical research, particularly the
> young men and new ideas, as well as the special efforts to
> exploit the promising leads of cancer.[23]

The appropriate framework for cancer was provided by the National Institutes of Health.

The AAMC was also deeply troubled by the effects a separate
cancer agency would have upon NIH. "If the cancer institute is
torn out of the closely woven fabric of research relationships
within NIH," Cooper warned, "the losses would far outweigh
such advantages as might accrue." An independent cancer authority, he continued, "may begin the destruction of the National Institutes of Health and their vital role in coordinating the scientific
attack upon disease. . . . It can only result in an unfortunate and
unwise dispersal of this great scientific strength. This fragmentation of the biomedical sciences, we feel, will not be in the best
interests of this Nation."[24] Senator Kennedy questioned Cooper
very closely on this point, and seemed far from persuaded of its
validity.

Kennedy then questioned Cooper on a more sensitive issue.[25]
He introduced fiscal 1969 federal budget figures for medical
schools. The total was $770 million: $100 million was for construction, $540 million for research, $70 million for noneducation
programs, and 80 percent of the remaining $67 million was for
education. "So we naturally gather," Kennedy asked, "that the
medical schools themselves, who are in the great financial crisis

. . . are very dependent upon . . . research grants from the National Institutes of Health, are they not?" That was true, Cooper responded, and support for research and research training should be increased. But "a proper balance" was needed and the support of "other programs" should also be increased so the outputs of health professionals could be maintained.

The senator asked: "Do you see the establishment of a separate agency in any way threatening that research resource in terms of the medical schools?" Not the level of support, Cooper replied, but the kind of research supported:

> The only thing we would be concerned about would be that the new authority would reduce the kind of support which the medical schools are now receiving, for the kind of research which they do best and would, in place of this, devote most of its efforts toward contract or directed research, which over the short run is important and necessary but which will not be effective over the long term unless we continually bring in new information from basic research.

Did he see anything in the separate agency approach, Kennedy asked, that threatened basic medical research? "No, sir," Cooper said. Did he have any reason to believe basic medical research might be threatened? "Not to our knowledge," the AAMC president responded, but that depended on how the program was organized and administered.

Kennedy was needling Cooper. If an independent agency for cancer would not jeopardize the research funds flowing to medical schools, why the AAMC opposition? There was another unspoken irony in the situation. Health manpower legislation, which would provide substantial financial support for medical education, was at that moment being considered by the Senate subcommittee. This legislation, far more than the cancer bill, was the top legislative priority for the AAMC and the medical school deans in 1971. Who should be championing the position of the association in opposition to the Nixon administration? None other than Senator Kennedy. The AAMC, obviously, was in a slightly awkward position in opposing him on S. 34.

The next witness was Dr. Howard A. Schneider, chairman of

the public affairs committee of the Federation of American Societies for Experimental Biology. Schneider, an enthusiastic and energetic witness, referred to those who had preceded him: "We have heard from the chiefs today, but I will now speak for the Indians. I come from the reservation to speak for 11,000 biomedical scientists who stand at the benches and the laboratories and do the work."[26] Though the biomedical scientific community applauded the goal of "the conquest of cancer at the earliest possible time," Schneider said his intent was to address the strategic flaws in S. 34.

Forcefully, Schneider asserted that the dismantling of the NIH would be "counterproductive" and might "leave us worse off than when we began." "My constituents," he continued,

> operate in a climate which is determined in large part now by . . . the National Institutes of Health. When [NIH] and their programs . . . are thriving, all of American biomedicine . . . thrives too.
>
> The reverse is also true. When the NIH shivers, American biomedicine . . . sneezes. Now the overriding scientific reality is that the whole scientific effort in cancer itself is similarly interwoven with the advancing knowledge of many fields so that the plucking of the National Cancer Institute from the . . . supporting cooperation supplied by the many diverse life sciences within . . . the National Institutes becomes a scientific blunder which can but hinder coordination rather than help. . . . The disappearance of the National Cancer Institute and its ostensible replacement by some kind of super agency, will be the beginning of the ultimate piecemeal destruction of the National Institutes of Health. This we deplore.[27]

Senator Kennedy responded that were he convinced that those Schneider represented were threatened in any way by S. 34, he would not have sponsored it and would have actively opposed it. Rather, the senator thought, S. 34 "could very well work to expand the understanding and knowledge of the biomedical area in terms of additional resources, energies, and creative efforts."[28]

The final witnesses that day were Philip R. Lee, chancellor of the University of California's medical school in San Francisco, and Mr. James F. Kelly, vice-president for administration of Georgetown University, Washington, D.C. Both men were former high-ranking HEW officials—Lee as assistant secretary for health and scientific affairs from 1965 to early 1969, and Kelly as comptroller from 1965 to 1970. Lee applauded Kennedy for his initiative in "reordering our national goals and priorities," in bringing "health to the center of the stage as an area of prime public policy concern," and in reversing "a perilous downward drift in the support of biomedical research."[29]

In his judgment, Lee said, the Panel of Consultants had raised three questions. First, was health really a major concern of the American people and were they willing to increase support for biomedical research? Health, he observed, had "moved close to the top in the list of national priorities" in the past decade, and he believed the public was prepared to increase its investment in biomedical research "if they had some reasonable assurance that this would, in fact, reduce or eliminate some serious disease problems." The second question was whether cancer was of special concern to the people, and whether existing research gave an expanded attack "a reasonable chance of success."[30] There was no doubt of its special concern, he indicated, and now the Panel of Consultants, the subcommittee, the president, his science adviser, and the officials of HEW and NIH concurred that a major cancer research effort had good prospects for success.

The heart of the matter, the third question, was "what is the best mechanism to achieve our objectives?" Many studies had examined the NIH over the years, Lee noted. The Wooldridge Report, probably "the most comprehensive" in the decade, had concluded that the NIH was "essentially sound" and its budget was "being spent wisely and in the public interest." His own personal experience with NIH over fifteen years, Lee said, as a grant recipient, as HEW assistant secretary, and as head of a major medical school, led him to "express my grave concern and my strong opposition to the specific proposal . . . to establish a National Cancer Authority and amputate the National Cancer Institute from the National Institutes of Health." A successful battle against cancer would require collaboration among the several NIH institutes like that which had been typical in the past. He was particularly distressed, Lee said, that if cancer was given autonomy, then "next year the advocates of heart disease would begin asking for a new National Heart Authority [and] the following year it might be arthritis, alcoholism, or mental retardation."[31]

The former assistant secretary had been instrumental in the reorganization of HEW health functions during 1966-68. He reflected that "major problems, loss of momentum, and other difficulties can arise with major organizational changes," and urged the subcommittee to give careful attention to this point. Kelly reinforced him: "we found . . . by trial and error . . . that an organization is a fragile thing . . . when you start to move boxes around . . . you [have] a disruptive influence."[32] Reorganization was appropriate, in Kelly's mind, only when the purpose for doing it was very strong and the existing organizational arrangements very weak. Neither condition was present in this case, he judged.

Not surprisingly, given his fifteen years as budget officer and comptroller of HEW, Kelly questioned the wisdom of budget autonomy. In his experience, he said, the NIH director had exercised greater independent status than any HEW agency head he had known, but not in determining his budget level.

> I just don't believe that you can create an organization that will determine its own budget level. You may establish one that . . . has greater influence over this determination, but I think that it is determining the level of funding, which is essentially a presidential recommendation to the Congress and a congressional decision, which is the greatest influence and deterrent over what an administrator can and cannot do.[33]

On budget matters, in short, higher authority always determined priorities.

Lee introduced two letters into the record. One from Charles J. Hitch, president of the statewide University of California system, to Senator Alan Cranston, characterized a separate cancer agency as "an ill conceived approach" that "would retard rather than aid the achievement of the goals we all seek."[34] The other was from James Shannon to the chairman of the Department of Medicine at the University of California, San Francisco.

Shannon, writing about the cancer initiative, indicated he was highly pleased with the apparent national rededication to a research attack upon the "as yet intractable" disease problems, and with the recognition that the "science base" in neoplastic disease was "far less broad and solid" than had been true in the space and atomic energy fields.[35] This satisfaction was offset somewhat by "uncertainties" over proper and proposed funding levels and "by deep concern about Congressional views on the authorities and organizational arrangements" for the cancer program.

Shannon was mainly concerned with the governance of biomedical research. "Above all," he wrote, NIH symbolized "a set of processes for the governance of the orderly growth and development of science," and was "an invaluable and irreplaceable guarantor to the nation that order, stability, sound judgment, balance, flexibility, responsiveness, and responsibility" would guide medical research's attack upon disease, disability, and death. A separate cancer agency would not accomplish anything that could not be done within NIH. On the other hand, such an agency would be "divisive" and threaten the integrity of NIH. "Orderly governance would be replaced by anarchy," Shannon predicted, and "program emphasis would be entirely determined by uncritical zealots, by experts in advertising and public relations, and by rapacious 'empire builders.' " These latter forces had played an "invaluable role" in the past in making the lay public aware of a real possibility to control disease through research. "As forces modulating the scientific judgment process," he wrote, "their contributions have been positive and important. As determinants, however, I would expect them to create chaos." The former NIH director clearly viewed Mary Lasker's most recent effort as a mixed blessing having prospects for genuine progress and for real mischief.

Shannon felt the panel's report reflected the frustration that occasionally overcame program people within NIH. "In my years as Director," his letter concluded,

> scientific judgments frequently got lost in, or unnecessarily and improperly diluted by, the bureaucratic machinery at higher levels in DHEW. Fortunately, on major issues we were usually able to place our case before the Secretary and thereby restore

proper perspective. If some way could be devised to facilitate access to the Secretary, DHEW, by the Director, NIH and the Director, NCI, something useful would have been accomplished. . . . The problem of access to the Secretary is, in my experience, the really critical one.

With Robert Finch gone from the secretary's position, and with Elliot Richardson now on the scene, Shannon was suggesting that a satisfactory arrangement might be possible. But the problem, quite clearly, was *not* access to the president.

Kennedy had spent much of the morning slouched in his swivel chair, listening to whispered comments from an aide. He had roused himself from time to time to challenge witnesses and ask pointed questions of them. Now, shortly before 1 p.m., he adjourned the subcommittee until the next morning when the case for S. 34 was to be made.[36]

ADVOCATES

Benno Schmidt, Mary Lasker, Luke Quinn, and several panel members had listened to the opposition testimony that first morning. They returned to the hearing room a few hours later to prepare for their own Wednesday morning appearance.[37] They were joined by the other members of the Panel of Consultants and by the representatives of the American Cancer Society who would give testimony the next day. Schmidt took the panel members through the carefully written twenty-four page statement they would present. Quinn, who had coached many citizen witnesses over the years, conducted much of the tactical discussion. Talk centered on ways to respond to the objections that had been raised to S. 34. The panel's distinct advantage, of course, lay in its opportunity to present its case and rebut its critics on the last of this two-day hearing.

Only Senators Kennedy and Javits were present when the subcommittee met at 9:30 a.m., Wednesday, March 10. They heard first from Senator Charles Mathias, Jr. (R., Md.), who proposed that Fort Detrick, formerly the U.S. Army's biological warfare research center, be converted to use for cancer research.[38] This proposal would surface from time to time throughout the hearings and would be affirmatively endorsed by the president later in the year.

The two senators greeted their many old friends among the panel members who were there that morning. The witnesses for the Panel of Consultants were Benno Schmidt, Dr. Sidney Farber, Dr. Lee Clark, Dr. James Holland, Dr. Henry Kaplan, Dr. Joseph Burchenal, and Mrs. Anna Rosenberg Hoffman. Dr. Mathilde Krim, also present, did not testify.

The congressional mandate, Schmidt began, included S. Res. 376, which had established the panel, and H. Con. Res. 675, which expressed the unanimous sense of Congress that "the conquest of cancer should be made a national crusade." So armed, the panel's report had reviewed the current status of cancer research, identified the areas of greatest promise, and indicated the steps to be taken to make the conquest of cancer a major national goal. But, the panel chairman indicated, since a number of the panelists had been present on the previous day, their intention was "to devote our time this morning primarily to the issues which were the subject of discussion here yesterday . . . to refer briefly to certain of our findings and recommendations which have raised questions and, in some cases, have invited opposition."[39]

Why the sudden interest in cancer at this particular time? Schmidt had been asked. It was the number one health concern of the American people, he responded, feared more than any other disease. Cancer struck at human life and human dignity, inflicting great suffering and financial hardship on individuals and families. There was little doubt, Schmidt said, that the American people were behind the determination of the Congress and the administration to give a higher priority to the cancer problem. Beyond this, "those most familiar with the cancer problem, the scientists and professional men who have spent their lives working in this field, feel that the time is especially right today for an intensified and accelerated effort."[40]

Schmidt asked Dr. Holland of Roswell Park Memorial Institute, and president of the American Association for Cancer Research, to discuss the scientific advances that had led cancer researchers to conclude that the time was right for an expanded effort.[41] The fundamental advances in molecular biology, Holland noted, related to the ways in which genetic information was stored and transmitted, had permitted investigation of the abnormalities of the cancer cell. Also, he said, viruses had been proved responsible for various cancers in frogs, chickens, mice, rats, guinea pigs, rabbits, cats, dogs, and monkeys, and it was expected that there would soon be definite proof that viruses caused certain human cancers. Further research had suggested that the immune response of the body bore a relationship to breast cancer, skin cancer, bone cancer, malignant melanoma, neuroblastoma, and acute leukemia, and a number of studies were pursuing the implications of this research.

In addition to these major developments, there was increasing evidence, Holland pointed out, that many chemicals in the environment could cause cancer in experimental animals and man, and this warranted further research. In chemotherapy, he said, thirty-two different drugs had been developed that demonstrated useful, reproducible anticancer activity, and these advances, together with new techniques in surgery and radiology, had led to new therapeutic strategies. Radiation research and therapy had progressed, both in terms of energy source and means of delivery, and cancers once thought susceptible to palliative treatment only were now being cured. A "broad and powerful wave of advances in cancer research" had occurred, Holland said, which meant that "a greater [research] effort can and should be sustained, one that can and should be successful."[42]

Schmidt then reiterated what the panel had felt were the three missing ingredients in the existing organization for cancer—"an effective administration," "a comprehensive national plan," and "the necessary financial resources." He asked Lee Clark to amplify on the need for effective administration. An expanded cancer program, Clark responded, would require a "simplification" of organization, and a "drastic reduction" in the number of people involved in administration.[43] The desired organizational efficiency did not exist within the National Cancer Institute, and the panel doubted whether a new expanded program could be organized within NIH and HEW. In the past, he noted, the federal government had given responsibility for scientific projects of comparable magnitude to an independent agency. This was the reason why the panel felt a National Cancer Authority, separate from NIH, was essential. Only through this means could conquest of cancer become a national goal.

Then, Clark pointedly addressed the relationship between program and administration. "We believe," he said,

> that it is important to get this program out from under the six tiers of bureaucracy . . . and have an Administrator responsible for cancer who is not subordinate to those responsible for 11 other health institutes and multiple health programs. Only in this way will cancer be given the kind of emphasis implicit in the congressional mandate. We believe that results should be more important to the administration, the Congress, and to the American people than preserving the apparent organizational symmetry which would seem to be preserved by leaving the cancer program in the National Institutes of Health.

Clark concluded that a separate agency need not isolate the cancer program from the rest of the scientific community. On the contrary, it would permit full mobilization of the relevant scientific effort, he said, "but at the will of those responsible for the cancer program and not of those who wish to preserve the status quo."[44] Organization, in short, was seen as a major determinant in the progress toward the conquest of cancer. That and control of the effort by the cancer research community.

This prompted Schmidt to respond to the criticism that the space program analogy was inappropriate to cancer. "The valid analogy," he argued, was not the scientific analogy but the organizational analogy. In order to succeed, he continued, the cancer program needed "the same independence in management, planning, budget presentation and the assessment of progress" that the space program had had.[45] In this respect, Schmidt said, the independent authority analogy was valid, but it did not mean physical relocation of the cancer program, nor fragmentation, isolation, or loss of grant-supported research. Happily, the space program analogy could be invoked where persuasive and ignored where questionable.

Dr. Henry Kaplan was asked to discuss the need for a coherent and systematic attack on cancer. However, the Stanford radiologist spoke principally of preserving and strengthening the role of basic science in the cancer research effort. The panel, he said, had been keenly aware of the need for a balanced, integrated set of recommendations that would maintain the quality of cancer research, especially basic science, while the program was being expanded. While some areas of promise in cancer research could not be exploited at the applied research level through the contract mechanism, many areas still lay within the realm of basic biomedical science. Basic research, which "must not be hampered and restricted," Kaplan asserted, flourished when two administrative features were used: "the grant mechanism of support, and . . . a peer review system to evaluate quality and relevance of research proposals." The panel had attempted to balance these considerations in their recommendations, he continued, but S. 34 could be improved, and "the scientific community . . . firmly reassured," if the language of the bill could be amended to restate the commitment to an increased emphasis on the grants mechanism and the principle of peer review.[46] Such action, he suggested, would result in the full and enthusiastic endorsement of the cancer effort by the nation's basic biomedical scientists whose contributions would be essential to the program's success. Schmidt chose Kaplan to discuss the "comprehensive national plan" in order to deemphasize the images of directed research that had been conjured up by the panel's report. But while Kaplan was seeking to allay the fears of the scientific community, he was also displaying his own latent anxieties, as would be quite clear at a later time.

Dr. Joseph Burchenal then testified about the financial resources required for the cancer program, simply restating the recommendation of the panel's report. He provided no further justification for the funds requested, but since the president had already asked for an additional $100 million, it was not considered necessary to devote much attention to this matter. Indeed, one of the intriguing aspects of the entire legislative debate was that no one felt compelled to justify the recommendation for more money and few thought it prudent to attack it.

The hearing, which had fewer observers in attendance than the previous day, droned on. The script was being closely followed, down to the thank you's and introductions. One physician dutifully read, "Members of the Subcommittee," even though Javits had left and Kennedy was the only senator present. But Kennedy was listening carefully. Where he had slouched in his chair the previous day, he was now sitting up straight and taking notes.[47] And there was no more attentive person in the room than Senator Kennedy when Benno Schmidt asked Sidney Farber whether the goal of conquering cancer was a realistic one.

"I have no question," Farber responded, "that this goal is a realistic one." Though cancer was not a single disease, and would probably not yield to a single form of treatment, Farber was certain that prevention of cancer would be extended due to "vast new insights" in science. Moreover, the cure rate for cancer could be increased from one-third to one-half of all patients if only "we could apply for the benefit of every patient . . . the knowledge and early diagnosis and treatment which is available in the centers of expertise." Summing up, with his infectious optimism, Farber said:

> One by one the diseases which we identify as cancer will yield. Based on the new insights with which I am familiar there is no question in my mind that if we make this effort today, and if we plan it, organize it, and fund it correctly, we will in a relatively short period of time make vast inroads in the cancer problem as we know it today leading to the eventual control of cancer.[48]

It was left to Anna Rosenberg Hoffman to deliver the forceful conclusion to the panel's testimony. "[W]hen I was Assistant Secretary of Defense," she recalled, "I appeared before . . . the Senate time and time again to ask for sums so great that the money we are asking for today for the fight against cancer seems infinitesimal." Comparing the $410 per person expenditures for national defense with the 89 cents per person spent for cancer research, she added, "Gentlemen, I believe that we are asking for too little, and personally I would have asked for more."[49]

Mrs. Hoffman thought of herself as an "expert in bureaucracy," and her Pentagon experience had taught her what it meant to go through "layers and layers of so-called authorities" to get a decision. Yes, she felt, the space agency was a good analogy and without it there would not have been a man on the moon. Then, in a peroration intended for an audience larger than the single senator present, she said:

> Senator, you and the members of the U.S. Senate have the opportunity . . . seldom given in the lives of men—even Senators—to turn on the power that eventually could save the lives of hundreds of thousands of men, women and children in the United States and pass on that knowledge all over the world and the name of America would be blessed. You and I have known some of your ablest colleagues who might have been saved and the many dear ones in our own families who still can be saved if we waste no more time and let S. 34 be our next "man on the moon."[50]

On that elevated note, the panel concluded its formal presentation.

Senator Kennedy wanted to know about the effect of the proposed new agency upon the NIH. The *Washington Post*, in a story on the previous day's hearing, had made much of the charge in Shannon's letter that emphasis in research priorities was in danger of being determined by "uncritical zealots, experts in advertising and public relations and rapacious 'empire builders.' " "Not guilty," Schmidt replied. He regretted the appearance of the statement in the press, thought it was shocking for Shannon to have made it, and expressed disbelief that he could have really meant it. Then, throwing back the challenge, Schmidt added: "I think what he really meant to say is that . . . I had too much to do with building the NIH to be in favor of anything that takes anything out of it."[51]

Why did the panel wish to disassemble the NIH? Senator Kennedy asked, since the charge was that this would be the effect of a separate agency. The panel harbored no such desire, Schmidt responded. On the contrary, he said, "we would have liked to have concluded that this program could be conducted right where it is because the last thing that any member of this panel wanted to become involved in, and the last thing I wanted to become involved in, was any kind of a fight over the details." But, he added, and this was the telling point, "[w]e want a national cancer program and we want it to succeed."[52]

Kennedy also asked Schmidt to identify the six levels of bureaucracy above the Cancer Institute. Schmidt pointed out that the director of NCI reported to the deputy director for science of NIH, the NIH director, the Surgeon General, the assistant secretary for health and scientific affairs, the under secretary, and the secretary of Health, Education, and Welfare. But it was the effects of layering—inordinate delays, ill-defined authority, and unclear responsibility—rather than the fact of layering itself, which the panel found objectionable.

The senator asked the scientists on the panel if a separate cancer agency would have a limiting and harmful effect upon basic research. Dr. Holland did not think the organizational setup would change the scientific activity of participating scientists. Further, reflecting the view of the cancer research scientist, he said: "There is nothing wrong with working on the problem [of cancer]. I think it is a myth that those who are working on cardiovascular disease are more likely to find the answer to cancer than those who are working on cancer."[53] Dr. Kaplan, while supporting this view, reemphasized that reliance upon grants, rather than contracts, and upon the peer review process, would insure that fundamental research would be adequately supported. Concurrently, new multidisciplinary cancer research centers would permit the rapid exploitation of new scientific leads.

Senator Hubert H. Humphrey (D., Minn.), back in the Senate only a few months, appeared later that morning to support S. 34.[54] He had seen a brother and a son go through the agony of cancer. As vice-president, he had been chairman of the Space Council and strongly supported the view that the space agency-type of organization was more suitable to meeting urgent national needs than was the regular bureaucracy. Characteristically, he attacked the "piecemeal and penny-pinching approach" of President Nixon, commended the panel and Senators Kennedy and Javits, and urged them to report a bill out soon.

Schmidt said, as the hearing ended, "The only regret I have in the situation as it has emerged is that there seems to have developed a conflict here between this panel and what is called the administration view." The panel chairman reviewed what he had done in seeing George Shultz, in attempting—though unsuccessfully—to see Elliot Richardson, and in working closely with Carl Baker. Schmidt did not want differences to become political: "The problem is bipartisan and nonpartisan." The lay members of the panel were about equally divided in partisan terms, and he had no idea what were the politics of the medical and scientific members. But all members were dedicated to the cause of the conquest of cancer. "So we want no political conflict in this thing," Schmidt said.[55] Noble sentiments, to be sure, but it was the fact that cancer had become a political issue that would greatly contribute to the realization of the objectives being sought by the panel.

On the afternoon of the 10th, two representatives of the American Cancer Society testified in favor of S. 34.[56] They were Dr. H. Marvin Pollard, the current president, and Dr. Jonathan E. Rhoads, the immediate past president and member of the Panel of Consultants. The session did not begin until 2:50 p.m., and was once interrupted by a vote on the Senate floor on enfranchising eighteen-year olds. The statements were brief, and supported the testimony of the Panel of Consultants. Senator Kennedy's questioning was perfunctory.

IF CANCER, WHY NOT HEART?

Dr. Campbell Moses, the medical director of the American Heart Association (AHA), was the final witness of the day. He endorsed the principle implicit in S. 34, that there should be "the implementation of a major effort coordinating and utilizing all of the strengths of the Federal Government to achieve the conquest of cancer at the earliest possible moment." A "dangerous cutback" was occurring in the support for categorical disease control in state health departments, in the abolition of the chronic disease program of the Public Health Service, and in the reduction of support for the regional medical programs in heart disease, cancer, and stroke. If the principles embodied in S. 34 were enacted, and the full Federal resources aimed at cancer control were coordinated in a massive effort, then "we reasonably could move much more rapidly toward the conquest of this disease."[57]

Moses made it perfectly clear, however, that the support of the AHA was contingent upon the prospective benefits that would accrue to heart disease research from such backing.

> We must emphasize, however, that if the state of research in cancer makes reasonable such a comprehensive effort to the control of cancer today, an exactly parallel effort is even more appropriate in the field of heart disease. In the cardiovascular field, we know that the expansion of our research effort, and the comprehensive full-scale application of the fruits of already available research and technology, would save lives now.[58]

The claim for parallel treatment was strong. Heart disease research was more advanced than cancer research, and heart disease was the number one killer disease while cancer was second some distance behind.

Moses acknowledged that the question of a separate authority outside of HEW had been "hotly debated in and out of Government." Speaking for his organization he said:

> If faced with the same opportunity for the cardiovascular field and with full recognition of the recent cutbacks in categorical support, we in the American Heart Association would opt in favor of a separate authority analogous to that outlined in Senate bill 34. Structuring all of the categorical programs at NIH as an authority reporting directly to the President via the HEW Secretary might be even more advantageous for advancing the health of the Nation.[59]

Would spinning off the separate agency from NIH lead to a dismantling of that organization? Senator Kennedy asked. Surprisingly, Moses thought not and said instead that it might provide a new opportunity and new resources to achieve the goals that NIH was set up to achieve. Kennedy thanked the AHA representative for his comments and observed, "There are those who say if you can't get a raise for yourself, the best thing that can happen is for the fellow next to you to get a raise."[60]

The opponents of S. 34 had argued that creating a National Cancer Authority would create pressures for similar treatment from other categorical research interests. The testimony of the

AHA's medical director was explicit confirmation of this argument. The threat of the fragmentation of the National Institutes of Health had suddenly become very real. It sent a cold shiver through officials at the NIH and through many of their constituents in the biomedical scientific community. Mobilization to encounter S. 34 was set in motion.

The initial Senate hearings on S. 34 suggested the main lines of the legislative debate that would follow. All parties agreed that cancer deserved a substantially higher priority than it had at that time. But the debate about scientific opportunities in cancer was so general as to be uninformative. The opponents of S. 34 made no explicit or detailed attack on the underlying assumption of the panel that major scientific opportunities existed for cancer. But the private views of many knowledgeable scientists, however, were that neither major scientific nor clinical opportunities existed. The silent acquiescence to the panel's argument by the opposition to S. 34, therefore, suggested a tactical accommodation to a view of political reality that this was not an arguable issue. Furthermore, on the question of the appropriate level of resources that might be required for an expanded cancer research program—a matter closely related to priority, there was also little attention given. In short, no challenge was made by the opponents of S. 34 to the claim that cancer deserved higher priority even though many had reason to doubt the claim.

Though repeated references were made to the importance of maintaining the grant mechanism and peer review, the philosophy and strategy of research management was hardly touched upon. The panel, it appeared, did not wish to make much of real management issues. The opponents of S. 34 had little to say beyond a defense of existing NIH practice.

The primary political issue that dominated the debate was organizational. The advocates of S. 34, including Senators Kennedy and Javits, basically argued that success in cancer research would be enhanced by organizational autonomy from NIH and by removing NCI from the layers of bureaucracy in the Department of Health, Education, and Welfare. But there was practically no questioning of the analytical support for the panel's position, nor any detailed examination of actual administrative problems by opponents of such drastic organizational change. Shannon was probably the most analytical, but the language of his letter was inflammatory in a way that negated the force, if not the logic, of his argument. Opponents of a separate agency concentrated their fire on their assessment of the probable adverse effects of reorganization on the entire biomedical research enterprise. The positions taken on the organizational question strongly support the view that conflict over organization is politics in another guise. The style of debate reflected the consistent, nonanalytical approach to these issues by both advocates and opponents.

7

Mr. Dominick's Number,
Mr. Kennedy's Bill

*'When I use a word,' Humpty Dumpty said in rather a scornful tone,
'it means just what I choose it to mean—neither more nor less.'*
*'The question is,' said Alice, 'whether you can make words mean dif-
ferent things.'*
*'The question is,' said Humpty Dumpty, 'which is to be master—that's
all.'*

Lewis Carroll
Through the Looking-Glass

The two days of March hearings were described by one report as
"strictly educational" for the Senate health subcommittee,[1] since
the eight Democratic members and two of the six Republicans
were among the co-sponsors of S. 34. But political education on
controversial legislation occurs in many ways and is hardly limited to formal congressional hearings. Nor is the audience limited
to members of Congress, but includes the president, his administration, the permanent bureaucracy, partisans for and against the
legislation, the press, and the public. Finally, the education is
never exclusively focused on the issues in conflict, but also involves identification of the areas of agreement between advocates
and adversaries, and clarification of the intensity, political resources, and willingness to compromise that each brings to the
debate.

The Panel of Consultants had worked behind closed doors during 1970 in the preparation of its report. The carefully laid
groundwork with the Senate health subcommittee insured that the
initial reception of its report would be favorable. But the hearings
on the Conquest of Cancer Act thrust the proponents of an expanded cancer program into a new phase of the policy process in
which the task of education was far more complex. They had publicly signaled the seriousness of their intentions, and this action
invited equally serious response.

THE ADMINISTRATION'S POSITION: APRIL

Even as Benno Schmidt was deploring the fact, the debate over S. 34 was coming to be seen as a struggle between the Kennedy subcommittee and the Nixon administration. The lines between the opposing camps were drawn more firmly during March and early April, and supporters of each group were lining up accordingly.

The administration had found much support from the scientific community as a result of Dr. David's February speech. It received further encouragement, while the position of the Panel of Consultants was significantly damaged, when Joshua Lederberg, one of the panel's members, withdrew his support for the separate agency proposal. On March 10, Lederberg, probably the most well known medical scientist on the panel, wrote Senator Kennedy and explained his change of view. He had favored a separate cancer agency initially, but several recent developments had led him to believe that such an agency "is not the best choice among new directions." Specifically, these developments included:

> the President's new-found but unambiguous commitment to health research in general, and to cancer in particular; evidence of a new balance, consistency and efficacy in the administration of HEW under Mr. Richardson's leadership, and the commendable and remarkably articulate commitment to the integrity of basic scientific work in the health field as expressed by the new science adviser to the President, Mr. Edward David.[2]

The Stanford geneticist and Nobel laureate had found David's arguments against separating cancer research from the NIH to be persuasive. The problems that had beset medical research, he wrote, were less "attributable to administrative weaknesses within NIH," than to "the hindrances that NIH itself had to face within the HEW bureaucracy." Since that issue had received well-deserved publicity, Lederberg continued,

> I would now advocate that the Cancer Agency *not* be separated from NIH. Rather all health research should be knit together within a single agency, specifically an augmented and strengthened NIH, administered by an Assistant Secretary for Health Research reporting directly to the Secretary. . . . [T]he

country would be better served by reinvigorating the NIH as an integral unit than by inventing new streamlined procedures for the benefit of an excised and transplanted cancer authority.[3]

The panel, whose unanimity had been frequently cited by Schmidt, now had a minority of one.

But the voice of this single scientist was influential. *Science*, in its March 26 issue, proclaimed, "Lederberg Opposes Cancer Authority." To insure maximum attention, the event was highlighted by a thick black line around the entire story. "The campaign to wrest cancer research from the National Institutes of Health," the account read, "suffered a setback last week. Joshua Lederberg, a member of the commission that first called for the separate National Cancer Authority, defected to the pro-NIH forces."[4] The scientific press lost no time in exploiting this crack in the position of the Panel of Consultants.

Also in early March, the General Accounting Office (GAO) issued a report, *Administration of Contracts and Grants for Cancer Research*, which had been formally requested by Senator Yarborough the previous September.[5] The study had found lengthy delays in the awarding and funding of both research contracts and grants by the National Cancer Institute. This lent support to the claims of the Panel of Consultants that NCI was hobbled with bureaucratic red tape. On the other hand, the recommendation that contract authority be delegated to the NCI from the secretary of HEW, Elliot Richardson had written on January 21, was already being put into effect.[6]

The GAO report also recommended that NCI program managers be permitted to award research grants up to a specified dollar limit "without review by study sections."[7] The secretary indicated that this idea would be considered in future evaluations of the project grant review system. But since the Panel of Consultants was then seeking to assure the scientific community that a separate agency did not mean an attack on the peer-review system, the appearance of this recommendation was rather untimely.

Supporters of S. 34 were also beginning to confront active opposition from the administration. Richardson, secretary of HEW since June 24, 1970, adopted a low profile and played essentially a behind-the-scenes role.[8] He and Edward David, for example, were the principal authors of the proposal contained in the president's health message of February 18, a less drastic organizational alternative than the proposed National Cancer Authority. The secretary also visited Senator Javits and urged him to support the administration's proposal, but the New York senator declined and indicated his intention to co-sponsor the Kennedy bill. It was also revealed that Richardson's failure to testify for the administration on March 9 was purely political. The policy of the administration was to resist, unless absolutely necessary, sending a cabinet member to testify before any potential 1972 Democratic presidential candidate. This reduced the possibility of extensive publicity for that candidate if a confrontation occurred.

Robert Marston was encouraged by Richardson to lobby actively on Capitol Hill, and he spent a good deal of time visiting key senators, especially those on the subcommittee.[9] He and Benno Schmidt wound up on one occasion providing one senator with an hour-long discussion of the pros and cons of a separate agency. Afterwards, Marston recalled, "we both stood around and kicked stones outside and thought it would be a good idea to have the debate in front of the committee."[10] That "good idea" never materialized.

Marston also visited Senator Warren G. Magnuson, on March 19, at the senator's request. Magnuson, as chairman of the Senate Labor-HEW Appropriations Subcommittee, wanted to talk about the president's proposed additional $100 million for cancer research. The senator, who favored more money for cancer research, was opposed to creation of a separate cancer authority even though he had been visited by Mary Lasker, Luke Quinn, and Mike Gorman.[11]

Dr. Leonard Laster, an aide to Dr. David in the Office of Science and Technology, did a good deal of work on the Hill for the administration along with Marston. Laster dealt mainly with the staff to individual senators, however, and when a senator expressed interest, science adviser David would follow this up with a personal visit. Senator Richard S. Schweiker (R., Pa.), who reportedly was "leaning" against a separate cancer agency, received such a visit from David.[12]

Beyond these direct approaches, Marston was also seeking to mobilize the scientific community to make their views known to their individual senators. Contacts were made with medical-scientific societies and to individual NIH grantees. They were asked to write and, if possible, visit their senators and urge them to oppose a separate cancer authority. Benno Schmidt complained in mid-March, "Marston is telling everybody getting grant dollars from the NIH, 'get in here and help us out.' Marston is trying to mobilize the NIH constituency against a separate authority."[13]

Inside the bureaucracy, however, things were in disarray from mid-February until early April. The president's health message had proposed the creation of a cancer conquest program whose director would be in the office of the NIH director, and who would oversee all cancer research done by any of the nine institutes. The officials of the National Cancer Institute, not surprisingly, were quite concerned about the prospects that NCI would be downgraded if the proposal were implemented. NIH officials were disturbed by the prospects of conflict between a "cancer czar" giving direction to individual institutes, and each institute director trying to run his own program and organization. Marston, in fact, held a retreat outside Washington on March 5 and 6 on the problems associated with the president's plan. "The whole purpose of the meeting," one participant said, "was to settle people down a bit, to try to keep the family together."[14]

There was good reason then, why the administration's testimony on March 9 had been a bit vague on details of the proposed new arrangement. When Dr. Egeberg indicated that administrative plans were being developed, Senator Kennedy asked him, "Could you then, Dr. Egeberg, submit to us in as great specificity as possible exactly how you expect to formulate the kind of priorities in the NIH that you have commented on here this morning?"[15] The senator repeated the same question in a letter of March 17 to Secretary Richardson.

A general HEW strategy session was held in Richardson's office on March 19.[16] The secretary, Edward David, Leonard Laster, Robert Marston, and Carl Baker were present, as well as several other HEW officials. A memorandum drafted by the HEW deputy assistant secretary for health and scientific affairs was the basis for discussion. The meeting led to a formal response by Richardson on April 2 to Senator Kennedy's letter.

The secretary's letter suggested that the objectives of the HEW were "identical in intent" with those of S. 34, that the issue was the best means of achieving "our mutual objectives," and that this could be done within the NIH "without compromising in any way the basic spirit and intent" of the bill or the recommendations of the Panel of Consultants. The proposed cancer conquest program had three objectives, Richardson indicated. The first was to provide a context for mobilizing scientific and managerial resources for the attack on cancer. The second objective was to assure that increased, continuing priority would be given to progress in cancer by DHEW and NIH officials. The final objective was to build upon the existing strengths of NCI and other NIH institutes in cancer research, and on NCI's "significant experience . . . in the adaptation of new management techniques to biomedical research planning."[17]

A complex plan of organization, developed initially and largely within NIH, was laid out for the new program.[18] First, a Bureau of Cancer Research would be established within the NIH, which would include the National Cancer Institute, a program management group, and a special cancer initiative planning group. The cancer program would be the responsibility of the bureau. The bureau director would also be a deputy director of NIH, a status that would assure him direct access to the NIH director. These arrangements would insure that the cancer program would receive special attention and emphasis within the overall management and operations of NIH. Within the bureau there would be a deputy director who, in turn, would be designated as director of NCI, thus bringing continuity of effort to current programs. All cancer funds within the NIH would be under the responsibility of the new bureau, including the proposed $100 million request submitted earlier by the president. The NIH director would approve all proposed plans and budgets, while the director of the Bureau of Cancer Research would manage all bureau activities.

A Cancer Conquest Advisory Council was also proposed to absorb the existing National Advisory Cancer Council. Its functions would be guidance in policy and program formulation, program management overview, and review of grants. Finally, an Executive Task Force on the Cancer Conquest Program would be established, consisting of the HEW secretary, the assistant secretary for health and scientific affairs, the NIH director, the director of the new bureau, and the science adviser to the president. This task force would provide further assurance of increased time and attention being paid to the cancer program by top federal officials.

This proposal constituted an explicit rejection of Marston's assertion in his December memorandum to Egeberg that the existing NIH framework was adequate for an expanded national cancer program. But it was an equally explicit rejection of the proposition that a separate cancer authority be created. The administration appeared to be drawing the line very firmly against taking cancer research out of NIH.

Firmness on policy was paralleled by an increased political firmness. Senator Dominick had written to Richardson immediately after the March hearings, expressing his concern with the independent agency recommendation. While the Panel of Consultants had recommended a NASA-type agency, he wrote, many in the scientific community had reacted strongly against this and had urged retention of cancer research within NIH. Furthermore, Dominick noted, when Senator Yarborough had introduced S. Res. 376 in April of 1970, he had specifically requested that the panel "direct particular attention toward the creation of a new administrative agency."[19] Consequently, the Colorado Republican was skeptical of the panel's objectivity on this point and wished to know the extent to which the panel members had consulted with HEW and NIH officials during their study.

Richardson responded on April 5. The panel staff, he wrote, had been quartered in the office of the NCI director from May 1970 through mid-February 1971. Neither officials nor employees of NCI, nor members of the panel or its staff, had interviewed HEW or top NIH officials during the study. Dr. Marston, the secretary indicated, had received a courtesy call at the beginning and at the end of the study, "but no substantive discussions on either scientific or management questions were held with him." Neither Dr. Egeberg nor Dr. Steinfeld had been contacted, even though Steinfeld had previously been at NCI. Though Mr. Schmidt and several other panel members had met with him for one hour on January 29, 1971, Richardson wrote, "no discussions of the study took place prior to that time between the panel, its staff or the Office of the Secretary." The secretary concluded with this fillip: "It is thus clear that, with the exception of the officials and employees of the NCI, members of the Panel did not consult with top management officials either of the Department or NIH with regard to the scientific and managerial aspects of cancer research."[20]

By early April, the administration had defined its position on cancer research. It had also begun a political counterattack—by direct lobbying of important senators, by mobilizing the scientific community, and by open criticism of the panel itself. The unresolved question was whether any counterattack could succeed without proposing alternative legislation.

THE SENATE MAILBAG

There was never any question in the minds of the members of the Senate health subcommittee that new cancer legislation would be adopted by the 92nd Congress. Nor was there any question of this in the minds of other close observers of the legislative scene. What was unclear, however, was whether this legislation would establish a separate cancer authority outside NIH.

In fact, there was much reason to doubt that a separate agency would be created. This was the case, at any rate, during the month and a half following the March hearings. Among the Democrats, only Senators Kennedy and Pell clearly favored a new agency. On the Republican side, Senator Javits had made his support for a new agency clear, but he was reportedly willing to listen to the administration's alternative proposal. Senator Nelson (D., Wis.), it was indicated, was prepared to oppose such an agency within the subcommittee. Nelson's Democratic colleague from California, Senator Cranston, had disclosed at the hearings that he was inclined against a new agency. Four subcommittee Republicans—Senators Dominick, Prouty, Packwood, and Beall—opposed taking cancer research out of NIH. In an undecided group were four Democrats—Williams, Eagleton, Hughes, and Mondale—and one Republican—Schweiker. The prospects of the subcommittee recommending a separate cancer authority were not promising in late March.[21]

The speed with which events moved from the early December report of the Panel of Consultants to the early March hearings of the Senate subcommittee had left the medical-scientific community little time to mobilize opposition to S. 34. In fact, the initial reaction of many within the scientific community was favorable to the report of the Panel of Consultants. If cancer benefited, they reasoned, then all of medical research would benefit. In this interpretation, they were trapped by prior experience with Lasker-inspired munificence. Increased funds for cancer would have been welcomed had they provided an answer to the funding problems of the rest of biomedical research. Only when the schismatic implications of the separate cancer agency recommendation were recognized, however, did the scientific community begin to regard the cancer initiative as a threat.

The task of mobilizing opposition to the Panel of Consultants, therefore, fell upon the permanent organizations of medicine, which more readily perceived the true significance of the panel's report. The American Medical Association (AMA) had written President Nixon in mid-February supporting his cancer initiative and urging that it be conducted within NIH.[22] On March 15, the AMA's executive vice-president wrote Senator Kennedy expressing the organization's opposition to S. 34.[23] The AMA feared that the abandonment of the peer-review system would create "a distinct and justifiable uneasiness" within the scientific community. Furthermore, the letter read, the separation of the National Institute of Mental Health (NIMH) from NIH in 1967 had been a "bitter disappointment" to both NIMH scientists and those concerned with mental health. Fragmentation of NIH, and slowdown of cancer research would be undesired results of reorganization, the AMA spokesman said.

Two organizations, however, the Association of American Medical Colleges (AAMC) and the Federation of American Societies for Experimental Biology (FASEB) did most of the work of mobilizing the academic medical-scientific community to oppose S. 34. The AAMC acted mainly through the Council of Academic Societies, which represents medical school faculty and is organized along lines of the medical-scientific disciplines.[24] FASEB, which consists of six scientific societies with a combined membership of over 11,000 scientists,[25] worked primarily through its public affairs committee,[26] and especially through its chairman and the full-time director of FASEB's office of public affairs.

In general, while AAMC and FASEB have different purposes and constituencies within academic medicine, they also have shared purposes, overlapping constituencies, and a multitude of personal, informal relations between officers, members, and staff. Together, in early and mid-March, they generated a number of letters to the Senate health subcommittee, some to Kennedy but many to Dominick, laying out the case against separating cancer research from NIH. Almost all of these letters voiced support, even enthusiasm, for an expanded cancer effort. None supported and many opposed S. 34. Most were strongly opposed to a separate cancer authority and equally in favor of keeping cancer research in NIH.[27]

The AAMC, of course, drew support from the resolution it had adopted at its February meeting. The American College of Physicians, a member of the Council of Academic Societies, supported this view in a letter to President Nixon in early March.[28] So did the dean of the New Jersey College of Medicine and Dentistry in a letter to Dominick a few weeks later.[29] Luther L. Terry, M.D., vice-president for medical affairs at the University of Pennsylvania, and Surgeon General under Presidents Kennedy and Johnson, wrote Senator Kennedy that an independent cancer organization would "create separate, conflicting, and overlapping national authorities" that would dissipate funds for administrative purposes.[30] Robert A. Aldrich, M.D., vice-president for health affairs at the University of Colorado Medical Center, wrote Dominick that a separate agency's impact would be "highly destructive to the cancer research base" in universities.[31]

Senator Kennedy heard from Harry M. Rose, M.D., the representative of the American Association of Immunologists to the FASEB public affairs committee: "Cancer will not be abolished by legislative fiat, nor by administrative control, nor by the mere expenditure of money."[32] A similar statement was submitted to the subcommittee by the American Society for Biological Chemists, an FASEB member organization.[33]

The subcommittee heard from other medical scientists. The president of the National Academy of Sciences, Dr. Philip Handler, wrote Kennedy on March 15 that the cancer effort should be used to strengthen the NIH rather than create a separate agency.[34] He indicated that this view was also held by the Institute of Medicine of the Academy, the President's Science Advisory Committee, and the chairmen of the medical schools' departments of biochemistry. The director of the Penrose Cancer Hospital in Colorado Springs, J. A. del Regato, M.D., wrote Dominick as a member of the National Advisory Cancer Council and objected to the proposed new council having half of its membership drawn from lay individuals.[35] An interesting letter came to Dominick from Frank Zboralske, M.D., director of the division of diagnostic radiology at Stanford Medical School, urging that expansion of the cancer program take place within NIH and NCI. The chairman of the Stanford department of which Zboralske was a member was none other than Henry S. Kaplan, M.D., member of the Panel of Consultants.[36] Seymour S. Kety, M.D., the research committee chairman of the Massachusetts General Hospital, wrote Kennedy indicating that the scientific community at the famous Boston hospital viewed "with alarm" the implications of S. 34 for the NIH.[37]

With the exception of three practicing physicians in Colorado, all of the letters came from individuals in medical schools. Harvard, Duke, Colorado, and Stanford were prominent among these. The writers were predominantly associated with the basic medical sciences—physiology, pathology, microbiology, biophysics, biochemistry, and genetics—and relatively few with the clinical sciences. But since many were known personally to Senate subcommittee members, and held in great respect by them, their influence was substantial. Moreover, the relatively few letters printed in the committee hearings represented only the tip of the academic science iceberg.

In addition to letters to the subcommittee, other more public means were used to express opposition. An editorial in *Science* concluded: "Increase in long-term support for cancer research is fully justifiable and should be implemented. However, the likely result of a hurry-up-and-wait crash program is wreckage of the nation's medical research enterprise without much counterbalancing progress in coping with cancer."[38] *Science*, with well over 100,000 subscribers, both informed and reflected the views of the scientific community. Mobilization was underway in earnest.

The supporters of S. 34 had pursued a strategy up to this point that involved a low profile but high impact. When it became clear in mid- to late March that the counterattack by the scientific community was having its effect, the proponents of S. 34 began to search for a public strategy to give their cause extensive publicity through the media.

In 1938, Congress had authorized the president to proclaim April of each year as "National Cancer Control Month," and the proclamation has been faithfully issued ever since. April is also the month in which the American Cancer Society conducts its annual fund-raising drive. So a publicity campaign to generate widespread support for S. 34 had the advantage of a prepared audience, since thousands of ACS volunteers were then being organized for door-to-door soliciting. In mid-March, the national headquarters of ACS sent a memorandum to all state and local chapters to inform staff and volunteers about developments related to the national cancer program. In answer to the question of "why change?" from NCI to a new organization, the memorandum said that a new agency was needed that would be "free of the administrative and financial restrictions imposed on the regular health agencies, and [capable of] using a new approach, suitable to a crash program."[39]

A related advantage for a publicity campaign was that the annual ACS science writers conference was being held in Carefree, Arizona, in early April 1971. Dr. H. Marvin Pollard, president of the society, said at the opening session that "to produce new and effective methods of cancer control will require the same kind of effort that went into the development of the atom bomb or the space program that placed a man on the moon."[40] He came down squarely in favor of a separate cancer agency, a fact reported in newspapers all over the country.

Dr. Solomon Garb, whose exaggerated claims were equalled only by his near-legendary letter-writing abilities, sought to mobilize cancer research scientists to offset the views of medical scientists he felt had little knowledge of cancer. Though a member of FASEB's American Society for Pharmacology and Experimental Therapeutics since 1951, he was unable to get his correspondence printed in the *FASEB Newsletter*.[41] Garb also wrote to many newspapers around the country. A typical letter, appearing in the upstate New York *Ithaca Journal* in April, urged readers to write Senator Javits in support of S. 34. HEW officials were opposing the bill, Garb indicated, because they didn't want the Cancer Institute removed from their authority and they were trying to get senators to vote against it. "The fate of the bill," he wrote, "the fight against cancer, and the lives of many present and future cancer patients hang in the balance. You can make the difference."[42]

A very moving letter appeared in the April 14, 1971, *Wall Street Journal*. A father described how he and his wife had been "plunged into numb agony" sixteen months earlier by the news that their five-year-old son had leukemia. "In the preceding days," the father wrote, "as Scott went through a series of diagnostic tests, we had begun to fear the worst—but hearing the worst was still a terrible experience. We endured constant heartache and shed many tears in the weeks that followed. Life just didn't seem worth living."[43] The writer made a heartfelt plea for more money for cancer research.

A number of stories related to cancer appeared in March and April,[44] but the most significant publicity occurred in mid-April, when Ann Landers devoted an entire column to urging support for S. 34.[45] "Dear Readers," she began, "if you are looking for a laugh today, you had better skip Ann Landers." But, "if you want to be part of an effort that might save millions of lives—maybe your own—please stay with me." She asked, "Who among us has not lost a loved one to cancer?" In 1969, she added, more Americans died of cancer than were killed in World War II. One of four Americans alive today, she wrote, would develop cancer, and two of three of those would die from it. "Cancer," she continued, "claims the lives of more children under 15 years of age than any other illness."

She put another question to her readers: "How many of us have asked . . . if this great country of ours can put a man on the moon, why can't we find a cure for cancer?" The reasons were simple: a national campaign had never been launched, funds for medical research were grossly inadequate, and the bulk of the tax dollar was going to national defense.

Senators Kennedy and Javits had sponsored legislation to establish a National Cancer Authority to be for cancer what NASA had been for space. "Today," she wrote, "you have the opportunity to be a part of the mightiest offensive against a single disease in the history of our country. If enough citizens let their senators know that they want the bill—S. 34—passed, it will pass." Each reader was urged to write his two senators, care of the Senate Office Building, Washington, D.C., 20510, with at least a three-word message: "Vote for S. 34." "And sign your name, please," Ann Landers concluded.

Ann Landers' plea produced an avalanche of mail.[46] Senator Williams of New Jersey received 11,500 letters by the end of the first week in May. Senator Pearson of Kansas and Senator Byrd of West Virginia received 7,000 and 3,500 letters respectively in that same period. Senator Beall of Maryland received 500 in two days. Senator Packwood of Oregon received "a couple of thousand." Senator Hughes of Iowa received "too many to count." Senator Nelson of Wisconsin received about 8,000 letters. Senator Bentsen of Texas said he had never seen such a deluge. Senator Humphrey reported that he had been swamped with letters. Senator Percy of Illinois had "never had such a tremendous response from any other issue." Senator Cranston of California, in November 1971, said, "Thus far, in nineteen seventy one, the biggest volume of mail has come in on the Cancer Act, S. 34. On that one bill alone, I received sixty thousand letters in a five week period."[47] The mail produced a nightmarish problem in Senate offices as they struggled to answer the flood of letters. "Impeach Ann Landers" signs appeared on the desks of secretaries throughout the Senate Office Buildings. As Elliot Richardson later told her, "You don't know your own strength."

THE PRESIDENT'S POSITION: MAY

The Citizens Committee for the Conquest of Cancer took out full-page advertisements on Sunday, May 2, in major newspapers throughout the country. The ads carried the text of Ann Landers' column, and named the two senators to whom the readers of the particular paper should write. The Cleveland *Plain Dealer*, for instance, urged people to write Senator Robert Taft, Jr., or Senator William B. Saxbe, the two Ohio senators. The advertisements listed three co-chairmen of the committee, Dr. Sidney Farber, Emerson Foote, and—new to the cause—Cleveland millionaire, Howard Metzenbaum.[48]

The one-two punch of the Ann Landers column and the full-page advertisements resulted in an estimated 250,000 to 300,000 letters to the Senate by mid-May favoring S. 34. Later estimates would place the total closer to one million! The Republican senators who had been opposing the bill became alarmed and increasingly fearful of being depicted as "in favor of cancer." They appealed to the White House to take them off the hook, pointedly noting that the White House was in the same predicament they were. The urgency of their appeal had already been underlined by quite active, independent efforts of the Panel of Consultants.

Benno Schmidt, Mary Lasker, Luke Quinn, and several others began in mid-April discussing ways to counter the objections to S. 34 then being raised by the academic medical community. They also discussed how they could communicate directly with the president. Almost nine months earlier, at the July 1970 meeting of the Panel of Consultants, Elmer Bobst had indicated that he had urged President Nixon to take an interest in cancer. Schmidt had then expressed the hope that Bobst would again speak to the president in support of the panel's recommendations at the proper time. Bobst had agreed to do this, and went to the White House on April 27 to talk to his friend, the president, about one of his life-long passions—the war on cancer.[49]

Bobst, who had served on the National Advisory Cancer Council from 1953 through 1957, in addition to his work in the anti-smoking campaign of the American Cancer Society, had long thought the research efforts of both the ACS and NCI had been "unfocused." "[T]he many layers of bureaucracy in the National Institutes of Health and its parent, the Department of Health, Education, and Welfare, led to such a scattershot spread of research money that it was doing little good where it really counted," he wrote in his autobiography. Consequently, when he visited the White House, he said:

> I urged the President to support a proposal that would make the Cancer Institute autonomous, thus stripping away the bureaucratic layers on top of it, give it an annual appropriation without any definite limitations, subject it to a carefully considered program of *directed* research in the most promising areas, and have it report directly to the President.[50]

Bobst was urging the president's support of the recommendations of the Panel of Consultants and the provisions of S. 34 in the strongest possible terms.

The panel met two days later in Benno Schmidt's office to hear a report on this visit. In addition to urging the president's support for a separate cancer agency, Bobst indicated, he had pointed out to Nixon that he could do this simply by executive order, thus making the cancer effort his program. The president, he said, had not taken a public position on a separate agency, but had "tested the market" through his science adviser, Dr. David. But, Bobst noted, the president's reaction was not negative and he had not dismissed the proposal out of hand. Bobst also reported that he had seen John Ehrlichman after talking to the president, and that Ehrlichman had later spoken about the matter to Richardson and Marston.

There was intense activity underway both at the White House and within the Senate health subcommittee. The outpouring of mail supporting S. 34 had caused a definite shift within the subcommittee. One report suggested that a bipartisan majority was prepared to vote the bill out of subcommittee. Prospects for passage were even better within the full committee, and better still in the full Senate. A "markup" session was scheduled by Kennedy for Friday, May 7, but Dominick requested a delay until the following Tuesday in order to give the White House time to act.

Over the weekend, the White House staff presented the president with four alternatives. He could do nothing on organization and simply add money. He could submit a reorganization plan that would either rearrange cancer research within NIH, or remove it from NIH entirely. He could seek legislation to establish the National Institutes of Health as an independent agency outside of DHEW. Or, he could create a conquest-of-cancer agency within NIH, with a director reporting to the secretary of HEW.

The first alternative was quickly discarded. The president had tried to add money without changing the organization, and that effort had become increasingly untenable. A reorganization plan was a theoretical possibility.[51] But this was a politically inappropriate response to an issue that had been spearheaded by the Senate health subcommittee. There was no way that Kennedy, Javits, and their colleagues could now be preempted from their share of the credit for the cancer initiative, regardless of White House desires. The idea of an independent NIH found little support among the White House staff. Politically, no one was promoting it. Organizationally, it ran counter to other efforts underway within the administration to consolidate rather than fragment the federal bureaucracy. The president chose the last option, but amended it: "We'll take number four," he reportedly said, "but the director will report to me."[52]

The president's decision meant that a message had to be prepared and a bill drafted before the scheduled session of the health subcommittee. On Tuesday, May 11, shortly before 11:00 a.m., the president announced to the White House press corps that he was sending legislation to the Congress to establish a Cancer Cure Program within the National Institutes of Health.[53] Final approval of his earlier request for an additional $100 million for cancer research was expected soon, he indicated, and the legislation was being submitted to implement that program. The Cancer Cure Program, the President added, would be "the one program within the National Institutes of Health and within the whole health establishment . . . that is independently budgeted and that is directly responsible to the President of the United States."[54]

Nixon stressed that he was assuming this responsibility "because of a very deep personal concern about this problem, . . . and also because I believe that direct Presidential interest and Presidential guidance may hasten the day that we will find the cure for cancer." The president recalled that his favorite aunt, Elizabeth, had died from cancer when only 38 years old. He also remembered his last visit to Senator Robert Taft, who died of cancer, and the many hours he had spent at Walter Reed Hospital when John Foster Dulles "was withering away with cancer."[55]

The program was being established within the NIH, Nixon said, because that organization had successfully fostered coordination and cooperation among scientists, and had earned the respect of the scientific community as well as the gratitude of those whose lives had been enriched by its successes. Within the NIH, the Cancer Cure Program could take "fullest advantage of other wide-ranging research." But this program, he said, was to be "one of our highest priorities, . . . not compromised by the familiar dangers of bureaucracy and red tape." Consequently, Congress was being asked to give it independent budgetary status and make its director responsible "directly to the President." Furthermore, Nixon said, "we must do a better job of tapping the Nation's administrative and organizational skills" in seeking the conquest of cancer. It was important to keep in mind, however, that "biomedical research is a notoriously unpredictable enterprise. Instant breakthroughs are few and the path of progress is strewn with unexpected obstacles. As we undertake this crusade we must put on the armor of patience, ready to persist in our efforts through a waiting period of unknown and possibly anguishing duration."[56] But though the cure for cancer might not come quickly, the president added, "it will not fail because of lack of money . . . it will not fail from lack of organization . . . it will not suffer for lack of . . . having special Presidential interest."[57]

At a press conference following the president's dramatic announcement, Elliot Richardson, Edward David, and Robert Marston responded to reporters' questions. "Mr. Secretary," one journalist asked, "what is the effect of this announcement likely to be on S. 34, which the Kennedy Subcommittee is supposedly marking up today?" Richardson hoped that the subcommittee would "regard this approach as meeting the need which we both recognize for a kind of advisable managerial thrust and focus for

the effort, while at the same time avoiding the problem created by
S. 34 of taking the cancer effort out of the National Institutes of
Health."[58]

The subcommittee met that afternoon at 2:30 p.m. Senator
Dominick drew attention to the president's announcement and
asked that no action be taken on S. 34 until the administration's
bill had received a hearing. Senator Kennedy expressed delight
that they were sponsoring legislation, but suggested that it was
rather late to do so. Dominick agreed, but hoped the administration could be heard. Senator Javits concurred, saying that they deserved their day in court. So a one-day hearing was scheduled for
the first week in June.

Among the Senate Republicans, the president's action brought
a sense of relief. But other reaction was critical. A *Washington
Post* editorial complained that:

> politics has begun to affect cancer research . . . President
> Nixon's proposal strikes us as somewhat peculiar, although it is
> understandable that he thought he had to do something dramatic
> to take the spotlight away from Senator Kennedy's even less
> acceptable proposal. . . . Although the President cast this pro-
> posal as one to eliminate red tape and focus presidential power
> and national interest on speeding up cancer research, it smacks
> of a proposal aimed at winning for the White House in advance
> any political glory that may result from the potential benefits of
> increased research funds.[59]

Forced to choose between the two proposals, the editorial concluded, it would take the president's, although even that seemed
unnecessary.

S. 1828: AN ACT TO CONQUER CANCER

The proponents of S. 34 had been stung by the criticism from the
scientific community and by the realization that it was having an
effect upon members of the Senate health subcommittee. As a result, the panel discussed how it might respond to such criticism at
its April 29 meeting in New York. Five amendments to S. 34

were considered which, it was felt, would be responsive to most, if not all, of the questions raised by the scientific community. These were then sent to Senator Kennedy, and an agreement was reached whereby Senator Williams, chairman of the full committee, would introduce them. Williams proposed them for consideration on May 6, in order to "accommodate specific needs and to overcome justifiable concerns" that had been expressed by many.[60]

One amendment specified that the new cancer authority should be physically located at or near the present site of the NCI. Another placed both the administrator of the new agency and the NIH director on the proposed National Cancer Advisory Board to insure coordination between cancer and the rest of biomedical research at the highest administrative levels. A third amendment provided that funds could be transferred from the proposed new agency to other NIH institutes for cancer-related research. Yet a fourth amendment authorized the administrator to take action so that "all channels for the dissemination and cross-fertilization of scientific knowledge" between NCI and the other NIH institutes were maintained under the new arrangement "to insure free communication between cancer and the other scientific, medical, and biomedical disciplines."[61] Finally, the administrator was directed to "insure proper scientific peer review of all research grants and programs," and the new board was directed to see that he performed this duty.[62] The clear purpose of these amendments was to reassure the scientific community that peer review of scientific research grants would be maintained.

The scientific community had been keeping the pressure on. The Association of Professors of Medicine, a constituent member of the AAMC's Council of Academic Societies, met in Atlantic City in early May. They released a statement signed by 79 professors and chairmen of medical-school and teaching-hospital departments of medicine opposing a separate cancer agency. "We strongly endorse current efforts to strengthen the fiscal and organizational structure of the nation's cancer programs. We believe that progress toward this goal can be best achieved through the framework of the National Institutes of Health because solutions will surely require cross-fertilization of many disciplines."[63]

The statement was sent to President Nixon and other government leaders.

The embarrassing fact was becoming apparent that no medical-scientific organization in the country, other than the American Cancer Society, had endorsed a separate cancer agency. Even the American Association for Cancer Research, the cancer research scientists' own organization, could not be persuaded to support a separate agency. On April 9, at their annual meeting, the AACR's board of directors adopted a resolution which noted that several alternative administrative proposals were being considered for cancer, asserted that any program "must recognize the complexity of cancer, the requirement for long-term sustained national commitment, and the need for expanded resources," and listed seven characteristics such a program should include.[64] Notable for its omission was support for a separate agency. Ironically, the president of the AACR was then Dr. James F. Holland, member of the Panel of Consultants, who obviously could not deliver the votes. Had he been able to do so, no time would have been lost in publicizing the fact.

Meanwhile, the president's request for an extra $100 million for cancer had moved steadily through Congress. On May 26, Nixon signed the Second Supplemental Appropriation Act for 1971 into law, in which was included the extra funds requested in January. Emphasizing that "money alone will not be enough," the president noted that that was why he had requested legislation to establish a new Cancer Cure Program.[65] He urged the Congress to be as prompt in enacting the newly proposed legislation as it had been in appropriating the funds.

The request for the extra $100 million had provoked a good deal of grumbling in some quarters. The request was for additional obligational authority to be made available for fiscal 1972, but the administration's announced intentions were that the money would actually be spent over a two-year period from 1 July 1971 to 30 June 1973. On April 23, Luke Quinn wrote that the net effect of this was to increase the fiscal 1972 budget by $50 million and the fiscal 1973 budget by the same amount. This was no more than the Congress had done in fiscal 1971, he continued, without prompting from the administration. With asperity, he concluded:

For HEW to call this feeble gesture a "new Cancer Conquest Program" is less than realistic. . . . This additional evidence of the futility of the existing bureaucratic structure makes it clear that we must look to the independent agency as the only hope for the solution of cancer. If we are really serious about the conquest of cancer there can be no compromise with the independent agency concept and no compromise with the terms of S. 34, the bill which will bring the independent cancer agency into being.[66]

This message went out to Quinn's cancer mailing list.

The suspicion generated by the $100 million episode helped set the stage for the events that followed the president's May 11 press conference. Though Nixon had stopped short of recommending a separate cancer agency, most observers thought that the president, in recommending a program whose administrator would report directly to him and which would have independent budget authority, had taken a long step toward accommodation with the proponents of S. 34. Lane Adams, executive vice-president of the Cancer Society, indicated immediately after the press conference that the ACS had long supported the concept of an independent cancer authority. "We are gratified," he said, "that the President also supports the advantages of this type of approach to the conquest of cancer and the establishment of an independent cancer agency directly responsible to him."[67] The *Blue Sheet* for May 12 described the Nixon plan as keeping the cancer program within NIH, "but just barely."[68]

The language of the bill that was actually sent up to the Congress that afternoon, however, contained some distressing surprises for those who had seen the president's announcement in a favorable light. Elliot Richardson's letter of transmittal, for instance, made reference to Nixon's State of the Union message and his Health message, but no explicit reference to the press conference of that very morning.[69] The bill itself was more distressing. The new program was to be "in the National Institutes of Health" and would "administer the authority of the Secretary [of HEW] . . . with respect to cancer."[70] Though the bill did specify that the administrator of the program "shall serve under the direction of the President," there was no language suggesting that he would report directly to the president. Finally, no provision was made in the legislation for independent budgetary authority for the new program.

Luke Quinn was very upset. He wrote Benno Schmidt on May 12, that "the provisions of the bill, as I scan them, do not fulfill the promise of the President's press release."[71] A program within the NIH and under the HEW secretary was "hardly an independent agency," Quinn noted, as he asked for Schmidt's reaction to the bill. Quinn also brought these matters to the attention of Elmer Bobst a few days later. The bill submitted, he wrote, was "a far cry from the independent agency sought by the Panel of Consultants and by S. 34." And there was "nothing in the administration bill providing for the budget independence that the President intended it to have when he made his press statement."[72]

Why this divergence between the president's statement and the bill actually submitted by the administration? Quinn, in his letter to Bobst, stated that the bill was being drafted at HEW when the president spoke and was not complete before the press conference ended. As a result, he wrote, "it may therefore not contain what the President wanted it to contain and what he conveyed to you it would have in it."[73] The bill, he continued, neither embraced the panel's recommendations nor the independent agency concept, but simply reaffirmed the existing situation where NIH reported to the HEW secretary and he, in turn, reported to the president. What was being suggested, and what had in fact happened, was that the White House and HEW were going in somewhat different directions on the cancer legislation.

Bobst, who had written Lane Adams on May 13 about how the president had "come out unqualifiedly in the support of a special authority to be devoted directly to the conquest of cancer," was also upset.[74] He communicated his distress to Ken Cole, one of John Ehrlichman's subordinates on the White House staff. Bobst was quoted as saying that the legislation violated "a clear, concise understanding I had with the President."[75]

The White House response was to dispatch Cole to meet with Benno Schmidt in New York and work out differences between the administration and the supporters of S. 34. There then began a complex series of negotiations involving Schmidt for the Panel of Consultants; Kennedy, Javits, and their aides for the Senate subcommittee; and Ken Cole and his aide, James Cavanaugh, for the White House. Neither HEW nor NIH were included.[76]

The Senate health subcommittee met on Thursday, June 10, at 9:35 a.m., in the New Senate Office Building. Senator Kennedy was in the chair, and Senators Nelson, Eagleton, Cranston, Javits, and Dominick were also present. The ostensible purpose of the hearing, as Kennedy put it, was to hear administration witnesses "discuss the merits of S. 1828."[77] But the opening remarks strongly suggested that a compromise had already been reached.

Kennedy expressed his gratification with the president's May 11 statement and his personal commitment to an independent cancer program directly responsible to him, having independent budgetary authority, and being within the NIH. Only the president, the Senator stated, "can mandate that level of priority and independence of action."[78] It therefore seemed possible, he said, to establish a cancer conquest agency within NIH, and that is what he hoped Congress would do. Senator Javits praised "the really extraordinary exercise of statesmanship by the administration and by the Chair," and observed that "[a] very great measure of agreement seems to be present." Senator Dominick congratulated the chairman for his efforts to work out "a mutually agreeable bill" with the administration. Jocularly, Kennedy suggested that "maybe we ought to just recess the hearing and pass the bill as long as we have everything on track."[79]

The chairman then turned, "with a warm sense of welcome," to Secretary Richardson who was there to present the administration's views. Richardson emphasized the two main features of the president's proposed program.[80] First, the program would have the full support of the president's power, prestige, and personal interest, and the director would report directly to the president. Second, the program would have independent budgetary status. The OMB would give budget guidelines to the program, not through HEW and NIH, and the program, in turn, would submit its budget request each year directly to OMB for the president's approval. The difference between the secretary's May 11 letter and his June 10 testimony was obvious to all.

How long would it take, Kennedy asked, to set up the new agency?[81] Both Richardson and Marston thought that work could begin immediately upon appointment of a program director. Would you feel "comfortable" about arrangements between the new agency director and your own responsibilities? Kennedy asked Marston. The NIH director replied that he had originally feared that the closeness of cancer research to the rest of biomedical research would be ignored, and that there would be a tendency to overpromise quick results. On both counts, he had been reassured since his last appearance before the subcommittee.

Javits, like a lawyer leading a friendly witness, asked Richardson if the autonomy needed for the conquest of cancer could be attained "through an independent agency within the National Institutes of Health?" The secretary responded affirmatively. "[B]ut do you really agree with this," Javits asked him, "and have your heart in it so that in that fantastic skill of bureaucracies you will really try to make it work?"[82] Richardson pledged HEW's "utmost efforts" to that end.

Senator Dominick had few questions, and Kennedy then turned to Senator Gaylord Nelson of Wisconsin, who had been absent from the March hearings. In the past two months, he had had extensive conversations with his good friend, Philip P. Cohen, professor and chairman of physiological chemistry at the University of Wisconsin Medical School. Nelson, in fact, had had dinner with Cohen and Marston very soon after the March hearings.[83] Though an initial co-sponsor of S. 34, Nelson had become persuaded that a separate cancer authority was undesirable.[84]

Nelson displayed none of the bonhomie that characterized the hearing that June morning. "This bill, as I read it," he said, "is S. 34 with minor amendments." It involved creating an independent agency, with a director two grades above the executive salary level of the NIH director, who reported to the president and not through the NIH director. The proposed agency was to have a separate budget, its own advisory board, and its own peer-review system. In effect, the senator said, it was an independent agency, housed within NIH but there "in name only."[85] Senator Javits intervened and there followed a most remarkable colloquy:

> *Senator Javits*: Would the Senator yield? Could the Senator identify the bill to which he refers?
>
> *Senator Nelson*: S. 1828, committee print.
>
> *Senator Javits*: Which committee print? I think we ought to know what the Senator is directing his attention to.
>
> *Senator Nelson*: I am talking about the bill which the Secretary is giving his testimony on.
>
> *Senator Javits*: As I understand the Secretary, he was testifying to S. 1828.
>
> *Senator Nelson*: That is what I am talking about.
>
> *Senator Javits*: Not a committee print.
>
> *Senator Nelson*: I have the committee print.
>
> *Senator Javits*: Could we know which it is? There have been a number of them. It would be useful so we are not confused. What committee print is the Senator using?
>
> *Senator Nelson*: It is the same. The June 4 print.
>
> *Senator Dominick*: I haven't seen any committee print at all. I would be interested in that myself. [Laughter.]
>
> *Secretary Richardson*: Neither have I.
>
> *Senator Nelson*: All right. I am addressing myself to S. 1828, June 8 committee print. It is the same as the June 4 print. It is the one the Secretary is testifying on.
>
> *Senator Javits*: He is not testifying to that. He is testifying to 1828 as originally put before us. If there are going to be any compromises, they have yet to be made. They may have been discussed, but the question—I think it is very important to understand that he is testifying in favor of the administration bill, period.
>
> *Senator Nelson*: He is not testifying in favor of the—he doesn't favor the Dominick—
>
> *Senator Javits*: I don't say he doesn't favor it, but that is not his testimony in chief. He will confirm that.

Senator Dominick: He favors the Dominick-Griffin bill.

Senator Nelson: Do you favor the Dominick-Griffin bill?

Secretary Richardson: Yes. I am testifying for S. 1828, as introduced. Senator Javits, it is correct that there have been changes in S. 1828 that have been discussed and have been reflected in various committee prints, but I am not testifying on behalf of any such variant of S. 1828.

Senator Nelson: Pardon me, Mr. Secretary. I didn't understand that.

Secretary Richardson: I said, Senator Nelson, I was confirming what Senator Javits said. I am testifying on behalf of S. 1828 as introduced, and not in support of any variant thereof that may have been the subject of intervening discussion.

Senator Nelson: So, you are not in favor of the Dominick-Griffin bill as of committee print, June 8, 1971?

Secretary Richardson: That is correct. I am not saying that there may not be provisions in it that the administration would be willing to accept, but I am here today not to support or to negotiate a specific variation of S. 1828 but to explain why the administration submitted it in the form in which it was introduced by Senator Dominick and why we think the provisions in it make sense, as we saw them at the time.

Senator Nelson: Well, are we going to have another hearing, then, Mr. Chairman, on the Dominick-Griffin print of June 8?

Senator Kennedy: I don't expect so. I hope that we won't. As I understand, there have been a number of prints, to reflect the alterations and adjustments in the bills. The most recent subcommittee print has not been introduced. It is different than the administration's proposal which is the bill the Secretary was asked to come up here and comment on. Now, since the administration's bill was introduced, a number of issues and questions were raised which culminated in a recent committee print which includes a number of different recommendations, adjustments and changes to the original bill. But to my knowledge this committee print has not been available to the Secretary, and has just been made available, as a matter of fact, to the members of the subcommittee.

Senator Dominick: Would you yield there? As a matter of fact, the June 8 Dominick-Griffin print, I haven't even seen.

Senator Nelson: Well, I was dealing with what I assumed was the reality. Let's face it, all the publications are saying that there has been an agreement and that S. 34 and the President's proposal have been put together and that is the political reality. There is no point in kidding ourselves. I want to discuss the current political reality, not some theory.[86]

The current political reality, of course, was that the negotiations had moved quite some distance from consideration of the original S. 1828 that had been introduced a month earlier. Benno Schmidt, after receiving Kenneth Cole, the White House envoy, had gone to see Senators Kennedy and Javits to explore ways out of the impasse. Schmidt thought agreement was very close and that what the panel and the subcommittee wanted was very close to what the administration would accept.

Kennedy had then made an amazing political move. News accounts throughout May had featured the difference between S. 34 and S. 1828 as one of personal political maneuvering between Kennedy and Nixon, each seeking political advantage out of the issue.[87] Furthermore, the administration's position on health issues in general had created very bad relations with the subcommittee. Disagreement had existed on practically every health bill. Even so, Kennedy suggested that one way out of the impasse was for the subcommittee to report out S. 1828—the administration's bill! All that was needed was to make it consistent with the president's press conference statement by substituting the language of S. 34.[88]

In short, though Javits was at pains to say that "nothing has been locked up," Nelson had exposed the fact that agreement between the subcommittee and the administration was close at hand. The elements of the agreement were that an *independent* Conquest of Cancer Agency would be created *within* the NIH. The director of this agency would report directly to the president, not through the NIH director. The agency would have its own independent budgetary status, and all of its funds would be spent for cancer-related activities. The subcommittee could claim it had created an independent agency. The administration could assert it had retained the effort within the NIH. Both could be appeased.

One account of the hearing indicated that a compromise cancer bill had been hammered out that carried the number of President Nixon's proposal but had the substance of Senator Kennedy's bill. Senator Nelson had taken an unusual step and discussed the committee draft that contained the main elements of the compromise. The report added: "Committee drafts of bills are not usually discussed in public. But as a result of Nelson's disclosure, yesterday's hearing gave a rare public exposure to the closed-door bargaining that goes on between congressional committees and administration officials."[89] Indeed it had!

Nelson created another surprise that day. Attacking the alleged unanimity of the panel, he claimed that within the Panel of Consultants "the vote was [only] 16 to 10 in favor of S. 34."[90] Since then, three distinguished scientists—Dr. Lederberg, Dr. Kaplan, and Dr. Rusch—who originally voted "aye" had changed their minds, and a vote taken today would result in a 13-to-13 tie!

Benno Schmidt lost no time in responding to this when he testified several hours later. He told Senator Nelson that he was unaware of the source of the figures on the panel vote, but he wished to make clear that the panel had been unanimous on the recommendations "at the time our report was delivered to the Senate."[91] As for Dr. Lederberg, Schmidt disclosed, he had attended none of the panel's meetings nor had Schmidt ever met him! When the report was presented, Schmidt continued, he had asked the Stanford scientist if he wished to be excluded since he had not had an opportunity to participate. Lederberg had chosen not to do so. Only after the panel made its report public had Lederberg indicated that he wished the cancer program to remain within NIH. The Nobel laureate had lent his prestige to the Panel of Consultants, which they had been quite willing to accept, but not his participation. Lederberg had obviously become a source of embarrassment, and Schmidt was finally prepared to jettison him before further damage was done.

Schmidt explained that the panel had met in New York on April

29 to discuss the proposals outlined in Secretary Richardson's letter of April 2, and to reconsider the idea of a separate agency. Thirteen members had been present at that meeting. Dr. Kaplan had indicated his deep misgivings about separating cancer from NIH and wished to go on record in favor of leaving it within NIH. The panel, therefore, was divided by a 12-to-1 vote on the issue at that point.[92]

Had any other votes been taken? Senator Kennedy wanted to know. Only at the time of the final recommendations, Schmidt replied. On the matter or organization, "I know there were variations in the strength of the conviction," Schmidt disclosed, "but at the time of our original recommendation the vote of the panel was unanimous."[93] Senator Nelson pressed his question again, "the vote was 16 to 10, wasn't it?"[94] No, Schmidt replied, and Lee Clark and Sidney Farber confirmed this. Nelson suggested that perhaps he had "misunderstood" two panel members who had suggested a 16-to-10 breakdown on the separate agency issue. But this exchange established Lederberg's nonparticipation in the panel's work, and revealed that he had been joined in his opposition to a separate agency by his Stanford colleague, Henry Kaplan, one of the most active of the panelists.

In addition to the administration and the panel, the subcommittee had invited the Association of American Medical Schools to testify on S. 1828. The AAMC president, Dr. John Cooper, reiterated that his organization could not condone "the piecemeal dismantling" of NIH, an organization that represented a unique joining of the nation's scientific community with "public aspirations for the conquest of disease."[95] He restated the AAMC's opposition to separating cancer research from NIH.

Cooper was familiar with S. 1828 as it had been introduced. He was also familiar with the amendments on peer review that Senator Williams had proposed, and the proposal of Senator Nelson to make NIH an independent agency. But he was not familiar in detail with the negotiations that had been taking place between the White House, Panel of Consultants, and the Senate subcommittee. He referred ominously to "two dark undercurrents of opinion" that appeared to be sweeping the national debate on cancer "into potentially treacherous waters." Bluntly he stated his anxiety: "An unfortunate misconception apparently is developing that the mere injection of additional Federal cancer research funds will produce somehow an instant cure for cancer. Its equally misleading corollary is that the key to the conquest of cancer . . . lies in the managerial efficiency and the capacity of the medical-industrial complex."[96]

Kennedy, obviously nettled, stressed that he, and Mr. Schmidt's committee, and the administration had all emphasized that there would be no quick and easy answers to the problem of cancer, and that they had sought to eliminate any false hopes on this matter. What was it about the "medical-industrial complex," Kennedy queried, that he found disturbing?[97]

Most important, Cooper said, the "whole scientific review process" was of primary concern to the biomedical scientific community.[98] The AAMC was not certain, he indicated, that the valuable, effective peer review mechanisms of the past would be continued in the new setting. Why, Kennedy persisted, did he think that? It had been Secretary Richardson's earlier response to Senator Nelson, Cooper said, that was disturbing. Be specific, Kennedy insisted. Cooper was: the secretary had suggested that new mechanisms might be sought, that this new undertaking might require a different approach, because with cancer the point had been reached where some large-scale activities had to be used rather than individual project grants.[99]

This fractious discussion continued. Nelson reread the language of the bill: "The Board shall insure that the Director by regulations maintain scientific peer review of research grants and contracts." Joseph Murtaugh, who was with Cooper, suggested that this left open the possibility that a peer-review system would be established separate from that of NIH. Kennedy, quite irritated, said he did not see how they could draw that conclusion from that language. Would they feel better, Kennedy asked the two AAMC officials, if the subcommittee's report indicated that the director of the new program follow "the time honored and tested" procedures used by NIH in the past?[100] That would be reassuring, they replied.

The mood was tense. Senator Schweiker asked whether the AAMC resolution of February, opposing the cancer authority, was representative of its assembly. Cooper responded that Dr. Jonathan Rhoads, a member of the Panel of Consultants, was a member of their assembly and that he had presented the panel's recommendations and the case for the separate agency. Even so, the resolution had been adopted overwhelmingly. And a number of individual medical-scientific societies had adopted supporting resolutions since then, most recently the professors of medicine meeting in Atlantic City.[101] The scientific community was not divided. They were drawing together.

Schmidt, Farber, and Clark testified at the end of the morning and sought to rebut some points raised by Senator Nelson and Dr. Cooper. Benno Schmidt did indicate to Senator Dominick, though, that the amendments which had been suggested to the administration's bill, and which were incorporated in the committee print and the most recent version of S. 1828, would essentially carry out the recommendations of the panel. Satisfied, Dominick left for another meeting.[102]

The June 10 hearing on cancer legislation ended with a few tactical embarrassments to the proponents of S. 34. But strategically, a compromise was near and the votes were at hand. That, after all, was what counted.

With the compromise bill hammered out, the Senate Labor and Public Welfare Committee met on June 16 to consider S. 1828. Senator Nelson moved that the Cancer Conquest Agency budget go to the director of NIH for review and evaluation before its submission to the president. Four other senators—Cranston, Hughes, Stevenson, and Beall—supported him, but the motion lost, 12 to 5.[103]

The sponsorship of the bill remained to be settled. Though the committee was reporting out a bill with the administration's number, Kennedy was entitled to be listed as the principal sponsor. But he leaned forward across the table and, looking directly at Senator Dominick, said, "Peter, why don't you report the bill?"[104] To be sure, it was an act of political shrewdness, but it was also one of political generosity. The committee unanimously reported S. 1828 to the full Senate. Dominick had been assuaged. Kennedy had been successful.

The Senate considered the legislation on Wednesday afternoon,

July 7. Dominick, Javits, and Kennedy were the principal spokesmen for the bill. Each praised the others, the Panel of Consultants, and the administration for an excellent compromise. Kennedy was careful to assure his Senate colleagues that nothing in the hearings or legislative record implied that this approach to cancer indicated an intention to apply it to other diseases. Prudence required that the cancer effort be properly evaluated before such extensions could be drawn.[105]

Senator Nelson rose to say that no one could quarrel with the objective of the legislation, but he had concluded that a separate cancer agency would hamper cancer research and jeopardize the entire biomedical research enterprise.[106] He found as the only reason for favoring the bill "a kind of emotional commitment" to doing something with an independent agency that, it was felt, could not be done within NIH. It was a dangerous precedent, the Wisconsin senator thought, and heart- and lung-disease advocates were sure to follow with claims for similar treatment.

Senator Cranston was the only other person to voice reservations about the bill. He thought the objections of the scientific community were "serious and valid," that the elevation of cancer outside of NIH would be "a virtually irresistable precedent" for other diseases, and that the proposed arrangement contained the seeds of much dispute between the cancer program and the NIH director. But, though he had "grave reservations" about the bill, he intended "with some reluctance" to vote for its passage. He indicated why:

> However, we must conquer cancer. This measure—considered with care by so many public officials and private citizens, many of them professionals with vast experience and the very highest of qualifications, others of them laymen like the indomitable Mary Lasker with the deepest dedication and concern and with truly remarkable experience in what it takes to launch a vital, new undertaking with requisite public, private, professional, and lay support—constitutes what all of us must hope is a start toward the end of cancer.[107]

Hope, and a large measure of political realism, had overcome Cranston's apprehensions.

Nelson, like Cranston, had grave reservations about the bill. A popular governor of Wisconsin before he entered the U.S. Senate, he had a maverick's reputation for voting his convictions against the popular views of his constituents. He had no fear of voting against the cancer initiative. Besides, that was precisely what some of his constituents at the University of Wisconsin Medical School were urging him to do.

Eighty senators were present when the vote came at 5:30 p.m. The vote was—Yeas, 79, and Nays, 1. Senator Nelson had cast the single negative vote.

The Panel of Consultants had enjoyed the advantage of surprise in the early months of 1971 in their advocacy of S. 34. As that advantage diminished, the debate was expanded to include the constituents of NIH and journals like *Science* that served them. The advantage began to shift to the opponents of the cancer initiative. Erosion of support for S. 34 ceased immediately, however, with the Ann Landers article and the resulting outpouring of mail. It was one of the great strengths of Mary Lasker and her associates that they could successfully move from an inside strategy directed to the Congress to an outside strategy aimed at the public and back to an inside strategy focused on the White House. The administration moved steadily toward the Senate position during this time, largely in response to public opinion. As it did, the degrees of freedom of the NIH were sharply curtailed and increasing reliance was placed upon the medical-scientific community to make the counter-arguments to the cancer initiative. But an accommodation had been reached between the Senate and the White House by early June and attention now turned to the House.

8

The Cup and the Saucer

Mr. Preyer: . . . *I do want to thank the Senator [Nelson] for a very hard-hitting statement. I would like to comment that I was a little shocked at the degree of fallibility of the Senate on this where they voted 79 to 1, but I am sure you could point to equally horrible examples from the House that you correct in the Senate.*

I think this may be a good example of the virtues of a bicameral legislature. . . . Somebody has said that Congress functions better as a brake than it does as an accelerator. So I hope after we do our thing as a brake, we can then proceed to press on the accelerator and get on with the battle here against cancer. . . .

Mr. Symington: *Senator [Nelson], it is a pleasure to see you here. I think it was George Washington who described the Senate as the saucer that cooled the legislation from the House. I take it you would like to see us play that role today.*

> Members of the Subcommittee on
> Public Health and Environment,
> U.S. House of Representatives,
> September 16, 1971

Senate passage of S. 1828 caused a number of actions and reactions. Robert Marston commented that he had, with the president's support, worked so strongly against the creation of a separate cancer agency that "the process of changing direction toward the compromise solution now supported by the Administration may cause me to leave behind a few skid marks."[1] The power that would be lodged in the administrator of the proposed new agency was so vast that it stimulated among many an intense interest in finding someone to head the new cancer effort.[2] The emphasis on a national plan led Dr. Baker and his NCI colleagues to initiate a major planning effort.

In the summer of 1971, the debate over the new cancer agency shifted from legislative chambers to the editorial pages of the country's leading newspapers. A July 11 *New York Times* editorial declared that "[the] weakness of Congress in handling a technical issue has rarely been more glaringly demonstrated than in the 79 to 1 vote by which the Senate approved the so-called Conquest of Cancer Act."[3] The "most disturbing element" of the debate, it continued, was the argument put forward by Senator Javits that scientists were close to "one final push" in a breakthrough for cancer. Javits, obviously stung, replied that his statements "merely paraphrased" the conclusions of the panel of experts and the Senate had acted "with wisdom and compassion on the basis of substantial scientific testimony."[4] Dr. H. Marvin Pollard also criticized the editorial as a "disappointment" that "seemed to address itself to shadows rather than facts." Not only would significant cancer results flow from the new bill, he wrote, "the upgrading of the cancer program . . . will have an equally stimulating . . effect on all . . . medical research."[5]

In an August 9 letter, thirteen scientists, five of whom were Nobel laureates, wrote of the Senate bill that they found it "hard to imagine a scheme with more potential for undermining the scientific integrity of NIH and the authority of its director."[6] They accused Senator Javits of being misleading in suggesting that scientific rather than political grounds were the basis for the new agency, that a scientific breakthrough in cancer was near, and that basic research would be stimulated rather than curtailed.

LEGISLATIVE OVERSIGHT

The rapid passage of cancer legislation by the Senate created widespread expectations that the House would also act quickly, especially since S. 1828 now had both the backing of Senator Kennedy and the support of the Nixon administration. There were signs within the House to suggest speedy action. The Rooney resolution of 1970, for instance, had been unanimously adopted one year earlier. The Gallagher resolutions of 1970 and 1971 had secured 85 and 100 co-sponsors respectively.

More importantly, on December 16, 1970, within two weeks of the submission of the Panel of Consultants report to the Senate, Representative Harley O. Staggers (D., W.Va.) introduced a bill that was identical to the Yarborough bill. Luke Quinn then rounded up over 100 co-sponsors before the 91st Congress adjourned. Staggers, chairman of the House Committee on Interstate and Foreign Commerce, whose subcommittee on public health would consider the cancer legislation, became even more active when the 92nd Congress convened in January 1971. He announced his personal support for the Kennedy-Javits bill. Then, on February 4, with Representatives Minish of New Jersey and Pepper of Florida, he introduced legislation identical to S. 34, having one hundred and three House co-sponsors.[7] Among these were sixteen members of the full committee, including three members of the health subcommittee. It looked as though the House was being mobilized for quick action. But all such assumptions tended to overlook Representative Paul G. Rogers.

Rogers, a Democrat from the Ninth District of Florida, was first elected to the House on January 11, 1955, having succeeded to his father's seat. He had been the second-ranking Democrat on the public health subcommittee for several years, but the chairman, Representative John Jarman (D., Okla.), had let him run much of the work of the committee. Not until 1971, however, in his seventeenth year in Congress, did he become chairman of the subcommittee in name also, and then there was some question whether two other more senior members of the full committee might take the position.[8]

Rogers was relatively unknown to the cancer-legislation proponents. He had not been invited to Benno Schmidt's luncheon of September 1970, since he was not at that time in a leadership position on health. Nor had Luke Quinn paid much attention to him, preferring to rely instead upon his old friend, Harley Staggers, chairman of the full committee.

In 1966, Rogers had chaired an extensive investigation into the organization of health activities within HEW.[9] The investigation had given substantial attention to the National Institutes of Health, so the Florida congressman had developed a respectable understanding of that organization and medical research. It was not out of character, then, when he said in mid-February 1971 that he had strong reservations about the Kennedy-Javits cancer legislation. "We already have a mechanism for cancer research," he said, and a special agency for cancer would increase administrative costs, disrupt ongoing work, and set a precedent for other health groups like heart disease.[10]

No one paid much attention to Rogers, at least not among the promoters of the national cancer crusade. The House strategy, to the extent one existed, was to work through Staggers. In March, Mike Gorman rather indiscreetly expressed the view that Rogers could be persuaded of the merits of S. 34. Rogers "always comes at these things cautiously," Gorman said, "but once you explain the reasoning he comes around. Besides that, Staggers is for it, and Staggers is his chairman. Chairmen are powerful men."[11] In July, *The Blue Sheet* quoted an unidentified lobbyist commenting of Rogers: "He'll come around—he doesn't want to be out-voted."[12] Luke Quinn, who detested Gorman's meddling in cancer matters, acknowledged in late July that getting a bill out of Rogers' subcommittee was the problem in the House, but he said nothing to aggravate Rogers.

Rogers became health subcommittee chairman as health issues were becoming high priorities on the national agenda. In 1971, the subcommittee would deal with health manpower legislation and drug abuse legislation, in addition to cancer. But these issues had a more narrow appeal. Cancer legislation posed a special problem for Paul Rogers. For one thing, though he had substantial experience with health, he was new and untested as chairman. Even subcommittee chairmen have great powers within Congress, but the exercise of such power is partly a function of the con-gressman's own skill in using the resources available to him, which in turn is critical to his reputation among fellow legislators. Rogers, as a new chairman of an important subcommittee, had need to establish himself with his House colleagues.

On cancer, Rogers knew perfectly well that House sentiment was sufficiently strong that legislation would be required. But he had three reasons for not wishing to pass the bill sent over by the Senate. The "independent agency within NIH" formula, in Rog-ers' mind, violated the integrity of NIH and thus ran counter to his personal convictions. Furthermore, he had no interest in being railroaded by any health lobby, regardless of how sophisticated or powerful it was. Finally, he had no desire to ratify a bill so closely identified with Senator Kennedy. To Rogers' irritation, Kennedy had already been dubbed "Mr. Health" in the Congress, though he had been subcommittee chairman no longer than Rogers and actually had less legislative experience on health matters. While he could have easily pushed S. 1828 through the House with a vote nearly as lopsided as that in the Senate, Rogers would have forsaken an opportunity to establish himself as a man of independence, thoughtfulness, and power with his House colleagues, members of "the other body," Mary Lasker's associates, and the Nixon administration. There was little to be gained in appearing to fold under pressure. The failure of the proponents of S. 1828 to appreciate Mr. Rogers' situation significantly affected the outcome of the legislation.

Rogers sought to assert some control over the situation by delaying the scheduling of hearings. In mid-July, he indicated that health maintenance organizations were next on the subcommittee's agenda and he announced no plans for cancer.[13] Soon thereafter, it was announced that no hearings could be held until after the August congressional recess was over.[14] Whether the subcommittee would deal with health maintenance organizations or with cancer would hinge on conversations between Rogers and Staggers. On August 8, Rogers announced that hearings would begin on September 8, and he reiterated his opposition to S. 1828.[15] Hearings were finally scheduled for September 15.[16] Though he had no intention of avoiding hearings, Rogers was seeking to exploit his control over scheduling to his own legislative advantage.

YOU CAN'T FIGHT SOMETHING WITH NOTHING

Mr. Rogers, however, fully appreciated that he had to bring out a respectable piece of cancer legislation. The pressure from his House colleagues was very strong, and supporters of S. 1828 were putting on additional pressure. The American Cancer Society wrote to selected state chapters to stimulate House support for the measure. The ACS was concerned about the prospects that the Rogers' subcommittee would consider amendments which would undercut the legislation and seriously interfere with the proposed new agency. The executive vice-president of the Tennessee chapter, for example, had it drawn to his attention by the national headquarters that Representatives Ray Blanton and Dan Kuykendall, both of Tennessee, were members of the Committee on Interstate and Foreign Commerce. It was suggested that these congressmen be contacted by individuals who knew them personally and be advised of the importance of the legislation.[17]

Mary Lasker went so far as to ask Representative Staggers to have the full committee conduct hearings on the legislation, thus preempting the Rogers subcommittee. This move would have been unprecedented for Staggers, and he was not prepared to violate the norms of congressional behavior. Though he strongly supported S. 1828, Staggers told Mrs. Lasker that he could not override Rogers' prerogative as subcommittee chairman.

Rogers also received a number of visitors. Max Geffin, publisher of *Family Magazine*, a good friend of Mary Lasker and Sidney Farber, and actually an emissary from them, came to see the Florida congressman. Geffin pointed out that since he had a home in Palm Beach as well as New York, Rogers was his congressman. The publisher expressed his dismay at the delay in House action, and asked if he could allay Rogers' fears about the Senate measure in any way. Rogers, a very persuasive individual, outlined the merits of what he intended to propose, and Geffin left Rogers with a promise to encourage his friends to support the chairman.

Benno Schmidt also went to see Rogers soon after action was completed on the Senate bill.[18] He and Rogers began a number of vigorous, lengthy, head-on debates that lasted over the entire course of action in the House. Though locked in an adversary position for several months, the two men nevertheless developed a large respect for each other. Yet neither would yield an inch to the other until forced to do so.

Rogers addressed himself to two principal tasks: drafting his own legislation and educating the members of his subcommittee to the issues involved. He had introduced the administration's legislation in early May, but this was simply a courtesy extended by the congressman in his official capacity.[19] Rogers had no commitment to this bill, and certainly none after it had been abandoned by the administration. He concluded, however, that he could not afford to find himself in the earlier position of the Senate Republicans of trying to oppose a bill with nothing but arguments about its weaknesses and adverse implications. Rogers determined he had to have his own bill.

Knowledgeable about the NIH and reasonably clear about what he wanted to propose, Rogers nevertheless needed help in drafting a bill. He sought this assistance from the Association of American Medical Colleges. The result of these efforts was H.R. 10681, "The National Cancer Attack Amendments of 1971," which was introduced on September 15, 1971, the day the hearings began.[20] This bill was distinct in several important respects from S. 1828.

One difference was that the scope of the legislation was broadened from cancer to include other major killer diseases, especially heart and lung diseases, and stroke. Collectively, these diseases were the leading causes of death, whereas cancer alone was only the second-leading cause. The national cancer attack program was to be coordinated by the NCI director, who was also to be elevated to be an associate director of NIH. Similarly, the directors of the National Heart and Lung Institute and the National Institute of Neurological Diseases and Blindness were to be made NIH associate directors. These three officials and the NIH director were to be presidential appointees. Thus, priority in medical research was not to be limited to cancer research.

A second difference was that H.R. 10681 proposed to retain the NCI and cancer research within the National Institutes of Health. The present state of scientific knowledge about cancer, heart and lung diseases, and stroke was attributed to "broad advances across the full scope of the biomedical sciences."[21] These advances, in turn, were a result of NIH programs that had created "the most productive scientific community centered on health and disease that the world has ever known." Specific provision was made for scientific peer review of all research grants and programs. The NCI director would be obliged to use the appropriate NIH peer-review groups to the maximum extent possible. He could establish other formal review groups as necessary with the concurrence of the National Advisory Cancer Council and the NIH director. Rogers' bill clearly rejected the separate agency status for cancer.

Rogers made his major concession to the opposition on the matter of the budget. The bill provided that the NCI director prepare and submit his annual budget request directly to the president.[22] The NACC, the NIH director, and the HEW secretary could comment upon the request, but could not modify it. The legislation also made specific authorizations for fiscal years 1972, 1973, and 1974 of $400 million, $500 million, and $600 million, while S. 1828 had only authorized "such sums as may be necessary." Furthermore, the president was enjoined to express to the Congress without delay the need for additional appropriations for developments in cancer, heart and lung disease, and stroke for which regular funds were not available.

Several other provisions were not matters of controversy. One was that existing cancer research centers would be strengthened and new centers established. These centers were to conduct "a multi-disciplinary effort for clinical research and training and . . . the development and demonstration of the best methods"[23] of cancer treatment. A consensus was developing on this issue. Another provision permitted the directors of the three particular institutes to award research and training grants of $20,000 or less after scientific review but without advisory council approval.

Rogers was active before hearings began in educating his fellow subcommittee members. He circulated to them a number of articles and editorials critical of the Senate legislation that had appeared in the national press. He provided the other congressmen with information on the critical issues involved in the legislation. This action paid major dividends. No member of the subcommittee expressed public support for S. 1828 during the summer. Moreover, when Rogers discussed his bill in a press conference on the eve of the hearings, five other members of the subcommittee were co-sponsors. They were Representatives David E. Satterfield (D., Va.), Peter N. Kyros (D., Me.), Richard Preyer (D., N.C.), William R. Roy (D., Kans.), and James F. Hastings (R., N.Y.). Only one Democrat, James W. Symington of Missouri, did not co-sponsor the bill, owing primarily to strong constituent pressure in favor of the Senate bill.

Representative Hastings, the single Republican co-sponsor, had received a telephone call at home on Tuesday evening, September 14, from Dr. John Zapp, the HEW deputy assistant secretary for health legislation. Zapp asked Hastings not to co-sponsor Rogers' bill and assured him he would get back to him the next morning. Zapp was out of town the next day, however, and an aide would not accept a call from Hastings' office. Not until the hearings were well underway that afternoon did Hastings hear from HEW, but by that time he had become a co-sponsor.[24] Mr. Rogers' bill had bipartisan support from the outset.

Of the three other Republican representatives, two—Ancher Nelsen of Minnesota and Tim Lee Carter of Kentucky—had acceded to the administration's request not to co-sponsor H.R. 10681. The third, Representative John C. Schmitz of California, an ultra-conservative from Orange County, thought the bill too expensive and refused to support it on those grounds.

Close observers could see the main elements of Rogers' legislative strategy emerging.[25] In introducing his own bill, the chairman was giving the subcommittee a wider scope of debate than was possible if attention were restricted exclusively to consideration of S. 1828. He did not wish the proponents of the Senate measure to define the terms of discussion. Furthermore, Rogers was clear that his primary audience was his own subcommittee. Given the situation in the House and in the full committee, the chairman would have to have near-unanimous support from his own subcommittee if he were to prevail.[26] Finally, in pointed contrast to the Senate, he indicated that the House subcommittee would hold hearings of three weeks or longer, not merely three days. Where the Senate was cursory, he would be thorough.

The House hearings began on Wednesday afternoon, September 15. Mr. Rogers welcomed Representative Staggers. The committee chairman had come to emphasize to the subcommittee his personal interest in cancer legislation. Staggers waved his greetings to Mary Lasker, whose presence made clear that the proponents of S. 1828 were in earnest pursuit of their objectives.

The supporters of the Senate legislation, however, lost an important strategist and tactician just as the hearings commenced. Luke Quinn, it was learned, had just been admitted to Massachusetts General Hospital. Ironically, the diagnosis was cancer of the liver, which had metastasized. The prognosis was pessimistic. Though hospitalized, Quinn would continue to play a role in the legislation, but his involvement would steadily diminish as his body waged a losing battle.

In his opening statement, Rogers said that all proposals aimed at developing a cure for cancer would be closely investigated and legislation would result. The threat of cancer, he pointed out, came from the disease itself and from the psychological fear it inspired. His conversations with scientists and doctors had revealed a cautious optimism about finding a cure for cancer. The situation was not entirely bleak, he continued. Two of every six persons contracting cancer were then being saved, and full implementation of known methods of diagnosis and therapy could increase that to three out of six. The challenge to the medical-scientific community, Rogers said, was "to find a cure for the 50 percent who suffer from a form of cancer which is presently without a cure."[27] This, he emphasized, would require the full resources of biomedical research, and that was why his bill reflected the intention to use NIH fully in the cancer effort. It would be unwise, he judged, to separate the cancer effort from NIH.

He turned then to the first witness, Elliot Richardson.[28] Richardson was accompanied by Stephen Kurzman, assistant secretary for legislation; Dr. Merlin K. DuVal, the new assistant secretary for health; Dr. Robert Marston of NIH; and Dr. Carl Baker of NCI. The resilient Richardson, defending the most recent official view of the administration, said that S. 1828 represented "the best means" for expanding cancer research. One essential feature was that the proposed legislation facilitated the direct access of the president to the director of the cancer agency, and provided the director with reciprocal access. This would be made possible by a more direct budget-review process than NCI had known.

The other essential feature was that cancer research was retained in a close, consistent relationship to the mainstream of biomedical research. This, the secretary explained, was to be accomplished by providing the new cancer agency with "autonomy in the development of its research program . . . in the framework of a mutuality of effort with other elements of biomedical research."[29] This rather tortured formula would soon be subjected to persistent skeptical questioning. Richardson concluded his testimony by noting that scientific accomplishments in the past fifteen years had created opportunities for expansion of research efforts, but warned that the dangers of over-promise should be kept always in mind.

Staggers asked Richardson whether S. 1828, as passed, was a better bill than the version introduced in May. Yes, the secretary responded. Did S. 1828 give "more impetus" to the program in its present form? Staggers asked. Definitely, came the answer. The committee chairman allowed that cancer was the most important legislation to come before the health subcommittee in 1971. He voiced his expectation that the subcommittee would "bring out a good, strong bill, even stronger, and better than the Senate version, which I think is a good bill in a lot of ways."[30] Staggers stopped just short of endorsing S. 1828 and thus openly challenging his subcommittee chairman.

Most of the questions to the secretary were directed to the organizational arrangements embodied in S. 1828. Representative Satterfield wanted to know how the independent agency would operate within NIH. Would it not fragment the cohesion now existing within the NIH? How, he asked, could the new organization be within NIH and operate as an independent agency? The secretary's convoluted answer was not satisfying to the congressman. Finally Richardson conceded that "[t]raditional organizational practice would counsel against doing this," but "the twin aims . . . of giving leadership and visibility to this effort and of preserving continuity and communication with other ongoing scientific efforts within the NIH have combined to produce this proposal now before us."[31]

Representative Kyros had a question about a different aspect of the new arrangement. Why should the director of cancer research become embroiled in politics? Why, he asked, should he be a presidential appointee confirmed by the Senate? Presidential appointments, Richardson replied, gave the president personal access to the program and insured that the new cancer director would be a man of both scientific stature and management capability. Senate confirmation was less important though normal practice for presidential appointments.

Representative Symington wanted to know if there were any precedents for an independent agency within another agency. The proposed organization was rather novel, Richardson conceded. Symington expressed the view that the proposed cancer research organization and the drug abuse reorganization supported by the White House were both inconsistent with President Nixon's larger effort to reorganize and streamline the federal bureaucracy. Perhaps the president was not that concerned about orgnizational efficiency, Symington suggested. As to the independent agency within NIH formulation, he found that "a distinction without a difference."[32]

The crash program implication was questioned by Representative Preyer. If it was a crash program being advocated, he asserted, it was for cancer research, not cancer cures. Moreover, he said, the cancer effort was "somewhat different from the NASA space program analogy." There "it was a question of just building it to get there and taking advantage of known laws [while] here we are looking for new theoretical breakthroughs."[33]

The matter of peer review was raised by Representative Roy, one of two M.D.s on the subcommittee. Marston's description of existing NIH practices and S. 1828 provisions apparently satisfied him. But then Roy wanted to know about the authority that the HEW secretary would exercise over the new cancer agency director. Richardson responded that his authority would be limited to collaboration between the Cancer Conquest Agency and the NIH. Rogers interjected that the bill stipulated that the new director would exercise the authority of the secretary, quite a difference from the original S. 1828 which vested authority in the secretary. All in all, Richardson's complicated responses were greeted with skepticism by many of the subcommittee members who reflected thoughtful concern for orderly processes of government.

Satterfield, Kyros, Symington, Preyer, and Roy were all Democrats. The Republicans manifested greater divergence of views. Representative Carter, the other M.D., was uncritically supportive of the secretary and the administration, and indicated his unqualified support for S. 1828. He was also rather critical of the Cancer Institute. In this respect, he was the best friend the supporters of S. 1828 had on the subcommittee.[34]

Representative Hastings expressed to the secretary his discomfort at being in disagreement with the administration. Why, he inquired, had the administration changed its mind from the version of S. 1828 submitted in May to that passed by the Senate in July? He had been convinced at the outset, Hastings said, that cancer research should be retained within NIH. He still felt that way, and found it difficult to accept statements that the agency would still be within "the confines of NIH." A significant change had occurred, Richardson conceded, but S. 1828 should be judged, first, on whether it was desirable to combine direct presidential responsibility for cancer research while retaining the effort within NIH, and, second, whether the bill accomplished this objective. All versions of the administration's position, the secretary noted, had been consistently affirmative on this initial point. The phrase, "within NIH," did mean something, and Richardson sought to indicate how cooperation between the cancer agency and NIH would take place. Clearly skeptical, Hastings hoped that the hearings would answer the question of "whether or not we are effectively removing the agency from NIH."[35]

The *Washington Post* had printed a letter on September 5, signed by ten scientists, of whom four were Nobel laureates in medicine.[36] Representative Schmitz drew attention to the letter, which claimed that the scientific community was overwhelmingly opposed to S. 1828. "I am not a scientist," he said, so "I can't make a judgment here." But it did seem "like a rather high-powered group, when you include four Nobel Prize winners in opposition to this bill." Describing himself as a "somewhat cynical" politician, Schmitz suggested that much of the legislative action might simply be "jockeying for position to take credit" for any cancer breakthrough. "Maybe if we kept all presidential candidates out of the cancer cure picture we might be better off," he added.[37]

Mr. Rogers, who had interjected questions throughout, reserved the end of the afternoon for his close questioning of

Richardson. He asked the secretary about the status of programs for the control of cancer: "Have you had a cut in those funds or not?" Richardson, caught off-guard, deferred to Marston, who also revealed momentary confusion. It was revealed that several cancer control programs had been phased out in 1970 and had not been reinstituted. Rogers was upset:

> [H]ere it is, where we have the answer to try to solve some cancer problems, and we talk about mounting a drive and the Department—and I think it is probably that OMB, too, has cut out many of the programs that could actually save lives right now with present knowledge, isn't this true? Don't we have present knowledge with early detection in many of these areas?[38]

It was a glaring contradiction in the administration's position that Rogers had identified: a large, expanded program in cancer research was being proposed a year after cancer control programs had been phased out.

Representative Ancher Nelsen, the ranking Republican subcommittee member, had said in his opening remarks that the task of the subcommittee was to examine all legislative proposals before it and "come up with the best and most effective measure."[39] At the end of the day, he noted that Richardson had been responsive to some improvements suggested by subcommittee members and suggested that working things out "in a quiet way" would prove to be an effective course of action.[40] Nelsen's efforts throughout the hearings would be conciliatory and directed to mediating differences between the administration and Mr. Rogers.

THE MINORITY OF ONE

The second day of hearings began with Senator Gaylord Nelson as witness. Nelson's sole dissenting vote in the Senate had amplified, rather than diminished, his influence with Mr. Rogers. S. 1828, the senator forcefully declared, was the "fundamental if not fatal assault on the organizational structure of NIH" and "the first giant step in the dismantling" of that organization. The statutory ties between the new agency and NIH would mean no more than the new director wished them to mean. Moreover, Nelson said, if the director and the deputy of the new agency were appointed by the president, with the advice and consent of the Senate, cancer research would be made political in a way unprecedented for medical research. Under S. 1828, Nelson pointed out, "not even the director of NIH is politically appointed; he is a career public servant."[41]

The administration, Nelson charged, had done an about-face on this issue. Neither the president, his science adviser, the HEW secretary, nor the scientific community originally wanted a separate agency for cancer research. The administration changed, Nelson said, "as a face-saving political compromise" when it became obvious that S. 34 would be adopted over their objections. But the compromise, the senator said, "simply changed the bill number from S. 34 to S. 1828, changed some language without changing the substance, and substituted Republican sponsorship for Democratic sponsorship."[42] In his judgment, it had been a compromise of a fundamental principle, since the original administration position had been right.

Nelson succinctly put the case for retaining cancer research within NCI and in the context of the National Institutes of Health.

> The enormous irony of proposing a moon-shot type of agency for cancer is that the breakthroughs to date have occurred because of the capabilities of the National Institutes of Health and its National Cancer Institute, not in spite of them. All of the major discoveries, including ones which fell out inadvertently from non-cancer research, have occurred largely because of the present broad-based, multi-disciplinary system of federally supported research embodied in NIH.[43]

The Panel of Consultants had charged that no comprehensive plan existed for a coordinated national cancer program, Nelson recalled, but this was "simply incorrect." Both Dr. Baker and Secretary Richardson had indicated that plans did exist for chemotherapy, viral oncology, and all research areas touching upon cancer. Moreover, experts were even then being appointed to draw up plans for other cancer research areas.

The senator's arguments came with drumfire rapidity. The biomedical science community was almost unanimous in opposing a separate agency, and only the Cancer Society supported it. Then, if cancer got a separate agency, heart research would soon seek similar status. Moreover, knowledge of cancer had not placed the country on the threshold of a moon-shot program. Furthermore, the senator said, ample funding was not insured by independent agency status, but the national commitment surrounding an issue gave it real priority.

The senator's greatest concern was the effect that "bypassing the director of NIH in formulating cancer budgets and programs" would have upon NIH. He reminded the subcommittee that thirteen prominent scientists had written, "it is hard to imagine a scheme with more potential for undermining the scientific integrity of the NIH and the authority of its director." The strongest argument in favor of retaining straight-line authority for NIH over the NCI, Nelson argued, was the little evidence that existed which showed that progress in cancer research had been significantly impeded by administrative problems.

Why, the senator was asked, had 79 of his colleagues disagreed with his views? For one thing, Nelson pointed out, there had been a major mail campaign. He had received over 6,000 letters in favor of S. 34, more than he had received on any other single issue except the 1964 civil rights bill. Also, he noted, each senator was busy, a prestigious outside panel had made a report, a subcommittee had considered a bill, and—lacking time to study the report—other senators had gone along favoring the general proposition of the conquest of cancer. Furthermore, the measure had been debated for only one hour or so on the Senate floor; had there been two or three days for debate, the issue might have been thoroughly aired.

Representative Preyer indicated that he was "a little shocked" at the Senate's vote. Representative Schmitz, who had been on the short end of some votes in the House, observed that "if you can get a good label attached to a bill, you have your job half done."[44] At the risk of alienating some conservative Wisconsin friends, he said, Nelson was to be complimented on "going beyond the label to the merits of the bill." Representative

Symington drew the inference from Nelson's testimony that the senator hoped for more reflection and deliberation by the House. The senator had successfully appealed to an underlying belief among many in the House that they could, and did, write better legislation than their Senate counterparts.

A number of members of the House presented testimony before the subcommittee on cancer legislation. Representative John Rooney was the first to do so. He was sponsoring House Concurrent Resolution 2, which was identical to his resolution of the prior year, to place the House on record once again in favor of providing the funds necessary for the cancer crusade. Rooney, the budget cutter, had no reservations about money for cancer research. Twenty-four years earlier, he explained, he had introduced a bill in the 80th Congress recommending a national drive to eliminate cancer, and had continued to do so over the years. He had been much heartened by the passage in 1970 of House Concurrent Resolution 675, and by President Nixon's request for an additional $100 million. The Brooklyn Congressman explained that he was a "competent witness" since he had been successfully operated on for lung cancer a few years earlier at the U.S. Naval Medical Center in Bethesda. It had taken time to come to the present situation, Rooney argued, because "scientific knowledge and technology grows in quantum jumps and has done so unheralded over the years until we are now at the point where we approach the very real goal of a major breakthrough on the cure of cancer."[45] The goal was within reach, the knowledge and resources at hand, and "Lord knows we have the desire," he emphasized. All that was needed was the vehicle, and that was S. 1828. Rooney was no less eloquent because his remarks had been written by Luke Quinn. If anything, the knowledge of his hospitalized friend's condition made him more forceful.

The next witness was Representative Claude Pepper, the Democratic representative from a district adjacent to Rogers', a former Florida Senator, and a long-time friend of Mary Lasker. The country was fortunate in having two pieces of cancer legislation to choose from, he said, but the desirable thing was for the House to adopt "as speedily as possible" the bill supported by the administration and enacted by the Senate with only one dissenting vote.

Pepper recalled his co-sponsorship in 1937 of the original National Cancer Institute Act, and his post-World War II support of more funds for cancer research. Though appropriations had grown slowly, they now reached more than $200 million annually, owing in large part to the persuasive urging of Albert and Mary Lasker. But, the congressman insisted, two billion dollars had been made available for the development of the atomic bomb without the Congress being informed, and no estimate had been made of the cost of going to the moon when that was set as a national objective. The purpose of S. 1828 was to establish a national commitment to conquer cancer. "[I]n the interest of this country and in the interest of those people who are suffering from cancer and their miserable loved ones," Pepper appealed to his Florida colleague to give favorable consideration to the Senate bill.[46] Pepper and Rooney made it quite clear to Mr. Rogers that there was much House sentiment favoring accommodation with the Senate on S. 1828.

The final witness that second day was Franz J. Ingelfinger, M.D., editor of the reputable *New England Journal of Medicine*, a citizen witness over the years for medical research, and a long-time associate of Sidney Farber and Mary Lasker. But today he was representing the American Medical Association as a member of the advisory committee on medical science to the board of trustees. The AMA view was that the cancer attack could be more effective through the NCI and within the NIH than through a separate, autonomous agency. Ingelfinger opposed S. 1828. The key issue was maintaining the integrity of NIH. He was asked by Representative Kyros if his statement was a personal one or one that reflected the AMA view. Ingelfinger responded that the statement had been approved by the board of trustees, and this was confirmed by a letter from the AMA five days later.[47]

DR. WHAT'S-HIS-NAME

The third day, Monday, September 20, opened with testimony by Representatives J. J. Pickle (D., Tex.) and Daniel Kuykendall (R., Tenn.), both members of the full Committee on Interstate and Foreign Commerce. Representative Pickle made a brief

statement supporting S. 1828. But Pickle, an old friend of Benno
Schmidt's, had advised the latter to get in and work with Paul
Rogers because, in the last analysis, something would have to be
worked out that was acceptable to the Florida congressman.
Pickle had told Schmidt to keep the pressure on and he would try
to help. He was doing his part that day. Kuykendall was also help-
ing by drawing attention to the financial plight of St. Jude Children's Research Hospital of Memphis, a center for treating cancer
in children. Representative Bob Eckhardt (D., Tex.), another
member of the full committee, came to hear the Panel of Consultants. Eckhardt, who had studied law at the University of Texas
when Benno Schmidt had taught there, was also friendly to the
Senate bill. The pressure was indeed being kept on Mr. Rogers.

That day the spotlight was on the Panel of Consultants. Benno
Schmidt appeared with Drs. Sidney Farber, Lee Clark, James
Holland, Mathilde Krim, and Wendell Scott. The presentation
followed the Senate pattern, with Schmidt making the major
points and his colleagues elaborating them.

The scientific and medical men who had worked with cancer
research, Schmidt began, felt that the time was "especially right
. . . for an intensified and accelerated effort" on cancer. Dr. Holland amplified this, indicating that developments in cell biology,
chemical carcinogenesis, virology and immunology, diagnosis,
chemotherapy and radiotherapy, and combination treatments constituted "a broad and powerful wave of advances in cancer research."[48]

The panel had concluded, Schmidt said, that three factors were
necessary for a national commitment to conquer cancer—effective
organization, a comprehensive plan, and adequate financial resources. Dr. Clark, widely rumored to be a leading candidate for
the directorship of the proposed Cancer Conquest Agency, was
asked to comment on the need for bureaucratic independence.
Clark reiterated arguments that had been raised earlier by the
panel, but spent some time seeking to allay the anxieties of the
scientific community. S. 1828 would not interrupt the current
work of NCI, he asserted, nor would it isolate cancer research
from the other biomedical sciences. The bill would not deemphasize basic research, nor threaten basic medical sciences, nor

216 • The Cup and the Saucer

diminish the use of grants, nor reduce support for medical schools. On the contrary, the Houston doctor argued, the legislation would provide authority and responsibility for management, planning, and budgetary development. It would provide that research scientists be involved in planning, and that scientific peer review be maintained. But Clark also made quite clear that cancer would retain the priority assigned it by Congress, and funds for cancer research would *not* be used for other programs, "however meritorious, such as unrelated biomedical research in other fields and medical education."[49] No one in the research community would suffer, Clark was saying, but funds for cancer research were going to be spent for cancer research. Period.

Dr. Farber was asked if he thought the cancer effort would fragment medical science. No, he thought, the exact opposite would be true. Fragmentation would be avoided through improved coordination of the relevant medical research, which was one reason why the panel had recommended an overall plan for cancer research. Refocusing on the point Clark had made a moment before, Farber said, "We want to see the cancer effort organized so that the mission can be clearly defined as the conquest of cancer, and so that the management, planning, and coordination can be directed to that end, and the progress of the effort measured against that standard."[50] Obviously, he implied, the existing NIH framework did not permit this to occur. Then, taking a swipe at the scientific community, he suggested that fragmentation had become "a favorite phrase" of those opposed to an independently managed cancer effort but that they had never done an accompanying analysis of existing fragmentation within NIH.

Dr. Mathilde Krim was asked to comment on the charge that a comprehensive plan would emphasize a systems or engineering approach to cancer research and downgrade basic research. Though applied research should be accelerated, she said, "it is equally important that basic research must not be hampered or restricted." With heavy emphasis on grant support, using peer review, and involving scientific participation in planning, there was no question, Dr. Krim argued, "that basic science related to cancer will be enhanced and enriched . . . rather than impeded" as had been charged by some who had "failed to grasp" the intent of the panel's recommendations.[51] She was careful to add, however, that she had no doubt that basic biomedical science beyond cancer would also prosper. More funds, better direction and planning, and increased results in cancer would make the case for similar improvement in the other biomedical sciences "increasingly irresistible."

Dr. Wendell Scott, who had been operated on for cancer during the work of the panel, and who had returned after missing several meetings to make a large contribution to its work, concluded the formal presentation. Though no one wished to raise false hopes, he said, "one by one the diseases which we identify as cancer will yield."[52] Based upon today's insights, he continued, there was no question but that undertaking, planning, organizing, and funding an expanding cancer effort would lead to "vast inroads" on the cancer problem in a "relatively short period of time." What better way than this, Scott asked, to begin "the reordering of our national priorities?"

Schmidt, unexpectedly, asked Mr. Rogers if Dr. Farber could make an additional statement, and the request was quickly granted. Farber noted that the NIH had made great strides over the years in basic research and medical education, but had been unable to make equal progress in the application of the results of basic research to the patient or in clinical investigation. The "crux" of the panel's recommendations concerned "the seat of decision making," Farber stated. Though the new cancer agency would remain "scientifically and intellectually" within the NIH, decision making would be vested in those dedicated to applying fundamental knowledge to the conquest of cancer. The Boston physician drove the point home with accustomed eloquence:

We cannot wait for full understanding; the 325,000 patients with cancer who are going to die this year cannot wait; nor is it necessary, in order to make great progress in the cure of cancer, for us to have the full solution of all the problems of basic research. . . . The history of medicine is replete with examples of cures obtained years, decades and even centuries before the mechanism of action was understood for these cures . . . all the way from vaccination, to digitalis and aspirin, where the mechanism of action is still not clear and yet benefits are obtained. . . . Finally, the main purpose of all cancer research and everything we are talking about is a prevention of cancer and the prevention of disability and death by a great expansion of clinical application of research in behalf of the patient with cancer. Everything that we have said finally must be directed and pointed clearly for the benefit of the patient with cancer and all of the basic research and the great expansion of clinical investigation which we have recommended will be the surest way of bringing that about.[53]

No one could match the aging patriarch of cancer research in arguing for more clinical research.

Paul Rogers began the questioning by suggesting to Benno Schmidt that the Senate legislation had not really carried out any of the panel's recommendations.[54] Unlike his own bill, it did not make specific funding recommendations though the panel report had done so. Furthermore, he pointed out, there was no provision in S. 1828 requiring the national plan recommended by the panel. Schmidt responded by conceding that the congressman's judgment on the first point was far better than his own and, after a see-saw argument, agreed that he had no objection to seeing the requirement for a plan spelled out in the law.

A moment later Representative Ancher Nelsen reported that individuals involved with cancer had expressed to him the concern that current progress might be disrupted by a new agency. In addition, other members of Congress had indicated their anxiety about NIH "that we are tearing down what we have already built."[55] The Minnesota legislator expressed a strong hope for accommodation between the two positions and noted with obvious satisfaction the two points on which Mr. Schmidt and the chairman had just agreed.

Mr. Rogers asked Schmidt whether any recognized scientific or medical society other than the American Cancer Society supported the independence of the Cancer Institute from NIH. Schmidt, on the defensive, expressed the view that any medical or scientific society would have a division of view on this matter. He noted that the Federation of American Societies for Experimental Biology had been listed for Thursday's forthcoming hearing as if "that organization were unanimously and unalterably opposed to a separate agency."[56] Dr. Krim and Dr. Farber, though members of FASEB, had never been consulted on the organization's views. Undaunted, Rogers suggested that those who went on record spoke for a majority and that was the only way to view these things. This was especially appropriate, he added, since he understood that several panel members no longer agreed on the independent agency argument. Nettled, Schmidt suggested that item be taken up in a moment.

Schmidt then pointed to the testimony of Dr. Ingelfinger, who had represented the AMA. Dr. Wendell Scott, he noted, was a member of the House of Delegates and had called the AMA president-elect two days before their testimony was to be presented. Scott was referred to the head of the legislative department who promised him he would call and read the testimony when it was drafted. Consequently, Scott had called that morning and asked whether the testimony had been read to the AMA board of trustees. It had not, nor had the trustees voted on it. Then Scott and Rogers got into a quarrel on whether the AMA supported the view expressed by Ingelfinger, or whether Ingelfinger had misrepresented the AMA viewpoint. Scott insisted that the testimony be regarded only as representing the advisory committee on medical science of the AMA, not that of the entire association.[57]

Rogers summed up by observing that the communications the subcommittee had received from the "vast overwhelming number" of medical and scientific organizations opposed taking NCI out of NIH. Obviously, some members of these organizations would disagree, but the overall position was one of concern. Schmidt rejoined, "I think a vast majority of those of whom you speak are not cancer scientists and I would suspect the intuitive response of people who are not cancer scientists to be that this gives cancer a priority, that it is divisive, and I think it takes a rather profound look at what is really happening to see that this will not be the impact."[58]

Representative Kyros commented a few minutes later that when many leading scientists and medical school deans were opposed to an independent agency for cancer, it made "laymen like myself pause." He asked how Schmidt explained the great amount of opposition. Schmidt responded by saying that it was quite easy to say that cancer belonged within NIH "until you have studied this subject quite deeply."[59] Furthermore, he said, "most of those deans and professors to whom you refer are not cancer people." Finally, a great deal had happened since the December 4, 1970, report of the Panel of Consultants, including the appropriation of an extra $100 million for cancer research. So the panel had been victimized by success in some measure. The refrain was the same: the opposition to cancer legislation came from those not closely associated with cancer research.

Kyros also indicated that he had been surprised by Senator Nelson's assertion that not more than a handful of senators had considered the implication of an independent cancer agency. Schmidt said he also had been surprised. The report had been presented in December 1970 to the full Senate Committee on Labor and Public Welfare. A luncheon had been held in February for all members of the Senate, hosted by Senators Mansfield and Scott. Hearings in March and May had closely examined the matter. He did not recall, Schmidt said, Senator Nelson's presence at either the luncheon or the March hearings. So, he confessed, "it came as rather a surprise to me that he felt that his colleagues had not fully appreciated this point."[60]

Senator Nelson had argued, Kyros persisted, that the administration changed its views on the separate agency only when it became clear that S. 34 would be adopted over its objections. "Do you know," the congressman asked, "what the administration wanted?" Schmidt knew that the NIH director had written the HEW secretary after the panel's December 4 report indicating opposition to a separate cancer agency. That had been the administration's position for a time, and was reflected in Secretary Richardson's letter to Senator Kennedy in early April. To the best of his knowledge, Schmidt continued, the president had not personally considered the organizational issue at this time. When Nixon did get involved, the panel chairman said, he concluded that cancer needed a degree of independence it did not have. So the change in the administration's position, Schmidt suggested, was simply a matter of the president "getting personally into the matter and concluding that the independent agency idea should be implemented to that extent while still saving the relationship with NIH to the extent he did."[61]

Kyros then asked about the reported 16-to-10 division on the separate agency, which Nelson had also mentioned in his testimony. Schmidt replied that he had no idea where Nelson got his figures, but that such a division had never existed. There were 26 members on the panel, Schmidt said, and there was never a meeting at which all were present. In fact, he reported, four members never came to any meetings, so a 16-to-10 vote could never have occurred. The panel did not proceed by voting, the New Yorker said, but by discussing and resolving differences. The only formal vote taken was when the recommendations were put in final form. At the final meeting, Schmidt said, "one by one, I read the recommendations and asked if there was any dissent from any of the recommendations."[62] There was none. The panel was unanimous.

Furthermore, the panel chairman continued, the four absentees had been offered an opportunity to withdraw their names from the report if they did not agree with the recommendation, but none did so. Dr. Lederberg, who had not participated in any deliberations of the panel, originally concurred and then later withdrew his support for a separate agency. Dr. Kaplan, at a meeting on April 29, indicated that he also had concluded that the cancer program should be retained within NIH. Reports had been widely circulated that Dr. Harold Rusch, of the University of Wisconsin, had also changed his position, but Schmidt indicated that he had a letter from Rusch that made clear this was not so. In short, a unanimous panel had come to have two, but not three, members who later dissented from the separate-agency concept.

Representative Tim Lee Carter, the Republican M.D. from Kentucky, compared what the panel was recommending—"this infusion of new blood"—to the revitalization of the Veterans Administration in 1945. He also revealed a strong antipathy for

Public Health Service and NCI personnel, which was somewhat baffling. Carter observed that PHS personnel became administrators, that they were not "deeply, scientifically knowledgeable," and that they appeared "a little bit arrogant" in dispensing money to those they wished to, even though they lacked the knowledge of where the money should be spent. "[W]e need an infusion of new blood from dedicated people such as you," he said to Sidney Farber, "to direct the research, the expenditure of funds, and the pinpointing of their allocations throughout the country."[63] It was his feeling, Carter said, that the NCI people were "really not sufficiently knowledgeable to manage this great campaign." Many scratched their heads at this line of thinking. It was not until later that Carter's underlying concern would become clear.

Rogers had initially expressed his concern with taking cancer research out of NIH. While S. 1828 read "within NIH," it also read "as an independent agency." That, he presumed, was the emphasis Schmidt gave the language. Farber amplified the panel's meaning later in response to a question from Kyros. The feature about Mr. Rogers' bill, he said, "that I would not like to see put into effect is the retention of the control of cancer by the director of the NIH." From his vantage point of sixteen years as a member of the National Advisory Cancer Council and four years as a member of the National Advisory Health Council, Farber was firmly persuaded that "the decision making in a given field should be in the hands of those who are devoted, who are experts, and who believe that the job can be done and who do not have to divide their decision making with decisions of 10 or 11 other institutions."[64]

Schmidt underlined the same point. "The crux of the matter," he said, "is whether you want the head of the NIH to run this program, or whether you want the head of the cancer agency to run this program." "The cry that has gone out," Farber interjected, "that the NIH must be preserved, that the administrative structure of the NIH must be preserved, is in error. The Congress appropriates one billion and a half a year not to preserve the structure of the NIH but to preserve the health of the American people, and so this will always be a changing structure." The full attack upon NIH direction of medical research was being made.[65]

On the prior Wednesday, Secretary Richardson had turned to Dr. Carl Baker, NCI director, for an answer. Sitting with Richardson at the witness table had been Mr. Kurzman, Dr. DuVal, and Dr. Marston, while Dr. Baker had been seated in the first row of chairs behind the table. After Baker's lengthy answer, Representative Nelsen had said: "The gentleman who just spoke, I don't have his name. Do you have a preference as to the bills that are before us, or are you a little cautious about answering that question?"[66]

Schmidt seized on that embarrassing incident very adroitly, and used it to drive home a telling point: "Figuratively, just observing Wednesday's testimony, we want to get the man who is running the cancer program at this table when your committee is having hearings on cancer, not in the front row behind the Secretary and Assistant Secretary and head of the NIH and so forth. You do not want 'Dr. What's-His-Name' running the cancer program."[67]

But Rogers was hardly prepared to yield to the pleas of the Panel of Consultants. Time and again he asked whether splitting cancer from NIH would not lead to a similar fragmentation by other diseases, especially heart and lung diseases, and stroke. The panel answers were never fully satisfactory to the members of the subcommittee. Rogers also pressed the panel about the provisions of H.R. 10681, but the line that had been developed was that they had not yet had time to study his bill. Thus, the panel sought to ignore Mr. Rogers' bill and focus attention on S. 1828.

When the panel concluded its testimony, Benno Schmidt said, "If we can work with you as this progresses, we would be available." Mr. Rogers thanked him but the panel was not invited back to present formal testimony. A skeptical subcommittee had listened attentively and that was enough. They were then into their second week of hearings and much work still remained.

In the House, as in the Senate, the issues of the debate were those that had been framed by the report of the Panel of Consultants. Representative Rogers, moreover, did not even question the need for new legislation; he accepted that need on the basis of political reality. Furthermore, the concession in his own bill on the budget process for the National Cancer Institute was a clear indication that he also accepted the arguments for an increased priority for cancer research. The continuing influence of the pre-legislative work of the Panel of Consultants was clear.

▪ 9 ▪

Mr. Rogers Builds a Case

In a review of the status of cancer treatment published in the New England Journal of Medicine *I said the following:*

> *I have attempted to point out what I believe to be hopeful portents, but it is clear that we are far from the moment when we shall cure or control all cancers. In my view, that goal will be achieved, but the voids in knowledge concerning normal and abnormal cellular mechanisms are so great as to make senseless an indiscriminate cancer-cure crash program. We have made progress in cancer therapy, and the prospects for the future are bright. However, the road will undoubtedly be long, the amount of basic information yet to be acquired is enormous, and the day of universal cancer control is, in my opinion, far off.*

Although that statement was written in 1966, I see no reason to alter it today.

Howard H. Hiatt, M.D.
professor of medicine, Harvard Medical School, and head of the cancer division,
Beth-Israel Hospital, Boston, Massachusetts, September 20, 1971

[W]e, as laymen, are so interested in doing something about cancer, and are caught up in this emotional rush to move forward. We must meticulously take the advice of everybody in order to accomplish the best possible means to achieve the goal.

Representative Peter N. Kyros
September 30, 1971

When he began the cancer hearings, Paul Rogers had a good idea of where he wanted to go. His problem was getting there. One necessity facing the subcommittee chairman was that he listen to his House colleagues whose interest in cancer legislation was very intense and clear, so the subcommittee heard testimony from 31 members of the House over the course of the hearings.[1]

Three senior members of the Committee on Appropriations testified to the need for substantially more money for cancer research. Fully twenty representatives supported an independent cancer research agency. Five members of the Committee on Interstate and Foreign Commerce who were not subcommittee members supported the Senate-passed legislation. Of these, Representative Pickle urged Rogers to bring S. 1828 to the House floor for a rapid vote.

The predominant sentiment echoed repeatedly by the congressional witnesses was support for an independent cancer research agency.[2] Mr. Rogers' House colleagues testified to their personal experience with friends and relatives struck down by cancer. They reiterated familiar arguments—the need for reordering national priorities, giving cancer research adequate financial support, eliminating bureaucratic red tape, and getting on with the job. But the urgency of the message was one that the subcommittee could not afford to ignore.

No more than four congressmen spoke in direct support of Mr. Rogers, but he sprinkled these supporters across the several weeks of hearings so as to avoid the impression that all congressional opinion was solidly behind S. 1828. Representative Don Fuqua (D., Fla.) indicated that the faculty from the J. Hillis Miller Health Center and College of Medicine at the University of Florida had persuaded him to change his mind from his earlier support for an independent agency to support for H.R. 10681.[3] When it became clear that Rogers and the subcommittee were intent on bringing out their own bill, several congressional witnesses began to urge compromise with the Senate legislation. Representative Gallagher, for instance, testified that the important thing was to get on with the job. "I want you to know, Mr. Chairman," he said, "that I am not wedded to any one particular approach," and he was certain that the subcommittee would report out a bill that would move cancer forward.[4]

THE CASE FOR S. 1828

The pattern of congressional hearings for any legislation is a product of the political forces at work and the way those forces are mediated by the committee chairman for his own purposes. Paul Rogers set himself apart from the Senate by holding extensive hearings. He gave his House colleagues ample opportunity to state their strongly held views. He created an alternative to S. 1828 by introducing his own bill. He established the subcommittee as a model of fairness in its willingness to hear all points of view.

Rogers also sought to control the manner in which the proponents of S. 1828 made their case. Equally important, he determined how he wished his supporters to make the case for H.R. 10681. Though short on support within the House when the hearings began, Rogers sought political strength and advantage in all the subtle ways available to a committee chairman.

The responsibility for making the case for S. 1828 fell to the administration, the Panel of Consultants, the American Cancer Society, directors of cancer research institutes, and others. Mr. Rogers scheduled things in such a way that these major proponents of the Senate bill testified early in the hearings.[5] Those arguing for that bill toward the end of the hearings were individuals of lesser stature and political force.

The president of the American Cancer Society, Dr. H. Marvin Pollard, and Dr. Jonathan E. Rhoads, the immediate past president, testified on the fourth day of hearings. They were more interested in refuting arguments raised by the scientific community than in advancing a positive case for S. 1828. Pollard, for instance, attempted to characterize the opponents of S. 1828 as fearful of change and uninvolved in cancer research. "It would indeed be unfortunate," he said, "if the honest, but inadequately informed judgment of distinguished scientists in other fields were to be given equal or perhaps increased weight over the well informed and experienced judgment of equally honorable, but specifically knowledgeable scientists in the cancer field."[6]

Rhoads regretted that the debate seemed to be losing sight of the central issue of how to save lives from cancer, and was becoming bogged down in questions about lines of authority, structures, principles, basic versus applied research. These were questions of means, not of ends, remote from the central issue.

Yet Rhoads argued that reducing cancer to a minor cause of

death would require numerous scientific breakthroughs, and implied that breakthroughs would come with the proposed reorganization. "There will have to be multiple breakthroughs," he said, which would require "the freedom [of] straight line administrative relationships between the Director and the President's office." The panel believed, he continued, that the present situation was unmanageable. So, returning to those allegedly remote questions of means, he said, "that is why we are calling for a new arrangement, a new organization, a new approach."[7]

Representative Satterfield pounced. Was it reasonable to expect a "breakthrough," or, he asked, was it not more likely that advance would be "one step at a time?" Rhoads backpedaled quickly. "I think there have been breakthroughs. [But] [t]here has not been a breakthrough," he said, "and I do not expect a breakthrough and I think we have really held up the case by leading the public to expect a breakthrough." Progress would indeed come a step at a time. Satterfield inserted an article by a science writer for the *Philadelphia Bulletin* that asked a dozen questions about promised breakthroughs in cancer which had been announced over the years at the annual ACS science writers seminar.[8] The congressman invited Rhoads to provide detailed responses to the questions, but none were forthcoming.

Pollard attacked the "narrow philosophy" of the NIH and suggested it had become obsolete because of rigidities in philosophy, mechanisms of grant application and review, and research emphasis. Representative Kyros pointed out that Dr. Rhoads had argued earlier for preserving the NCI as the core of the new organization and suggested that it was a contradiction for Dr. Pollard to attack the old system as inadequate. Rhoads, in a rambling answer, suggested that the old system should be preserved to maintain momentum while new admininistrative machinery was being established. Kyros pounced on this: "I thought the statement made was that the status quo of cancer research had become obsolete, yet you say there is impetus."[9]

The subcommittee members were being tough on the ACS representatives. Schmitz, the Orange County Republican, and a former college professor of philosophy, noted that both Ingelfinger and Pollard had cited the same article in support of diametrically opposed positions. Upon reading the article, he said, he could find no basis for inference either way.[10]

Mr. Rogers asked about the testimony of a Harvard medical scientist who had pointed out on the previous day that the death rate from cancer had leveled off in the past twenty years. Pollard conceded that this statement had been based upon ACS figures. Rogers, a very able interrogator, had been "startled" to hear that. "No one had brought that to the attention of the Committee until yesterday," he said, "and I think it is well for us to verify that fact, and perhaps let the American people know that."[11] While the overall death rate from cancer had leveled off, Pollard responded, that was not true for all types of cancer. But Mr. Rogers had made his point. Indeed, the subcommittee had made quite clear to the Cancer Society spokesmen and to the supporters of S. 1828 that they faced a challenge on practically every major argument they might raise.

Memorial Sloan-Kettering Cancer Center and M. D. Anderson Hospital and Tumor Institute had been well represented on the Panel of Consultants, so neither institution presented independent testimony. But Roswell Park Memorial Institute had been without a permanent director from March 1970 until November, when Gerald P. Murphy, M.D. and D.Sc., was appointed to that position. Consequently, Dr. James Holland, a member of the Panel of Consultants and a Roswell Park scientist, did not speak as the authoritative representative of the Buffalo institution. Murphy, relatively unknown on the national scene, had not testified before the Senate. He made his first congressional appearance before the Rogers' subcommittee on September 21. Essentially he reiterated the arguments of the Panel of Consultants and came down strongly in favor of S. 1828. Repeating the general line, Murphy said that he had not reviewed Mr. Rogers' bill in detail and would be prepared to submit a supplementary statement later. An open and friendly dissension on several points of difference occurred between Murphy, Rogers, and Representative Hastings, in marked contrast to the sharpness of exchange with the ACS that morning. As the questioning concluded, Rogers said: "We look forward to seeing your operation because we have heard a great deal about it from Mr. Hastings and we are hopeful that the

Committee can work out a time in its schedule."[12] Preparations were underway for a visit by the subcommittee to Roswell Park Memorial Institute.

The next witness was Howard Metzenbaum, whom Rogers welcomed as one of the national co-chairmen of the Citizens Committee for the Conquest of Cancer. The Cleveland millionaire, who indicated that his was a personal statement, emphasized the need to expand clinical research. While NCI did some clinical research, it was not "a first priority" for them at that time. "The NIH . . . does an excellent job in investigative and test tube-oriented research. [It] is a science-oriented body but it doesn't zero in on basic problems and basic diseases about which we are concerned. NIH, in the main, does not consider it a priority matter to deal with the human element." A separate authority for cancer research, he thought, would have more clinical centers and "the research that will be done [there] will be a kind of human-oriented research that will come up with the answers."[13] ·Only Mr. Rogers was present as Metzenbaum stated one of the underlying concerns of the proponents of S. 1828.

The Association of American Cancer Institutes, with 20 institutional members, was only able to persuade the directors of three smaller cancer institutes to testify in behalf of S. 1828. Dr. Donald Pinkel, the medical director of St. Jude Children's Research Hospital, came on September 28 to say that enough scientists had spoken on behalf of various research disciplines, but he came to speak "on behalf of our children who have cancer." "You have to hold them in your arms when they are pleading and they have high fevers and they are dying; and I have had literally hundreds of young children die in my arms, and I have had hundreds of experiences in notifying young parents that they are going to lose their child or they just lost their child."[14] No one could question his sincerity as he evoked the horrors of childhood cancer. Rogers asked Pinkel if he had read H.R. 10681 and Pinkel replied that he had not.[15]

Dr. John S. Spratt, Jr., chief surgeon and director of the cancer research center of Ellis Fischel State Cancer Hospital in Columbia, Missouri, testified on the following day. Spratt argued, "cancer right now is probably the most curable of all chronic diseases, but yet, frankly, it takes the systems management approach with good equipment, good organization, good business know-how, to pull this together and make it available to the public."[16] He did not think this kind of approach was possible within the NIH. But a quorum call on the floor of the House took all subcommittee members away save Mr. Rogers, and there was no more than ten minutes for Spratt to present his testimony. The subcommittee did not return to pursue the secrets of systems management.

Dr. John W. Yarbro, from the Institute for Cancer Research of the American Oncologic Hospital, Fox Chase Center for Cancer and Medical Sciences, Philadelphia, was the third cancer institute director to testify in behalf of S. 1828. He urged that NCI be provided a large measure of administrative and fiscal autonomy within NIH, but sharp questioning by Representative Kyros revealed that he had only an impressionistic view of differences between the Senate bill and Mr. Rogers' bill. Yarbro did provide evidence that he felt demonstrated a lack of concern with cancer by the medical establishment. The University of Kentucky Medical Center, for example, had been without a radiologist for almost one year, he said, owing not to a lack of money but rather to a shortage of personnel. He pointed out that the Masonic Cancer Hospital at the University of Minnesota had been initiated by the Masons and that the people of that state deserved more credit for it than the existing medical establishment. M. D. Anderson, Roswell Park, and Memorial Sloan-Kettering, he observed, owed more "to enlightened state legislators and private sources than to organized medicine."[17]

He turned to an analysis of the clinical research community to further illustrate his point. From 1966 through 1971, Yarbro said, 151 papers had been presented to the American Society for Clinical Investigations, the highly prestigious clinical research organization. Forty-five percent of these dealt with heart disease or related cardiovascular diseases, though heart disease and nephritis accounted for slightly less than 40 percent of the causes of death in 1967. By contrast, only 5 percent of the papers dealt with cancer, though it was the cause of death for nearly 17 percent of the population. "I am not suggesting their priorities are wrong,"

Yarbro said, adroitly understating his point, only that "their priorities are different."[18]

Three proponents of S. 1828 appeared before the subcommittee on October 4. Dr. Michael J. Brennan, president of the Michigan Cancer Foundation, testified that NCI had not been "a satisfactory instrument for the development of large-scale developmental programs." He emphasized the importance of programmatic research, while the subcommittee members stressed the need for basic research. Dr. Thomas C. Hall, director of the Division of Oncology and professor of pharmacology and medicine at the University of Rochester Medical Center, argued that "a network of national categorical institutes" like that advocated in S. 1828 was long overdue.[19] Hall also argued that NIH should not be preserved as it was. "Parenthetically," he added, "the best way we know of preserving it is to pickle it, dead, in formalin and that is now what the NIH deserves." Representative Symington's questioning revealed that Hall had an imprecise grip on the details of the two pieces of legislation under consideration. Dr. Emil Frei, III, the president of the American Association for Cancer Research, and associate director of M. D. Anderson, repeated the arguments that had been advanced by the Panel of Consultants. His testimony was clearer and more direct than that of the two preceeding witnesses, but no new arguments were introduced.

The case of the proponents for S. 1828 came into clear focus as these witnesses appeared before the subcommittee. If there were to be a major national commitment to the conquest of cancer, with a national priority comparable to that given the space program, then an independent agency had to be established that had considerable independence from the NIH. Neither the medical establishment nor the NIH was much concerned with cancer. If the NIH director were left with substantial authority over cancer research, he would seek to direct funds for cancer to the other NIH institutes in accordance with his natural organizational bias and on the grounds that unrelated basic research could contribute much to cancer research. Only by removing cancer research from NIH control was there any hope for rapid progress against the dread disease. Only in this way could emphasis and resources be reallocated from fundamental research to much-needed clinical research. Responsibility for cancer research had to be placed in the hands of those dedicated to the conquest of cancer. The basic argument, in short, was that the key determinant of scientific and clinical progress against cancer was organizational. It was hardly a universal view.

THE VIEWS OF THE SCIENTIFIC COMMUNITY

The testimony for S. 1828 concentrated mainly on the merits of organizational autonomy for cancer. The ambiguous formula—"an independent agency within the National Institutes of Health"—led some to stress that NCI was being kept within NIH, but greater weight was attached to independence. Little attention was devoted to the funds needed for an expanded cancer program, and to the nature of a comprehensive national cancer plan. Organization was the subject, autonomy was the issue, and this aroused a great deal of apprehension within the scientific community.

Paul Rogers countered the witnesses for the Senate bill by interspersing among them representatives of the biomedical scientific community. As the scientists went on record in opposition to an autonomous cancer agency, they supplied Rogers with a refutation of the Senate bill and a justification for his own. The witnesses he called to support his position included representatives of medical-scientific organizations, cancer scientists, and basic scientists. The net effect was to suggest that the Panel of Consultants, supported only by the American Cancer Society, was isolated from the mainstream of biomedical thinking.

The panel had suggested that opposition to S. 1828 from the scientific community came from those not involved in cancer research. The first witness to follow the panel, on the afternoon of September 20, was the living refutation of that argument. Dr. Howard Hiatt, professor of medicine at Harvard and physician-in-chief, was also head of the cancer division of the Department of Medicine of Beth-Israel Hospital. He had been one of the organizers of the statement issued on May 2 by the chairmen of departments of medicine opposing S. 34. Referring to that resolution, he stressed that maintaining the integrity of NIH had been uppermost in the minds of the signers.

Hiatt expressed gratitude to the president and the Congress for proposing resources that would permit "a more rapid search for solutions" to cancer. But, he cautioned, rapid progress would come only by involving all sectors of NIH and the biomedical community. Such a broad approach was needed because in cancer "we are dealing with so many conditions" and there were "very large gaps in our basic knowledge of these conditions."[20] Stomach cancer, he noted, was decreasing as a cause of death, but not because medical science understood why. Furthermore, many treatments that now benefited cancer patients—advances in surgery, anesthesia, and blood transfusion—stemmed from work originally unrelated to cancer.

Mr. Rogers asked Hiatt about the arguments that cancer should be taken out of NIH, that NIH was inefficient, and that the director of NIH was trying to run the cancer program without knowing anything about it. Hiatt responded by noting that there were nine other divisions besides the cancer division within the Beth-Israel department of medicine. All played a role, he said, in the management of the cancer patient. While more immediate and even long-range relevance might come from the cancer division, it was impossible to say where the critical data would originate in the quest for basic information. What about dividing funds between basic and applied research? the chairman asked. It was difficult to draw the line between them, Hiatt replied. Many in clinical departments had not stressed applied research enough. In cancer, "with the obvious desire to find solutions," there was the opposite temptation to move too rapidly.

What did Hiatt think, Rogers asked, about the charge that some within the biomedical scientific community opposed an independent agency because it might affect their own research funding? Hiatt did not think this was the case and noted that, as head of the cancer division, though it was in his short-term interest to support S. 1828, he did not think it in the long-term interest of his division. Though he had heard his Harvard colleague, Sidney Farber, argue that morning that increased support for cancer would lead to increased support for other research, Hiatt was skeptical. "[W]hen resources are finite . . . it is well to have an overview of where the emphasis should be."[21] If a situation arose where cancer research support caused the neglect of other research areas, he surmised, he was certain that the panel members would not regard that as in the national interest.

Hiatt drew attention to the most dramatic recent discovery in basic cancer research, an enzyme called reverse transcriptase, discovered by Dr. Howard Temin at the University of Wisconsin and Dr. David Baltimore at the Massachusetts Institute of Technology.[22] This scientific advance was already having wide promise of determining whether viruses caused cancers in man. Interestingly enough, Hiatt noted, Baltimore's work had been supported mainly by the National Institute of Allergy and Infectious Diseases.

Dr. Baltimore was the next witness. He had been working in virus research for a decade when laboratory results suggested that some cancer-causing viruses might contain an unsuspected property of a new type of protein. Drs. Temin and Mitzutani at Wisconsin had reached the same conclusion at approximately the same time, and the two papers were published simultaneously in June 1970. The speed of the dissemination and application of results could be seen by the fact that only two months elapsed from initiation of the work to its publication, and that within six months of publication the results were being applied in clinical research on humans.

The MIT scientist's position was that aspects of S. 1828 might endanger the programs and institutions which had made medical research so strong in the past. There was a need for new funding for medical research, training new scientists, establishing new research centers, and continuing productive research efforts, he said, but it alarmed him that the programs which had provided his own training and support were in jeopardy or had been abolished. To facilitate the elimination of cancer as a threat to health, Baltimore said, only those changes in NIH that were "absolutely necessary" should be made. Cancer should not be separated from the rest of biomedical research, and a crash program atmosphere should not be created, because "it will not speed up and may slow down progress and because the American people should not be misled into thinking that a cure for cancer is imminent."[23]

Though Baltimore did not think he would personally have any difficulty getting research funds, he did not like the thought of having to go to NCI for money for virus research. Virus research had originally been stimulated by the interest in polio, but the National Institute of Allergy and Infectious Diseases had continued supporting it after polio had been eliminated as a major disease because of an interest in the basic biology of viruses, not because of its practical application. Baltimore feared that a strong, dominant cancer agency would become "the arbiter of what is done and what is not done," that it would assume functions then being served in a "more flexible and less well organized way," and that everything else at NIH would become secondary to cancer.[24] This basic scientist, who had unexpectedly contributed to cancer research from unrelated work, was an outstanding illustration to Mr. Rogers of why the existing NIH should be preserved.

Rogers continued to mix representatives of medical-scientific organizations, cancer scientists, and basic scientists throughout the hearings. The third week of testimony began on Monday, September 27, with the appearance of three representatives of the American Society for Pharmacology and Experimental Therapeutics, a FASEB constituent organization. The president of this society, Robert F. Furchgott, M.D., expressed the view that enactment of S. 1828 would disrupt the NIH and that this would have "a deleterious effect on progress in biomedical research . . . and the rate at which useful clinical applications . . . from this research are realized." This society, at a business meeting one month earlier, had debated whether its opposition to S. 1828 should be presented by the public affairs committee, the executive council, or the entire organization. One hundred and sixty-three members had been present, approximately 10 percent of the total membership, and there were only six dissenting votes on the issue of speaking for the entire society.[25] Dr. Robert E. Handschumacher and Dr. Frederick S. Philips accompanied Furchgott and also testified against S. 1828 and for H.R. 10681.

Philips was professor of pharmacology at the Cornell Graduate School of Medical Science and chief of the pharmacology division at the Sloan-Kettering Institute for Cancer Research. He apologized for not bringing a prepared statement, but only after much "soul searching" had he been able to write down his thoughts. Even though his research had been funded for many years by the NCI chemotherapy program, and though he was in complete agreement with the spirit of S. 1828, he was opposed to any major administrative changes within NIH. NIH had fostered "the forward motion of biomedical research on a broad front" and the fragmentation resulting from an independent agency would be "a turn in the wrong direction."[26] The impact of a senior Sloan-Kettering scientist testifying against the views of the Panel of Consultants was lost on no one.

Dr. Henry I. Kohn, Gaiser Professor of Radiation Biology at the Harvard Medical School, and director of the Shields Warren Radiation Laboratory at Boston's New England Deaconess Hospital, testified next that H.R. 10681, though not perfect, was certainly better than S. 1828. He was one of the few witnesses concerned with planning for cancer research, which he felt should be a primary function of professionals throughout the country, not those "centralized in Bethesda" or any other single location. The failure to find cancer cures, he stated, was due to scientists and clinical investigators, not to the investigative creativity of the NCI directorate. But success would depend upon "the men who carry out the investigations, not the executives."[27] Officials of NCI should facilitate planning, Kohn said, but the planning process should be largely an extramural effort, since a comprehensive plan would require extensive participation by the scientific community to be successful.

Specifically, Kohn suggested three steps that might be taken. First, the NCI director might appoint a scientific planning committee to develop 5- and 10-year plans for limited fields. Then, as much as half the budget might be allocated for unplanned research under the present study-section system. Finally, planned and unplanned programs should be reviewed every 5 and 10 years by an ad hoc commission of experts, not by the NACC or the NCI.

Kohn was followed by Dr. Maclyn McCarty, vice-president and physician-in-chief of Rockefeller University, who testified on behalf of the American Association of Immunologists, a FASEB constituent. He addressed himself to immunology, one of the fastest-growing fields of biomedical science in the previous quarter century. This growth had occurred because new basic research findings had demonstrated that the implications of the immune process reached well beyond infectious diseases to cancer and organ transplantation. Furthermore, he said, the rapid growth coincided with and was nurtured by the unprecedented support for biomedical research provided through the NIH. A separate cancer agency would impair the effectiveness of NIH, and would stimulate demands for further fragmentation. An intensified cancer effort should be wholly organized with the existing National Institutes of Health.

The next day, September 28, Dr. Howard A. Schneider, chairman of the public affairs committee of the Federation of American Societies for Experimental Biology, testified that FASEB had opposed S. 34 in the Senate hearings, and was not impressed by the compromise legislation. FASEB was apprehensive that the proposed administrative changes threatened "great harm" to the "governance of biomedical research" and the NIH, Schneider said, harm that could delay or defeat the conquest of cancer. H.R. 10681 offered better prospects for the redirection and reinvigoration of medical research than did the Senate bill.

The American Society for Microbiology arranged for the testimony of two basic scientists. Dr. Harold S. Ginsberg, professor of microbiology at the University of Pennsylvania School of Medicine, argued that the nature of cancer was not well understood. He was seeking to counter the claim by proponents of S. 1828 that scientific knowledge of the nature and etiology of cancer was available and all that was needed was "a large-scale managerial-directed attack . . . to apply the basic information." "Only scanty data are available," he said, "on the differences between a normal and a malignant cell, and most of the evidence is at a descriptive rather than molecular level." On the organizational question, from the perspective of 20 years as a government consultant, Ginsburg observed that interagency cooperation was minimal and "protection of one's power base is maximal."[28] S. 1828, he thought, would result in "a horrendous duplication" of facilities and efforts wasteful of both resources and expert manpower.

Salvador Luria, professor of biology and Institute Professor at MIT, and a Nobel laureate in medicine, appeared with Ginsberg. He had lent his name to criticism of S. 1828 in letters to the *New*

York Times and the *Washington Post*, but his concern that day was "strictly scientific." Cancer research, he said, was not ready for a frontal attack on cancer like that made on poliomyelitis in the 1950s. An "effective, expanded, rationally and soberly conceived" research program could be worked out, he thought, but crash-program visions of a cancer cure in three to five years were "a self-delusion" and would be "a dangerous misleading of the public."[29] An expanded program should emphasize basic research and, in addition to money, should provide for large numbers of highly trained biomedical scientists.

Representative Kyros asked Luria about the reputation of NIH. The MIT scientist responded that it was very high overall, though it varied by institute. "[S]ome institutes," he continued, revealing a lightly guarded secret among biological scientists, "specifically the National Cancer Institute, have probably not achieved in some areas a great reputation because cancer research is an extremely frustrating thing and it was not accepted as a popular area of research until about 8 or 10 years ago."[30] It had not attracted the brightest scientific people, Luria implied.

Mr. Rogers was careful to hear from cancer researchers as well as basic scientists. George Nichols, Jr., M.D., was the scientific director of the Cancer Research Institute at the New England Deaconess Hospital, and a professor of clinical medicine at Harvard. His knowledge of cancer came from his experience as a physician, from seeing how subtly cancer developed and spread, from witnessing the pain and suffering it caused, and the terror it brought to victims and their families. But he also understood cancer from a different perspective, that of a researcher:

> I have come to know how complex the matter of cancer is; that it is probably not one but many diseases; that despite intense effort by a large number of highly skilled people, we are still today as much in the dark regarding its ultimate cause or causes as we were 50 years ago when Peyton Rous first showed that a virus was the cause of a particular malignant tumor in chickens.[31]

Nichols favored more money for research, longer research grants, and more well-trained people. Since he was the only cancer research institute director to testify against S. 1828, his remarks had a special force.

The president of the National Society for Medical Research, Dr. Maurice D. Visscher, testified on September 29 in favor of more resources for cancer research but against the creation of a separate agency. NIH as then constituted, he argued, could relate basic research to disease entities more effectively than if every disease category were independently organized to support basic research related to its particular objective. Retain NCI within NIH, he recommended, fund cancer research more generously, but avoid the dangers of waste and duplication.

The subcommittee heard representatives of several major medical-scientific organizations in the first week of October. Dr. John R. Hogness, the president of the Institute of Medicine of the National Academy of Sciences, argued on October 4 against overpromise to the American people. The scientific knowledge simply did not exist, he said, that would permit anyone to say with certainty that an organized attack on cancer would result in cures for cancer in 5 or 10 years.[32] Furthermore, he noted, many of the important scientific discoveries responsible for increased understanding of the malignant cell had not originated in cancer research but elsewhere in biomedical research.

On October 7, Dr. John A. D. Cooper, president of the AAMC, accompanied by Joseph Murtaugh, testified that the AAMC's executive council had endorsed the provisions of H.R. 10681 on September 17.[33] Cooper's remarks were brief, though they reflected no lessening of opposition to the Senate-passed bill. But since the AAMC had worked with Congressman Rogers on the drafting of H.R. 10681, had suggested witnesses who might be called, and had held numerous conversations with him during the hearings, they now had little need to argue in public the case they were successfully making in private.

Also on the seventh, Robert L. Hill, professor and chairman of biochemistry at Duke University, testified on behalf of the committee on public policy of the American Society of Biological Chemists. He made a plea for increased understanding by the public on the emotional issue of cancer.

We also believe that the citizens of this country should be

properly informed about the realities of the cancer problem. Many people tend to believe that a cure for cancer can be found within a short period of time. It is clear that many years of basic research and clinical experience underlie that progress which has been made to date in the treatment of cancer. Our citizens should be clearly informed that much remains to be done, and although past progress warrants a renewed and vigorous attack on the cancer problem, further progress may well require many more years of hard work. It is erroneous to believe that we presently have sufficient knowledge of the processes of living things, and that cancer can be cured simply by applying this knowledge. This false belief must not be allowed to dominate our thinking.[34]

Consequently, Hill concluded, his society supported H.R. 10681.

Hill represented the fourth of the six scientific societies constituting the Federation of American Societies for Experimental Biology to testify in support of Mr. Rogers and against S. 1828.[35] But this was only part of a yet larger number. In addition to the four official representatives, nine other scientists who spoke out against the Senate bill were members of constituent organizations of FASEB. Cooperating closely, the AAMC and FASEB had supplied Mr. Rogers with the witnesses he needed and they, in turn, had provided him with the arguments he wanted.

INVADING ENEMY TERRITORY

Paul Rogers, during the course of the subcommittee hearings, had adroitly drawn supporting testimony for his bill from the constituencies of his opponents. On September 21, for instance, Dr. Campbell Moses, the medical director of the American Heart Association, testified. Whereas he had supported S. 34 in the Senate hearings, Moses now spoke in favor of H.R. 10681. He endorsed the principle implicit in Mr. Rogers' bill, namely, the use of all the resources of NIH in implementing the conquest-of-cancer program. He did reflect the view of the heart disease research community, however, when he said "we emphasize particularly . . . that if the state of research in cancer makes reasonable such a comprehensive effort for the control of cancer today, an exactly parallel effort is even more appropriate in the field of heart disease. In . . . heart disease, we know that the expansion of our research effort and the comprehensive full-scale application of the fruits of already available research and technology would save lives now."[36]

Boston was the geographic center of Edward Kennedy's political constituency. The intellectual community of Harvard and MIT, moreover, had been an important part of that constituency since John Kennedy had been senator from Massachusetts. But Rogers had heard from Dr. Ingelfinger of the *New England Journal of Medicine*, from Drs. Hiatt, Kohn, and Nichols of the Harvard Medical School, and from Drs. Baltimore and Luria of MIT. Dr. Francis Moore, the renowned Harvard surgeon, also testified. Seven major medical scientists from the Boston area had been heard. Rogers had successfully invaded the Boston-Cambridge intellectual community and had found the natives friendly to a South Floridan.

The subcommittee had also heard from Dr. Philips, a senior scientist at Sloan-Kettering. It listened to Dr. Ginsberg, from the University of Pennsylvania Medical School, the home institution of Dr. Jonathan Rhoads. Mr. Rogers had poked in everywhere.

At no time did he do so with greater effect than on October 7 when Dr. Henry S. Kaplan, the Stanford radiologist, testified. Kaplan was introduced as a former member of the radiation study section at NIH, and a former member of the National Advisory Cancer Council, as well as a current member of the board of chancellors and chairman of the commission on cancer of the American College of Radiology. But it was with the relish of a skilled legislative debater that Paul Rogers noted, "I think it would be well to say at this time that Dr. Kaplan was a member of the Panel [of Consultants] and we do give great weight to the action of the Panel."[37]

Kaplan began by saying that a clear and convincing case had been made for an expanded cancer research program and that the only real question was how to do the job. The Panel of Consultants, he reminded the subcommittee, had called for adequate financing, a comprehensive national plan, and had laid great emphasis on "creating an effective administrative framework." The panel, he said, had heard Carl Baker, NCI director, and his staff identify a number of administrative delays confronting the NCI. Some of these involved needless requirements for approval of high-level personnel appointments well above the level of NIH director, which made the recruitment of skilled administrative and scientific personnel quite difficult. There were apparently unnecessary requirements for contract approval at the level of deputy director for science of NIH, even after contract proposals had been cleared through NCI channels. The reported "manifestations of bureaucratic complexity and inefficiency," Kaplan said, strongly influenced the attitudes of the panel and had an important influence on the final recommendation that the expanded cancer program be housed within an independent agency.

But, the Stanford researcher noted, the panel had been under severe time pressure. It first met in June 1970 and submitted its final report less than six months later. "It is perhaps not surprising," he observed, "that today, with the clarity born of hindsight, one can appreciate that there were certain gaps in the thoroughness with which various options were explored." For one thing, Kaplan suggested, it would have helped to hear from Dr. Marston about prospects for remedial action and reform of the inefficiencies identified by the NCI staff. Consequently, he continued, he thought that the panel reached the decision to recommend a separate agency—"a conclusion in which I concurred very reluctantly at that time"—without fully exploring the possibilities of leaving cancer research within NIH while upgrading and streamlining its operation.

The original Kennedy-Javits legislation had called for a completely independent agency, while the compromise formula in S. 1828 was for "an independent agency within the National Institutes of Health." Kaplan had concluded that this was a logical contradiction:

> The agency, if it is truly independent, is not in any functional sense within the National Institutes of Health; conversely, if it operates within the guidelines of the National Institutes of Health, it is not independent. I am convinced that this provision will continue to be open to serious differences of interpretation and will thus constitute an open invitation to potentially serious administrative conflicts. Accordingly, I was obliged to dissent

publicly from the position taken by other members of the Panel
of Consultants, with respect to S. 1828, primarily because of
my concerns on this specific point.

Kaplan found H.R. 10681 to have many commendable features, the most important being the retention of NCI within the
NIH. He made a number of suggestions on how to improve it, was
questioned briefly by the subcommittee members, and hurried
from the hearing room to catch a plane. But his testimony was
damaging to the cause of S. 1828, because the chief dissenter
within the Panel of Consultants had devastated the panel's work.

A VISIT TO BUFFALO

The 11 days of subcommittee hearings stretched into a fifth week
on Monday, October 11, and concluded with a day-long visit to
Roswell Park Memorial Institute. "[T]his committee came to
New York," Rogers said that day, because "our member from
New York, Mr. James F. Hastings . . . felt that [we] would gain a
great deal of knowledge" from such a visit.[38] Hastings, whose
western New York district bordered on Buffalo, had had a good
deal of contact with Roswell Park officials during the hearings,
and the suggestion for a visit had arisen out of those conversations. Paul Rogers, moreover, was pleased to return the favor
done him by Hastings in being the only Republican co-sponsor of
H.R. 10681.

It was not clear to everyone why the trip had been scheduled,
though it was obvious that political reasons made it impossible for
Rogers to visit either Memorial Sloan-Kettering or M. D. Anderson at that time. Benno Schmidt, however, lost no time in urging
Dr. Murphy, the Roswell director, to attempt to sell Rogers on the
need for some independence from NIH. A national priority for
cancer, Schmidt thought, required someone in charge who believed in that priority and who had the organizational autonomy
necessary to do the job. Murphy also heard from the American
Cancer Society, which offered help to publicize the visit.

Murphy and his colleagues had assessed the needs of the moment somewhat differently. They saw the situation polarized between supporters of S. 1828 and supporters of H.R. 10681, and
felt this threatened prospects for any legislation. Rogers, after all,

had said there would be no legislation before there would be the Senate legislation, and the Roswell Park people had recognized him as a determined man.

The need, in the eyes of Murphy and his associates, was for pragmatic compromise. Representative Thaddeus J. Dulski (D., N.Y.), in whose district Roswell Park was located, sounded that theme as he welcomed the subcommittee to their "combination hearing and on-site inspection of a great cancer research facility." Addressing himself to Rogers, Hastings, and Kyros, the three who made the trip, the Buffalo Congressman noted that, while he was a co-sponsor of the Conquest of Cancer Act, he was "well aware that compromises will be necessary—that's really the name of the game in legislating."[39] His concern was that an all-out fight against cancer be moved along as fast as possible.

Representative Jack F. Kemp, a Conservative/Republican from an adjacent district, reinforced the views expressed by Dulski. Kemp, who had been elected to Congress in November 1970, was a former professional football player, and he recalled men and women from the annals of sports who had been "cut down in the prime of their careers"—Ernie Davis, former All-American from Syracuse, Babe Zaharias of Olympic track and field and women's golf fame, Freddy Steinhorn of the Texas Longhorns football team. Even this conservative legislator felt that the time had come "for a reappraisal of priorities," and that there was a real need for more Federal funding for the cancer crusade "so that Roswell Park Memorial Institute and others can carry on and expand their vital work."[40] Support for cancer research reached clearly across party differences.

Dr. Murphy and his staff decided to adopt a neutral posture toward the two differing pieces of legislation. "It is not my purpose to come to you this morning to provide testimony," he said, "having had that opportunity previously."[41] Deftly, then, he removed himself and Dr. James F. Holland from appearing before the subcommittee, and thus finessed a potentially awkward situation, since Holland, as a panel member, would have strongly supported S. 1828. Murphy's staff then got on with the task of indicating the nature of and the case for comprehensive cancer research centers.

Dr. Edwin A. Mirand, the institute's associate director for educational affairs, outlined the needs for the training of additional personnel for an expanded cancer program. He reviewed educational programs at Roswell Park, noted that they had been hurt by cutbacks in federal training funds, and made a plea for "steadiness and continuity of policy."[42] Highest priority in training should go to pre- and post-doctoral research training in oncology, senior clinical traineeships, and training for residents wishing to specialize in cancer. Short-term training in diagnosis and patient care should be made available through cancer centers to physicians, dentists, and other health professionals. Training for allied health professionals in cytotechnology, radiotherapy, and nurse anesthetists should be reactivated. Mirand hoped that the subcommittee would "provide the legislative framework" for continuing current training efforts and also "the broader spectrum" of training programs in cancer which were so desperately needed for an expanded national program.

The questioning shifted to the preparation of grant proposals for the Federal government and how this might be improved. Mirand suggested that "a large institutional grant in specific categories" might be given to a comprehensive cancer center. Rogers liked the idea and thought that comprehensive clinical centers might have a peer review of their programs which would be "kind of an accreditation."[43] On the basis of this review, then, a certain amount of funding might go directly to approved centers and their programs.

The Roswell Park staff made a major effort to demonstrate the nature of a comprehensive cancer research center. Dr. David Pressman, the associate director for scientific affairs, described how the "closely knit operation" dealt with various aspects of cancer. The immune response of an individual to his own cancer, for example, involved cooperation among the departments of dermatology, medicine, experimental therapeutics, pathology, viral oncology, experimental biology, and biochemistry research. "Much of the progress here in cancer," he said, "has been made to keep the research efforts directed toward application to the cancer problem."[44]

Dr. Edmund Klein, chief of the dermatology department, voiced the same view:

Our success so far and our present and future work depends on the integrated efforts of our Cancer Institute. The physical as well as the intellectual proximity of experts in various fields is of cardinal importance for translating ideas and leads into new treatment methods. This environment can only exist at an institution where the administration, the professional and educational staff, and the technical and ancillary personnel have one major goal—to give the most effective care to the cancer patient.[45]

Throughout the day, the subcommittee heard the consistent theme of the necessity for informed balance between laboratory research and clinical application depending upon the particular aspect of the cancer in question. Dr. Samuel Freedman, chairman of the medical advisory board of the Cancer Research Society of Canada, and a visiting scientist at the Buffalo facility, stressed that "basic and applied research are not mutually exclusive." Dr. L. Siminovitch, another Canadian, stressed the need for "intensification of efforts to encourage those trained in the basic science disciplines to turn their hands to problems in cancer research."[46]

Dr. Murphy, in his concluding remarks, reemphasized the main point he and his staff had sought to make; "whether one calls this an agency or an institute," he said, "it is still a comprehensive program." Mr. Rogers thanked the director and his associates for giving the subcommittee a conception of what a comprehensive center should be like. The session had been useful, Rogers said, "because we have gotten to see things in action . . . and that is what we want in the cancer field."[47] The Florida congressman concluded by saying that what had been observed that day would help the subcommittee write the type of legislation which would help bring about such activity across the nation. With that, he and his colleagues headed back to Washington to write a bill.

When the house hearings began, it was assumed that Paul Rogers would not be able to stand up against the momentum generated by the proponents of S. 1828. That assumption changed as the hearings developed and extended over several weeks, and as a distinct polarization of views developed with regard to S. 1828 and H.R. 10681. It was clear that Rogers would not permit the passage of S. 1828, and he had marshalled powerful support for his position during the hearings. Compromise was obviously needed, but what was the basis for compromise?

▪ 10 ▪

The Politics of Compromise

"Don't Be Taken In By Them Scientists — Leave
Everything To Old Doc Politics"

It had been widely assumed that the House hearings would end in
the first week of October and that the subcommittee would go directly into executive session to write a clean bill. While the October 11 visit to Buffalo had come as something of a surprise to
many, it did little to alter the basic schedule of activity. And since
Mr. Rogers had used the hearings to effectively buttress his case,
the partisans of S. 1828 shifted their efforts to influencing the successive stages of the legislative process. Cancer legislation would
have to be considered by the health subcommittee, the full com-

mittee, the entire House, and, if a bill different from S. 1828 resulted, there would have to be a joint House-Senate conference to resolve the differences. Proponents of the Senate-passed legislation thought that their chances of prevailing increased at each successive stage. "When the 9th inning comes, we'll win," Luke Quinn insisted.

The Citizens Committee for the Conquest of Cancer became active again. Solomon Garb personally traveled to Mr. Rogers' home district and sought to organize people to bring pressure on the Florida congressman.[1] He wrote various papers seeking their editorial support, contacted a number of doctors, and, sought to mobilize the substantial Jewish population in Rogers' constituency in support of the Senate bill.

Howard Metzenbaum, throughout August, tried to secure the endorsement of Ohio's House delegation for the Conquest of Cancer Act. A press release issued just before the House hearings began listed 19 co-sponsors out of the 24-member delegation. On October 8, owing to Metzenbaum's efforts, the Cuyahoga County unit of the American Cancer Society held a luncheon to honor seven northern-Ohio congressmen who were co-sponsors, and to hear Charles Ebersol, vice-chairman of the ACS board of directors, talk about recent legislative developments.[2]

The American Cancer Society initiated a letter-writing campaign to influence members of Congress.[3] Dr. H. Marvin Pollard wrote more than one hundred congressmen on October 4. Luke Quinn, from his hospital bed in the Clinical Center at NIH, sent a similar number of letters to individual representatives on October 5 and 6, many of whom were old personal friends. Dr. A. Hamblin Letton, president-elect of the ACS, followed up with letters to many in the House on October 12. The message was the same in every case, vote for S. 1828. Congressional responses to Pollard, Quinn, and Letton generally indicated strong support for cancer legislation, substantial concern for the National Institutes of Health, a clear recognition of differences between the House subcommittee's bill and the Senate bill, and a legislator's recognition that judicious withholding of judgment was necessary.

The Citizens Committee and the ACS, though supporting the same objective, took independent action in efforts to promote S.

1828. On Saturday and Sunday, October 9 and 10, full-page advertisements were run in the 21 daily newspapers published in the congressional districts of the ten members of the House health subcommittee.[4] These advertisements were sponsored by the Citizens Committee for the Conquest of Cancer. Readers were invited to "let your Congressman . . . know that you want [S. 1828] passed." The ailing Luke Quinn, though adamantly opposed to these advertisements, was unable to persuade Mary Lasker not to take this action. Quinn thought that such heavy-handed tactics would only backfire.

The Cancer Society, believing that they had been unable to get fair treatment in the press, took out full-page advertisements in the *New York Times*, *Washington Post*, and *Washington Evening Star* on October 12. "An open letter to the 435 members of the House of Representatives from the president of the American Cancer Society," was an effort to go over the heads of the subcommittee and bring pressure on them through the opinions of their fellow legislators.

"We are for the Conquest of Cancer Act—S. 1828—unequivocally," the letter began, because it provided for "an unprecedented mobilization of brains, science, technology and money to smash the barriers that have stalled us" in the fight against cancer. The bill preserved the "essential scientific relationships" between NCI and NIH, but by cutting red tape represented "an advance in the mechanics of administration." The president, the Senate, the people, and the cancer experts on the Panel of Consultants favored this legislation. Those who favored the status quo did "not fully understand the situation," did "not have expert cancer knowledge," and had "never made a parallel study" to that of the Panel of Consultants. Without mention of Mr. Rogers' bill, the ad concluded by saying, "The people are desperate for action against cancer. . . . They are counting on you, their representatives. And so are we. Vote for S. 1828."

The Rogers subcommittee met on Tuesday morning, October 12, for its first day of executive session. The members had not been together since Thursday, with the exception of the three who had traveled to Buffalo on Monday. When they met, they compared notes on the weekend attack by the Citizens Committee and that morning's message from the American Cancer Society. They deeply resented the heavy-handed effort to stimulate constituent pressure through the local district newspapers. They also resented the implication of Dr. Pollard's letter that, after eleven days of hearings, they did "not fully understand the situation." Nor did Mr. Rogers appreciate the fact that neither group would even concede that he had sponsored alternative legislation. They saw the advertisements as a one-two punch by Mary Lasker and her associates designed to strong-arm the subcommittee into support for the Senate bill. But the actual effect of the advertisements was quite the opposite.

A "CLEAN BILL"

During the hearings, H.R. 10681 had been ignored by its opponents but had drawn support from the scientific community. Even witnesses who found the Rogers' bill preferable to S. 1828, however, had made specific suggestions for its improvement. Though the subcommittee had no intention of reporting S. 1828 to the full committee, it was clear that H.R. 10681 would have to be improved and rewritten. The task before the subcommittee was to report out a "clean bill" that would contain the basis for acceptance by the House, compromise with the Senate, and agreement with the Nixon administration.[5]

The immediate priority before Paul Rogers as he began four days of executive session was to achieve unanimity within the subcommittee. Five others—Satterfield, Kyros, Preyer, Roy, and Hastings—had co-sponsored H.R. 10681, so a long step had already been taken toward a united subcommittee. Symington, the only Democrat who had not joined Rogers in mid-September, strongly favored the chairman's approach and became an early supporter of it in executive session. Rogers was then left with the job of persuading the other three subcommittee Republicans—Nelsen, Carter, and Schmitz—of the merits of the House bill.

The subcommittee worked through the text of the original H.R. 10681. They reduced its scope from "cancer and the major killer diseases" to an exclusive concern for cancer. But they retained those provisions of the bill that elevated the directors of the National Cancer Institute, the National Heart and Lung Institute, and the National Institute of Neurological Disease and Stroke to be associate directors of NIH. These two steps were logically inconsistent but legislatively logical. If Rogers could restrict the scope to cancer, the Senate might wish to reserve new bureaucratic status to the NCI director. The House subcommittee was signaling a potential bargaining point to the Senate conferees. Paul Rogers had always understood that the bill he finally reported out would be a cancer bill.

New language was added which stressed that cancer research was to be retained within the NIH.[6] To provide for "the most effective attack on cancer," the subcommittee wrote into the new bill, "it is important to use all of the biomedical resources of the National Institutes of Health, rather than the resources of a single Institute." The NCI director was authorized to implement this general principle by taking whatever action was needed to insure that "all channels for the dissemination and exchange of scientific knowledge and information" were maintained between NCI and all other medical and scientific organizations. The earlier bill had granted this authority to the NIH director, so this reflected a general upgrading of the NCI director.

The general principle of making full use of NIH resources was incorporated through the provisions for "proper scientific review" of all research grants. The NCI director was to use existing NIH peer review groups to "the maximum extent possible." Where appropriate, he might establish other peer review groups with the approval of the NIH director and the National Advisory Cancer Council. The maintenance of quality in research grant award decisions was to be preserved.

The major concession of H.R. 10681 was retained in the new bill. The NCI director was granted substantial budget autonomy from NIH and HEW. He was to submit his annual budget request directly to the president. While the NACC, the NIH director, and the secretary of HEW could "comment" on the proposed budget, they could not change it. In addition, Rogers added language to insure that NCI received from the president and the Office of Management and Budget "all funds appropriated by Congress for obligation and expenditure." The NCI director was also authorized to make research grants up to $35,000 after scientific review but without NACC approval. So, while insisting that NCI remain an organizational part of the National Institutes of Health, Rogers and his subcommittee also provided it with a large degree of autonomy within the NIH.

The bill that the subcommittee was drafting amplified the provisions of a national cancer program. The NCI director, for instance, was directed to support "appropriate manpower programs of training in fundamental sciences and clinical disciplines." In this way, the subcommittee sought to reverse by statute the Nixon-administration policy of reducing, if not eliminating, the research training programs of NIH. The subcommittee judged such programs essential to insuring an adequate supply of scientists, physicians, and allied health professionals for the basic research, clinical research, and treatment needs of an expanded cancer effort.

The original Kennedy-Javits bill, and the compromise version of S. 1828, had called for the cancer program to strengthen existing comprehensive cancer centers and to establish new ones. The Rogers' bill had used this language but had defined centers to include the laboratory, research, and patient-care facilities required to develop and demonstrate the best patient treatment methods, but not to include the financing of extensive patient-care facilities. The clean bill specified that 15 new centers be established for "clinical research, training, and demonstration of advanced diagnostic and treatment methods" related to cancer. The new legislation authorized the director of NCI to enter "cooperative agreements" with public or private nonprofit agencies to pay for planning, development, and basic operating expenses of new or existing centers. Federal funds could be used for construction of facilities, patient-care costs associated with research, training of personnel, and the demonstration of treatment methods. Authority was provided to give $5 million support to a center for each of three years, with successive three-year renewals possible after scientific review. Core support for cancer centers, clearly, had been one outcome of the visit to Roswell Park Memorial Institute.

Rogers, in executive session, also insisted upon language that provided statutory authority to NCI for "the prevention, control, and eradication of cancer." Cancer control was further authorized $20 million, $30 million, and $40 million for fiscal years 1972, 1973, and 1974, respectively. The subcommittee was attempting to reinstate a program that the Nixon administration had terminated a year earlier.

Representative Nelsen had a legislative reputation as a compromise-seeker. "When I saw the storm clouds brewing," he said, "I started to search for ways to make the House version acceptable to both the White House, which had already endorsed the Senate bill, and to Senator Kennedy, the chief sponsor in the Senate."[7] His search began on the first day of the hearings when he suggested to Elliot Richardson that close liaison between HEW and the subcommittee might be the best way to work out differences. He had also expressed to Senator Nelson the hope that differences could be resolved "by sitting down together and working it out."[8]

On October 7, Nelsen questioned Representative Brock Adams (D., Wash.), an advocate of the Senate-passed legislation, on prospects for a compromise.

> In the Rogers bill you have your National Cancer Advisory Council and they sort of sit in an overview position and I was wondering if there is some way that we can find a sort of middle road in this proposition.
>
> There are those of us who time after time have seen some new thing come along so we lay another layer on top of an existing layer; nevertheless, I think there is merit in having somebody with a little muscle in here to see to it that this activity is stimulated properly.
>
> Now, if there was a little review group of two or three, appointed by the President and with authority of umpiring this activity, maybe we could accommodate the wishes of the so-called Senate bill. At the same time many of us don't want to fragment past activity but still wish to earmark funds, and a small group could sit in and watch and see the job is properly done.[9]

Nelsen's embryonic idea had developed sufficiently by the following week for him to propose that the House legislation incorporate a provision establishing the President's Cancer Attack Panel. The panel would consist of three citizens, of whom at least two would be distinguished scientists or physicians. These three individuals would monitor the cancer program. Most importantly to a Republican seeking a way to gain White House acceptance of the bill, this proposed panel would report directly to the president, thus meeting his oftstated desire to have direct responsibility for the cancer program.

No sooner had Ancher Nelsen proposed the idea of the president's panel than Paul Rogers accepted it. The idea had some problems but also a certain logic. For Rogers, accepting the panel idea, gained him the support of the ranking minority member for the subcommittee's bill. Only two Republicans now were outside the fold.

Representative Tim Lee Carter had been the only supporter of the administration's position on the subcommittee. Throughout the hearings the Kentucky M.D. referred to the time when the Veterans Administration, in 1945, had brought in Dr. Paul Magnuson, an eminent Chicago physician, to run their medical program. Magnuson, Carter would point out, had infused new blood into the program. The physician-congressman was also very critical of PHS personnel who became administrators and who were not "deeply, scientifically knowledgeable." In fact, this was Carter's veiled way of expressing his intense dislike for Carl Baker, the NCI director. Though a decision on who would run the cancer program could not be made until legislation was passed, Baker's own future prospects waxed and waned throughout this time and Carter may have had reason to think a new program would require new leadership.

But Carter had also been softened up by the extensive hearings. One of the last witnesses to appear before the subcommittee had been Francis Moore, M.D., professor of surgery at Harvard Medical School and surgeon-in-chief at Boston's Peter Bent Brigham Hospital. Moore's testimony against S. 1828 was impressive to Carter because the surgeon referred to example after example from 19th- and 20th-century medical history to reinforce his arguments. Had there been a diabetes institute in the late 19th century, Moore argued, it would not have supported the work of

Langerhans on the pancreas, work which led to the discovery of insulin as a means of controlling diabetes. The relation between insulin and diabetes was simply not recognized at the time. Had there been a heart institute in 1915, he doubted whether it would have put its money behind two people attempting to extract a fluid from the liver, a fluid now known as "heparin," which prevents blood clotting and which is essential for open heart surgery. Would a government institute on poliomyelitis have supported the work of Dr. John Enders in the early 1950s, work that had at the time more apparent relationship to kidney disease? Moore asked. He thought not.

Moore listed five medical research advances of the past quarter century—tissue transplantation, open heart surgery, poliomyelitis, oral contraceptives, and tuberculosis. In each instance, he said, support had gone to people—"often young people, often unheard of people"—in universities.[10] A recent informal discussion among a group of clinicians and scientists, Moore said, had not produced one example of a major advance that had come from a targeted institute.

Carter asked Moore what might have happened if there had been an NIH in 1870 and it had supported Langerhans. The Harvard surgeon responded that Langerhans might have done his work a bit faster, but since he didn't understand the hormone-creating function of the pancreas he probably would not have come up with the Banting-Best discovery. His support could have come, Moore added, only as bright young people were supported, not through targeted, directed research efforts. Carter thanked Moore for his testimony. It "has so many things that I remember," the Kentucky physician said.[11]

When Ancher Nelsen joined Paul Rogers as a supporter of the clean bill, Tim Lee Carter found himself under substantial pressure within the subcommittee to throw in his support. Whatever the reasons, whether an expectation that Carl Baker would not be running the new program, or the persuasive effects of testimony, or the internal pressure from within the subcommittee, Carter decided to support the rewritten House legislation.

At the end of four days of executive session, Paul Rogers emerged buoyantly to announce at a press conference that the subcommittee was unanimously reporting out H.R. 11302, "The National Cancer Attack Act of 1971."[12] Even Representative Schmitz, Rogers noted with obvious pleasure, had joined in the subcommittee decision to report this clean bill to the full committee. Nine members of the subcommittee, the chairman pointed out, Schmitz being the single exception, were now co-sponsors of the new bill.

Rogers had parlayed his considerable legislative skill into a unanimous position within the subcommittee. He added that he expected little or no opposition to the new legislation from within the full committee. The *New York Times* commented that "It would be unusual for the full committee to raise serious objections to a bill that had the unanimous agreement of that subcommittee, particularly after the subcommittee devoted four weeks of hearings to the measure."[13] Paul Rogers had passed a critical milepost.

REAR GUARD ACTION

Rogers had been deeply angered by the full-page advertisements sponsored by the Citizens Committee for the Conquest of Cancer and the American Cancer Society. On Monday evening, October 11, he accused the supporters of an independent cancer agency of trying "to stampede the Congress."[14] He was still smarting from these pressure tactics when he emerged on Friday to report on the unanimous action of the subcommittee on the legislation. "It is most unfortunate," he said with emphasis, "that some people are attempting to bring the issue of finding a cure for cancer into a political setting."[15] Despite these efforts, Rogers predicted that his legislation would be supported by the full committee and the entire House.

Senator Gaylord Nelson appeared with Rogers at the Friday press conference and also blasted the Cancer Society and the Citizens Committee. One was moved to ask, he said, why it was necessary to pressure the House through "a massive advertising campaign," alleged to have cost $53,000, when everyone knew a major cancer bill would be passed and signed by the president. The answer, Nelson said, was simple enough:

This group does not want the bill to be modified or changed in any way to correct what the scientific community considers a major organizational defect in the Conquest of Cancer proposal as it passed the Senate. Because they cannot win the scientific argument, they have launched a propaganda campaign at the grassroots level aimed at stampeding Congress by making a political issue of an important scientific matter.[16]

The American Cancer Society, which had hoped for "some positive impact" on the Congress with the advertisements in the *Times*, *Post*, and *Star*, was thrown on the defensive by this counterattack. Though it issued a press release on the fifteenth that was critical of the bill that Rogers' subcommittee had just reported, the ACS representatives were preoccupied with responding to Rogers and Nelson. Rather lamely they denied any attempt to pressure the House in favor of an independent cancer agency and declared that they had nothing to do with the advertisements in the 21 local newspapers.[17]

Ironically, the Op-Ed page of the *New York Times* for October 15, 1971, carried a piece by Dr. Emil Frei, III, president of the American Association for Cancer Research, entitled "The Conquest of Cancer." Frei, who was also an associate director of M. D. Anderson, argued a straight S. 1828 line in the article, which the ACS had worked hard to get printed. Coming when it did, the article was simply lost in the shuffle.

The administration was also active. On October 14, the White House announced to the press that President Nixon would travel the following Monday to Fort Detrick, Maryland.[18] On Monday, the eighteenth, the president announced the conversion of Fort Detrick to a cancer research center. He hoped that as the nation moved from a wartime to a peacetime era it could convert many of its military facilities "to meet pressing domestic challenges." The Fort Detrick scientists, Nixon said, could now devote themselves "toward saving life, rather than destroying life." The president referred to the Senate bill, which provided "an independently budgeted program with the director . . . responsible directly to the President," as "an important step" in the campaign against cancer. He urged the House "to act promptly on this matter, so that we can get on with this important work," The pressure on Paul Rogers was being maintained.[19]

With Luke Quinn in the hospital, Mary Lasker began to rely increasingly upon Mike Gorman for advice on how to deal with Paul Rogers. Coordination among the proponents of the Senate bill began to break down as a result. Mrs. Lasker's support of the advertisements in the local newspapers had been urged by Gorman, though Luke Quinn had strongly opposed the action. Moreover, those advertisements were sponsored independently of Benno Schmidt.[20]

Gorman stubbed his toe badly, however, in the week after the subcommittee completed its executive session. On October 18, he wrote each member of the Committee on Interstate and Foreign Commerce.[21] The establishment of the president's Cancer Attack Panel in the new bill was "a considerable improvement" over the original Rogers' bill of September 15, though it had "no really direct input" to the formulation of NCI programs. "Over the weekend," Gorman wrote, "I talked to many members of the National Panel of Consultants on the Conquest of Cancer, to top officials of the American Cancer Society, and to the key officials at the White House who have been working on the bill for the President." All found the panel desirable, but thought it was "floating somewhere off in space" under the provisions of the clean bill. A simple solution, he suggested, would be to add language that required the NCI director to submit the institute's broad policy and program plans to the president's panel, and to delete language that had the NCI director reporting directly to the director of NIH.

Gorman's letter suggested that this "very modest" proposal could be handled satisfactorily in the full committee markup session. "We are willing to go along with the revised Rogers' version," he wrote, making the NCI director an associate director of NIH, if the president's panel were given the proposed powers. Then, anticipating the joint conference committee, he wrote: "We have reason to believe that the Senate will back down from its insistence upon almost total independence for the Cancer Institute Director and accept this modified language." Gorman's boldness was only reinforced in the postscript he added for Representative

Brock Adams. "Luke Quinn," he wrote, "is in complete agreement with these changes."

Three days later, on the twenty-first, Dr. Pollard and Alan Davis of the American Cancer Society met with Mrs. Lasker and Mike Gorman in Washington. Furious at Gorman's purported representation of them, the ACS representatives demanded to know to whom he had spoken on the Panel of Consultants and at the White House. It turned out that he had not talked to anyone!

Pollard and Davis visited Representative Ancher Nelsen that afternoon to unofficially tell him they thought his three-member panel was a good idea. At Nelsen's suggestion, they also went to see Paul Rogers. The Cancer Society's advertisement, Pollard said, was not related to the advertisements in the subcommittee member's local newspapers. This was the first time Rogers had realized this fact. Then, referring to the October 18 letter, the two ACS representatives said that Gorman and the ACS had never had any relationship, never would, and that Gorman had absolutely no authority to speak for them. Rogers, his eyes flashing, leaned forward and said, "I'm glad you told me that." He, too, was furious at Gorman's crass attempt to steal credit for the compromise.[22]

The full-page ads reactivated the editorial page battle between the proponents and opponents of the Senate legislation. The *New York Times* chided the proponents for their apparent panic at Paul Rogers' efforts to win support for "a sensible approach to cancer research." "Such panic," the *Times* editorial said, "is the only explanation for the high-powered advertising campaign launched by the American Cancer Society in support of the Senate measure."[23]

Schmidt felt compelled to respond to the *Times* editorial. In a letter that appeared on October 25, he complained that the *Times* persisted in dealing with a complex organizational issue as if it were "an issue of right and wrong." The *Times* view was in opposition to the "vast majority" who had spent their lives in the cancer field, the president, 79 of 80 U.S. senators, the HEW secretary, and "an objective panel of consultants." Contrary to the implication of an earlier *Times* editorial, Schmidt said, the Senate acted after "a great deal of deliberation and study." The paper had also made a "hero" of Senator Gaylord Nelson, the lone opposition vote. Finally, Schmidt noted, the *Times* had "an unhappy tendency" to regard disagreement with its views as politically motivated. "As one who has no interest in the political side of the matter," he concluded, after weighing the considerations and arguments carefully, he found the views of the president and 79 senators "overwhelmingly preferable [for] the American people" to the point of view of the *Times*.[24]

The following day the *Times* printed a letter signed by thirteen heads of tumor clinics, experimental laboratories, and departments of medicine and surgery at Harvard Medical School. It took issue with Pollard's assertion in the full-page ad that the objections to S. 1828 had come from those without expert knowledge of cancer. This was not true at Harvard, and the thirteen, "long interested in cancer research," were also concerned with the treatment of "some thousands of cancer patients a year." S. 1828, though reflecting "a notable objective," was "poorly written legislation." The writers urged that Congress consider the Rogers bill as a basis for "a heightened attack on cancer" and attempt "in a critical and constructive spirit to make that bill the best one possible."[25]

A similar letter appeared in the *Washington Post*, signed by seven prominent scientists. Dr. Howard Temin and Dr. David Baltimore, co-discoverers of reverse RNA transcriptase, were signers, as was Dr. Charles Huggins of the University of Chicago, the only person to ever receive the Nobel prize for treatment of cancer patients. "[E]ach of the undersigned," the letter said, "is a bench scientist working in some area of cancer research, or a clinician working with cancer patients. Each receives or administers funds from the National Cancer Institute or the American Cancer Society. Each opposes the organizational provisions of S. 1828."[26]

The real action, however, was about to take place within the House Committee on Interstate and Foreign Commerce, which was scheduled to meet on Wednesday, November 3. A letter signed by the ten subcommittee members went to all committee members on November 1.[27] Everyone, it read, shared the common goal of controlling and eliminating cancer, and the subcommittee believed this effort should have "a unique priority." The subcommittee thought that its legislation represented "the most effective approach" to accomplishing this goal, and pointed out that "a vital, perhaps the most vital, element is retaining the National Cancer Institute within the National Institutes of Health." Rogers and his colleagues urged the committee members to support their legislation—"possibly the most comprehensive attack ever mounted against a single disease in the history of this nation."

Paul Rogers and his subcommittee members went into the full committee in a very strong position.[28] In addition to their own unanimous position, they drew strength from the limited Republican support for S. 1828 within the committee. Months earlier, Representative William L. Springer (R., Ill.), the ranking minority member of the full committee, had responded to a letter from Dr. John W. Moore, a professor at the Duke University Medical Center: "I note you feel fairly strongly," he wrote, "that whatever effort is initiated should be kept within the present National Institutes of Health. I am definitely inclined to agree with this approach, and unless there is some overpowering evidence to the contrary I will continue that opinion."[29] Springer, on March 19, 1971, wrote Dr. Pollard of the ACS. Noting that "new governmental entities" were among the "several suggestions" being proposed for an increased effort to conquer cancer, the Illinois Congressman expressed the view that "it does seem logical to me that existing agencies of government should be used to carry on the work."[30]

In late April, the minority counsel to the full committee had circulated copies of the correspondence between Senator Kennedy and Elliot Richardson to the Republican members of the health subcommittee in order to give "some foresight into the responses we can expect when the Committee considers the problem."[31] The minority counsel followed this with a memorandum which indicated that the Kennedy bill (S. 34) would create a new governmental entity to handle cancer research. "It is the stated intention of the Administration," he wrote, "to put an extra $100 million into cancer research in fiscal 1972 through the machinery of the existing Cancer Institute." Beyond this, the memorandum pointed out, "HEW proposes to make internal organizational changes in the Cancer Institute to streamline and intensify its effort in both in-house and contract research."[32] None of these actions, it was noted, required new authorizing legislation or reorganization plans.

The Republicans, very simply, had concurred with the administration's initial position that the cancer program should be kept within the NIH. The changes in position by the administration, however, had put the Republicans on the spot. They did not like the prospect of differing with the White House on legislation wanted by a Republican president. But Springer and his colleagues had concluded that the early position of the administration was much better than the later compromise position.

The committee met on November 3 and 4 to consider the revised Rogers bill. Normally, a full committee would require only a single day to deal with legislation brought to it from a subcommittee. Mike Gorman, at the end of the first day, interpreted the need for two days to mean that the subcommittee's bill was in trouble. He reflected the hope of the Lasker forces that Rogers could be overridden in the full committee. This hope was misplaced.

Rogers encountered two formal challenges from those who favored the Senate bill. Representative Pickle, reflecting Benno Schmidt's suspicion of the proposed President's Cancer Attack Panel, proposed an amendment to H.R. 11302 that would have made the members of the panel ex officio members of the National Advisory Cancer Council, and also would have made the chairman of the panel the chairman of the council as well.[33] This amendment would have fused the responsibilities of panel and council, reduced the proposed monitoring role of the panel, involved the panel more fully in policy development and even program management, and made the chairman of both bodies an extremely powerful figure. Representative Nelsen objected. The panel had been proposed to provide a regular review and evaluation of the cancer program, he argued, but the effect of the amendment would be to enmesh it with policy responsibility in a way that would vitiate the panel's ability to objectively evaluate the program. The amendment was put to a vote. There were 7 votes in favor, 20 opposed, and one "present."[34]

The direct challenge to Paul Rogers' position was made by

Representative Brock Adams (D., Wash.). Adams, in testimony before the subcommittee on October 7, had expressed concern about "a legislative conflict" that might prevent immediate passage of effective cancer legislation.[35] He suggested that one compromise might be to adopt a bill similar to S. 1828 and then establish a one-year commission to study how the NIH could be "reorganized and reinvigorated." Experience on the Science and Astronautics Committee, which authorized funds for NASA, Adams said, had persuaded him of the great advantages of an independent agency. While he had indicated a willingness to accept compromise if that were needed "to satisfy the biomedical community," the Washington congressman left few doubts that the subcommittee would be challenged in committee.

Ironically, Adams was emissary for the Panel of Consultants, Mary Lasker, the ACS, *and* the Nixon administration. HEW Secretary Richardson had written each member of the committee, urging them to report out the Senate bill rather than the one developed by Mr. Rogers. "Unlike H.R. 10681," the secretary wrote, "the unique organizational structure proposed by the Senate-passed bill would reflect the President's personal commitment to an intensive campaign to find a cure for cancer and facilitate direct White House interchange with the Director of the new agency in the development of necessary programs."[36] That this case should be made in committee by a liberal Democrat, however, merely underlined the absence of support for the president's position among members of his own party.

Adams proposed to strike out all the language of H.R. 11302 and substitute for it the language of S. 1828. Had the committee accepted this amendment, its action would have denied the importance of subcommittee specialization, repudiated the precept that doing one's homework was essential to writing good legislation, and deferred to the wisdom of the Senate, which few committee members were prepared to acknowledge. There were only 4 who voted for the measure—Adams, Pickle, Eckhardt, and Tiernan, all Democrats—and 24 voted it down.[37] The proponents of the Senate legislation had been decisively beaten.

The administration had been ineffective in securing Republican support for S. 1828. Only 10 of the 18 Republican committee members showed up for the session. One supported and one voted "present" on the Pickle amendment. No Republican supported the Adams amendment.[38]

Benno Schmidt, Mary Lasker, and the ACS were practically as ineffective. Chairman Staggers, on whom they had once pinned so much hope, chaired the committee session. He did not vote, however, wishing to offend neither Rogers nor the Senate bill proponents. Since the two challenges to Rogers were so overwhelmingly turned back, the chairman was completely unable to exercise the power that a closely divided committee would have given him.

The committee came to the matter of formally adopting H.R. 11302, which it did by a 26-to-2 vote. Only Adams and Tiernan voted against the bill.[39] "The committee's action," the *New York Times* reported, "halts a trend in which it had seemed likely that the Senate bill would sweep through Congress almost unopposed."[40]

Paul Rogers had passed another important milepost. The next step was to get through the House of Representatives. On that score, as a reporter for the *National Observer* wrote: "It looks now as though Rogers will win and Nixon and Mrs. Lasker will lose. Representative Rogers has the momentum and by most estimates he'll sail through the House and win the Senate."[41]

SUSPENSION OF THE RULES

The Committee on Interstate and Foreign Commerce brought H.R. 11302 to the House on November 10. The accompanying report detailed the purpose of the legislation and its projected costs, reviewed the hearings, summarized the provisions of the bill, gave a brief history of the National Cancer Institute, and presented the committee's views on the bill.[42]

The proposed President's Cancer Attack Panel received specific attention. The three-man panel, the report said, would have the duty to "monitor" the new cancer program and "report directly to the President." Since the president had many responsibilities and could not follow the progress of the cancer program on a day-by-day basis, "direct oversight" of the program would be accomplished by the panel. The panel, the report continued, would "evaluate the program and make suggestions for improvements," and would also report to the president "any delays or blockages of rapid execution of the program."[43]

The additional views of five Republicans on the committee— Nelsen, Carter, Hastings, Brown, and Frey—explained that the panel was an effort to respond to "the President's desire to have a direct line of authority over the cancer program," while retaining NCI within NIH.[44] The three-man body would give the president "the muscle he needs" to insure an effective, accelerated cancer research program. House Republicans were thus assured that the bill, in principle, was acceptable to the president.

Representative Pickle's "additional views" in the report indicated his intention to offer an amendment on the floor of the House. To the language which specified that the NCI director would report to the director of NIH, Pickle proposed adding "except to the extent that the President may from time to time otherwise direct."[45] Though suggested as a compromise between the Senate and House bills, the proposed amendment would have provided discretionary authority to the president that he could not have failed to exercise. If adopted, the amendment would give the NCI director immediate access to the president, circumventing the NIH director. It was this bureaucratic advantage that the proponents of the Senate measure were seeking. The Republican's provision for giving the president access to the program through the three-member panel was not their objective.[46]

Representative Brock Adams took eight pages of the report in which to develop his "minority views."[47] He discussed both S. 1828 and H.R. 11302, explained why he had offered the Senate bill as a substitute in the committee, and made a plea for support when the same amendment would be brought to the whole House. Representative Tiernan added further support to the Adams proposal. The proponents of an independent cancer agency were girding themselves for a direct attack on the Rogers' bill when it came to the floor. Since more than 100 members of the House had co-sponsored the equivalent of the original Kennedy-Javits legislation in early January, and many had co-sponsored the equivalent of S. 1828 in July and August, neither Pickle's proposed amendment nor Adams' could be regarded as an idle threat.

The 58th annual meeting of the American Cancer Society was held at the Waldorf Astoria Hotel in New York City on November 6. It was too early for celebration, and not too late for tactical planning. Even so, there was an air of celebration. Mrs. Julie Nixon Eisenhower was the president's personal representative to the meeting. She brought a message from her father, calling cancer research "one of the greatest endeavors we are undertaking in this decade."[48]

A special citation was given to Ann Landers for "extraordinary service" to cancer through her syndicated news column. Representative Daniel J. Flood of the House Appropriations Committee also received a special citation for his "outstanding leadership and consistent legislative support" for major health measures and the "vital programs" of the National Cancer Institute. Luke Quinn, still in the hospital in failing health, received a citation for his substantial contributions to cancer through two decades of work for the ACS. The society presented its National Awards for 1971 to Dr. Wendell G. Scott, former ACS president, member of the Panel of Consultants, and a distinguished radiologist, and to Laurance S. Rockefeller. On hand to present the first National Founder's Awards was Elliot Richardson, and the two recipients were Mary Lasker and Elmer Bobst.[49]

The legislative committee of the ACS, which had met on November 3, sent a resolution on the cancer legislation to the board of directors. The key paragraph contrasted S. 1828 with H.R. 11302:

> The Senate version contains three provisions not included currently in the House bill, namely, a separate Cancer Agency within the National Institutes of Health; a newly created position as head of that agency responsible directly to the President, and sole cancer agency responsibility for program plans and budget.[50]

Drafted by Luke Quinn, this statement clearly spelled out the objectives that were being sought in the Pickle and Adams' amendments.

In early November, President Nixon passed the word along to Republicans in the House that he wanted a cancer bill on his desk for his signature before Christmas. Even though the legislation had moved along rather quickly, this schedule posed some problems. The committee did not report the bill to the full House until November 10, a Wednesday, and the Congress was due to adjourn for Thanksgiving recess on the nineteenth, the following Friday. It would not return until November 29.

Any bill coming from a committee to the full House must go through the House Rules Committee, which schedules the flow of legislation to the floor, and specifies the "rule"—closed, open, or limited—that governs the opportunities for amendment.[51] Since the first session of the 92nd Congress was nearing its close, a good deal of legislation had reached the Rules Committee, of which the cancer bill was only one measure. Under normal scheduling procedures there was little likelihood that debate on the cancer bill could occur until after the Thanksgiving recess. This made it unlikely that the joint House-Senate conference committee could meet, negotiate differences, report back to both houses of Congress, and have a bill to the president before adjournment for the Christmas recess.

Consequently, an agreement was reached among Mr. Staggers, Representative Colmer (D., Miss.), the chairman of the Rules Committee, and the Speaker of the House, Carl Albert, to bring H.R. 11302 to the floor of the House on Monday, November 15, under a procedure known as "Suspension of the Rules." "Suspension of the Rules" can occur only on Monday. The formal motion is "to suspend the rule and pass the bill." Under this procedure, a single vote by the House circumvents the Rules Committee, brings a bill to the floor, and passes or defeats it. Since it is an extraordinary procedure, though, a two-thirds majority of those present and voting is required for the motion to carry, and no amendments to the legislation may be considered. Any measure brought to the floor by this means must be voted up or down.

This was the ideal parliamentary situation for Paul Rogers. The House could be responsive to the president's desire for a bill before Christmas. But those who were preparing amendments to H.R. 11302 were now placed in an impossible situation. If they wished a bill before Christmas, they would have to vote for Mr. Rogers' bill. If they wanted a chance to amend H.R. 11302, they would have to defeat the motion and wait perhaps several months to bring the bill to the floor under normal procedures. Representatives Adams and Tiernan, in a letter of November 12 to all House members, urged a "No" vote on suspension of the rules on H.R. 11302 the coming Monday. "Otherwise," they wrote, "the bill will be considered with no chance to offer amendments." But Rogers' position, derived from the advantageous combination of procedure and timing, made the position of Adams and Tiernan tenuous indeed.

The House met at noon on Monday. Yet another sticky problem had been dealt with by Rogers just as the House went into session. Section 410 of the bill contained language that would have authorized the NCI director to appoint an additional 50 scientific or professional experts without respect to normal civil service grade and salary limitations. These "supergrade" positions throughout the federal government were under the jurisdiction of another committee, the House Post Office and Civil Service Committee. Representative Thaddeus Dulski, chairman of this committee, told Rogers in the week before the vote that he and his committee would object to the suspension of rules as long as the supergrade positions remained in the bill. This threat was serious, since the House leadership would probably oppose suspension if a question of committee jurisdiction was at issue.

Rogers, however, told none of his subcommittee colleagues about the threat. Consequently, no approaches were made to Dulski to reach an accommodation. Rogers, in this way, placed Dulski in a potentially very embarrassing situation. Having made the threat to Rogers, he could not easily withdraw it and consent to this unsanctioned invasion of his committee's jurisdiction. On the other hand, since Roswell Park Memorial Institute was in his district, Dulski could not go on record as opposed to cancer legislation just to protect some congressional turf.

Rogers made no move toward accommodation until the fifteenth. Dulski, as a result, left the country and the responsibility for protecting his committee's jurisdiction with other colleagues. The health subcommittee members were informed of the situation a few hours before the scheduled vote. Much scurrying about led to agreement to delete the offensive provisions, announced by Mr. Staggers in his opening remarks.[52]

Representative Hale Boggs (D., La.) was Speaker pro tempore in the absence of Carl Albert. Soon after the opening prayer and preliminary business, he recognized Mr. Staggers. The commerce committee chairman rose and said: "Mr. Speaker, I move to suspend the rules and pass the bill (H.R. 11302) . . . to strengthen the National Cancer Institute and the National Institutes of Health in order to conquer cancer as soon as possible." Boggs allotted twenty minutes of debate to "each side," meaning the Democrats and the Republicans, not the pros and the antis.[53]

Staggers noted that the bill had emerged from 11 days of hearings, four days of consideration by the subcommittee, and two days by the full committee. The two negative votes in committee reflected disagreement over how to achieve the conquest of cancer, not disagreement with the objective itself. Pointedly addressing the Republicans, and any others who would invoke the president in support of their position, he said that the administration had taken four different positions on cancer legislation during the debate: "At first they were opposed to any legislation; then they proposed their own bill; then they appeared before our Committee and supported the Senate bill; and the last of their positions is support of the Senate bill with a substantial number of amendments adopted by the Subcommittee on Public Health and Environment."[54] The implication was that the president obviously wanted a cancer bill but not a particular bill.

In his remarks, Mr. Rogers drew the attention of his House colleagues to the fact that the approach of H.R. 11302 had been "endorsed by an overwhelmingly united scientific and medical community." Practically every significant scientific body in the U.S. supported the concept of maintaining the federal cancer research effort within the NIH. Several Nobel laureates in medicine, he said, including one whose award was based upon treatment of cancer patients, supported this concept. So did the heads of 72 departments of medicine in medical schools throughout the country, as did 13 faculty of Harvard Medical School long associated with cancer research or patient treatment. The legislation, Rogers concluded, was "a balanced innovative approach . . . conceived after rational study and close scrutiny by the hard working members serving on the Subcommittee."[55] The implied contrast with the Senate was clear.

Representative William L. Springer controlled the Republican time in the "debate." It was his "sincere judgment," he said, that H.R. 11302 was "the proper approach" to the cancer problem and would "hasten the day when this dread disease can be conquered." Tiernan challenged him:

> *Mr. Tiernan.* Mr. Speaker, will the gentleman yield?
>
> *Mr. Springer.* I yield only for a question.
>
> *Mr. Tiernan.* Is it not true the President approved the Senate bill as passed?
>
> *Mr. Springer.* He approves this bill.
>
> *Mr. Tiernan.* But the President also approves the Senate bill.
>
> *Mr. Springer.* I said the President approves this bill.[56]

Springer had returned to Washington from Chicago in early November with President Nixon on Air Force 1, and the two had discussed the legislation on that occasion.[57] His authority was unassailable.

Representative Ancher Nelsen expressed his pride in H.R. 11302, and considered it "one of the most important single pieces of legislation" dealt with by the Congress this session. Dr. Tim Lee Carter described the bill as a compromise arrived at "after much discussion and debate," which "by and large, [was] an effective piece of legislation."[58] Dr. Roy, Carter's fellow M.D. on the subcommittee, associated himself with the Kentuckian's remarks.

A great deal of strength was being mustered in favor of H.R. 11302. Nevertheless, Representative Adams took the floor to ask his colleagues to vote down the suspension of the rules motion and call up S. 1828 later. The medical community, he said, contrary to what Mr. Rogers had suggested, "is split as how to best approach this problem."[59] Those engaged in treatment of cancer and in clinical research favored S. 1828.

Representative Tiernan complained that Mr. Springer would not acknowledge the fact that the president actually wanted S. 1828. Representative Hastings replied: "The President, unfortunately, in consideration of the Senate bill, did not have the advantage of knowing that this House would come up with this superior piece of legislation."[60] It was an interesting legislative situation.

Liberal Democrats were making an appeal for support for the Senate bill on the grounds that President Nixon wanted legislation that was still identified with Ted Kennedy. Republicans, with a bill that was now acceptable to the White House, were not inclined to vote against the Republican leadership on the committee and subcommittee. Nor were they inclined to vote against Paul Rogers, the respected, conservative Florida Democrat who had crafted H.R. 11302.

When the allotted time was up, the Speaker pro tempore directed the Clerk of the House to call the roll. The House, by a vote of 350 in favor and 5 opposed, approved H.R. 11302. Paul Rogers relished the victory.[61]

CONFERENCE

The joint House-Senate conference committee met on Wednesday, December 1, to resolve the differences between H.R. 11302 and S. 1828.[62] The speed at which the conferees were moving meant that agreement was quite certain.

The House conferees were led by Harley Staggers, and included Mr. Springer and the 10 members of the health subcommittee. The Senate conferees, led by Senator Kennedy, included all 14 members of his subcommittee. Attention, though, was focused on Paul Rogers and Senator Kennedy. Kennedy had been impressive in pushing a cancer bill through the Senate that had President Nixon's personal approval and that drew only a single dissenting vote. Rogers, however, had been equally impressive in maneuvering his bill through the House legislative maze.

If anything, Rogers was in a stronger position in the conference than Kennedy. He had checked with all of the House conferees the previous morning. No one had any problems with the House position as they went to conference and they agreed to stand together in the negotiations. Kennedy, by contrast, had experienced some erosion of support for S. 1828 among the Senate conferees. Gaylord Nelson communicated rather openly with the House conferees during the conference, advising them on tactics in light of probable Senate moves. Senators Cranston and Schweiker had both informed Rogers beforehand that they favored the House bill. Finally, the administration took a hands off posture toward the work of the conference. Though still formally committed to S. 1828, the White House had already communicated through Springer to the House Republican leadership that H.R. 11302 was acceptable.[63]

The principal element of controversy between the Senate and the House had been the organization of the cancer effort. Where the Senate bill would create a new cancer agency "as an independent agency within the National Institutes of Health," the House bill provided that the NCI director report to the NIH director except on the matter of budget submissions. Rogers and the House conferees expected this issue to be the focus of argument and discussion, though they were not prepared to negotiate on it. To their complete surprise, the Senate barely gave a passing nod to the matter.[64] Kennedy did suggest changing the name of NCI to the National Cancer Authority, and the House members indicated a willingness to accept this if all other NIH institutes were changed accordingly. The matter was dropped.

The composition of the House-proposed President's Cancer Attack Panel dominated two days of conference deliberations. The Senate proposed to add an additional three members to the panel—the NCI director, the chairman of the NACC, and the NIH director. Their intent, which reflected Benno Schmidt's thinking, was to give the NCI director immediate access to the president and to establish organizational power for cancer outside the NIH. It was the same purpose that had been central to Congressman Pickle's efforts to amend H.R. 11302 in the full Commerce Committee. Rebuffed on one formulation, the Senate conferees responded with a new variation, all of which added the NCI director to the panel. The House was equally adamant in opposing any arrangement that included the NCI director. The panel was to be an oversight body, a watchdog, Rogers and Nelsen argued. If problems arose, slowing the cancer program, the panel was responsible for bringing them to the attention of the president for his resolution, but its role was not that of direct program management. The most telling House argument was that putting the NCI director on the panel was like having the fox guard the henhouse: an impartial review of the program would be impossible if those responsible for running it were also asked to pass judgment upon it.

The conference was deadlocked.[65] Senator Kennedy, at the end of the second day, proposed a weekend visit to President Nixon to resolve the difference of view. The House conferees rejected this idea, but the Senate conferees were ambivalent. If Kennedy went to the president, the House members feared, and if Benno Schmidt had the president's ear, as they believed he did, a nod by Nixon toward the Senate view would lock up the Senate position. It would also mean that the administration would bring pressure on the House Republicans to accept the Senate proposal. Presidential intervention in the conference could substantially alter the bargaining power of the two sides.

The conference adjourned on Thursday and agreed to meet again the following Tuesday. Whether Kennedy would visit the president was left in the air. He did not, but he and eight colleagues wrote the president on Friday, December 3, inviting his "specific guidance" on "the principal function and requisite membership" of the cancer attack panel. The House version, the letter noted, saw the panel performing oversight functions and consequently its membership was to consist of three nongovernmental experts responsible for advising the president on the program. The Senate view was that the panel ought to be enlarged to include three "key program experts" so they would have "a suitable channel for direct access to you."[66]

Benno Schmidt, who had actively promoted the Senate view of the panel, had been acting largely on his own in this matter. Over the weekend, though, Dr. Lee Clark and Dr. Gerald Murphy, reflecting long-standing suspicion of NCI on the part of the cancer research institutes, told Schmidt that they did not favor the addition of the NCI director to the panel. If the same official responsible for program management were in a position to review his own program, the cancer research institutes would be left without an appellate body. Clark and Murphy also passed their views along to the White House staff who were preparing the response to the Senate letter. Schmidt, who quickly saw the logic of this as it applied to Memorial Sloan-Kettering, accepted their argument.

The president's letter of Monday, December 6, said, regarding the cancer attack panel, "I could work effectively with either version." In vintage Nixon prose, he wrote:

> The Senate version would have the advantage of bringing those who are directly responsible for the Government's cancer effort into all discussions with the President concerning its progress. On the other hand, I can understand the feeling of the House conferees that the activities of the National Cancer Institute can be more effectively monitored if the heads of that program are not monitoring their own performance. In any event, the normal powers of the executive could still be used to include the agency heads when desirable if the House view is adopted, or to exclude them when appropriate if the Senate view prevails.[67]

There was no reason the president found to take a position on this matter when he could live with either outcome.

When the conference committee met on Tuesday, the Senate immediately accepted the House version of the President's Cancer Attack Panel.[68] Confronted with unyielding resolve by the House conferees, no support from the White House, and a negative opinion from key cancer institute directors, there was no other course of action open. The way was clear to resolve other less serious differences.

The House bill would have retained the National Advisory Cancer Council as established in the 1937 legislation. The Senate bill, on the other hand, would establish a National Cancer Advisory Board with 18 members, no more than 12 of whom would be physicians and scientists and no more than 8 would be laymen. In his letter, President Nixon expressed a preference for the Senate version, which provided for "expanded outside assistance." The House accepted the Senate's proposal with the addition of five ex officio members.[69]

Where the Senate bill had focused exclusively on cancer, the House legislation also gave attention to heart and lung diseases and stroke. The Senate view prevailed and all references to diseases other than cancer were deleted. The Senate also recommended deletion of all references to the directors of the National Heart and Lung Institute and the National Institute of Neurological Diseases and Stroke, and this was accepted by the House. This left only the House proposal for elevating the NCI director to be an associate director of NIH. Since this provision had not found support among the scientific community, the House yielded to the Senate and eliminated this provision.[70]

The final difference was on the manner of appointment of the NCI director. The Senate bill provided for appointment of the NCI director by the president with the advice and consent of the Senate. The House bill would have both the NCI director and the director of NIH appointed by the president. Representative Staggers was blunt on this point: "The Senate has never had advise and consent power over health matters and it will not have it now." The House view prevailed.[71] The remaining language of the bill was that of the House. At the end of the third day, the conference emerged to report agreement on "The National Cancer Act of 1971."

When the House received the conference report on Thursday afternoon, December 9, Mr. Staggers averred that the compromise bill was superior to the bills previously passed by either body. Mr. Rogers, differing slightly from this view, saw the conference result as "basically the bill which the House approved." "Mr. Speaker," he said, "this report represents a substantial victory for the House of Representatives, the biomedical community, and the American people." By insuring that the well-integrated NIH research program would be involved in the cancer fight, Rogers said, the American public were guaranteed that the attack on cancer would involve a marshalling of all national resources. "I can think of no better Christmas present for the American people," he concluded, "than to have this bill passed by this body and signed by the President without delay." The House of Representatives agreed to the conference report moments later.[72]

The Senate had met that morning and Kennedy submitted the conference report. The two pieces of legislation, he said, had been similar with "one key exception": "Many of us had felt that in order to develop a sufficient program to attack the problems of cancer, we ought to create a type of program modeled on the Space Agency. There were those in the scientific and biomedical research field who felt it should remain within the National

Cancer Institute, and the House provided accordingly."[73] In most other respects, the senator said, the two bills were consistent.

The Senate met on the morning of December 10 and accepted the conference report. The roll call vote was 85 in favor and none opposed.[74] The measure was sent to the president for his signature.

A CHRISTMAS GIFT TO THE AMERICAN PEOPLE

The National Cancer Act of 1971 was signed on December 23, 1971, in the State Dining Room of the White House. One invited guest later described the scene this way:

> Some hundred people . . . were escorted into a large room half filled with television cameras. . . . The doors in the front of the room were theatrically flung open and the President of the United States was announced. The audience rose, the television cameras whirred, still photographers ran around to vantage points, and the President, obviously fresh from a television makeup technician's chair, turned on a contrived smile. He came to sign the Cancer Act of 1971. Cancer research had entered the political arena. The Congressmen and Senators who guided the law into being smiled broadly as the cameras focused on them. Most of the scientists in the audience did not smile; many were worried. The hoopla surrounding the Cancer Act implied the conquest on cancer in the near future because a couple of hundred million dollars a year more were to be channeled into cancer research. Those of us there who knew the "state of the art" had cause to worry.[75]

Whatever the anxieties of the scientific community, however, this day belonged to the politicians and especially to President Nixon.[76] "We are here today," he said, "for the purpose of signing the National Cancer Act of 1971. I hope that in the years ahead that we look back on this day and this action is shown as being the most significant action taken during this administration. It could be." The president remarked on the season and the occasion:

> Hope and comfort, the relief of suffering and the affirmation of

life itself—these are the qualities which have traditionally been associated with the Christmas season. There could be no more appropriate time than this to sign into law the National Cancer Act of 1971. For this legislation—perhaps more than any legislation I have signed as President of the United States—can mean new hope and comfort in the years ahead for millions of people in this country and around the world.

While biomedical research was "a notoriously uncertain enterprise," the president said, there would be no uncertainty about the government's role in the cancer effort. "I am determined," Mr. Nixon continued, "that the Federal will and Federal resources will be committed as effectively as possible to the campaign against cancer and that nothing will be allowed to compromise that commitment."

The president had great confidence in that pledge, he said, because immediately prior to the ceremony he had asked Benno Schmidt to serve on the President's Cancer Panel for the three-year term and to be chairman in its first year. Schmidt, the president announced, had accepted this invitation. Mr. Nixon described Schmidt as "an effective leader of men and a dedicated community servant," and expressed the view that the nation was fortunate to have such an individual "heading this important panel in its critical first year." The president then gave Benno Schmidt the pen with which he signed his first name. It was a great moment of personal recognition for the chairman of the Panel of Consultants.

Richard Nixon wrapped himself in the mantle of leadership in the nation's new commitment to an attack on cancer. He reminded his audience that he had asked for such a program in his January State of the Union Message and his health message of February, and that he had submitted very specific proposals for a cancer-cure program in May. Beyond this, the Congress had appropriated the additional $100 million for cancer research he had requested, which brought the fiscal 1972 appropriations for NCI to $337.5 million. In addition, the president noted, he had announced the conversion that October of the biological warfare activities of Fort Detrick, Maryland, into a leading center for cancer research.

The president noted that the new act allowed him to appoint the director of NCI, and provided that the NCI budget be submitted directly to him. It also provided the three-member President's Cancer Panel and a new National Cancer Advisory Board, which bodies, made up of presidential appointees, reported directly to him. "The important result of all these provisions," Mr. Nixon said, "is to place the full weight of the Presidency behind the National Cancer Program" and to enable him "to take personal command of the Federal effort to conquer cancer so its activities need not be stymied by the familiar dangers of bureaucracy and red tape." "Having asked for this authority—and this responsibility," he said, impressing the presidential seal even more firmly on the cancer program, "I now pledge to exercise it to the fullest."

President Nixon emphasized that a "total national commitment" meant more than government. "It is essential," he said, "that an organization such as the American Cancer Society—which has raised so much money for this cause and which has done so much to promote research and education in this field—continue to play its full effective role." When he signed his last name to the bill, he gave the pen to Dr. A. Hamblin Letton, the new president of the American Cancer Society. Letton thanked the president for "a wonderful Christmas present" on behalf of the two and a quarter million volunteers of the ACS.

Senators Kennedy, Dominick, Williams, Schweiker, and Beall were present, as were Representatives Staggers, Springer, Rogers, Nelsen, Satterfield, Kyros, and Hastings. "I deeply appreciate the months of hard and careful effort," the president said to these legislators, "which so many of the Members of Congress gave to this cause." He was pleased, he said, that the Congress was "totally committed" to providing what funds were necessary for the conquest of cancer. The president posed for pictures with Senator Kennedy and Representative Rogers, which were transmitted by the wire services to newspapers throughout the country.[77] It was one of the most prominent displays of bipartisan cooperation seen in Washington during the first Nixon administration.

Among the one hundred thirty-seven guests, representing the Panel of Consultants were Lee Clark, Joseph Burchenal, Mathilde

Krim, Jonathan Rhoads, Emerson Foote, Anna Rosenberg
Hoffman, and William McC. Blair. Mary Lasker, of course, was
present.[78] The NIH was represented by Robert Marston, John
Sherman, and Kenneth Endicott, then director of the Bureau of
Health Manpower Education. The National Cancer Institute was
represented by Dr. Carl Baker, its director, and ten staff members.

President Nixon sought to capture the significance of the occasion in this way:

> As this year comes to an end, cancer remains one of man-
> kind's deadliest and most elusive enemies. Each year it takes
> more lives in this country alone than we lost in battle in all of
> World War II. Its long shadow of fear darkens every corner of
> the earth. But just as cancer represents a grim threat to men and
> women and children in all parts of the world, so the launching
> of our great crusade against cancer should be a cause for new
> hope among people everywhere.
>
> With the enactment of the National Cancer Act, the major
> components for our campaign against cancer are in place and
> ready to move forward. I am particularly happy that the year
> 1971—at the beginning of which I issued my call for a new
> campaign against cancer—can end with the signing of this
> landmark legislation.[79]

The National Cancer Act of 1971

The policy problems that confront us now, with the nation's declared commitment to conquer cancer, are somewhat like those involved in the artificial-heart question. There will be, for a while anyway, a running argument between two opposing forces. There will be on the one side those who believe that cancer is a still unsolved but eminently approachable scientific puzzle, requiring only enough good research by imaginative investigators on a broad enough biological base, and that, provided with enough financial support and enough time, we will, in one way or another, find ourselves home and dry. On the other side there will be those who believe themselves to be more practical men of the real world, who feel that we have already come about as great a distance toward understanding cancer as we are likely to come for some time and that we should give the highest priority to applying on a much larger scale what we know today about this disease—that with surgery, chemotherapy, and radiation we can now cure or palliate a considerable number of patients and that what we need at this time is more and better technology of essentially today's model. I do not know how this argument will come out, but I believe it to be an issue of crucial symbolic significance; whichever way it goes may possibly indicate the drift of biomedical science for the next decade.

Lewis Thomas, M.D.
"Reflections on the Science and
Technology of Medicine," 1972

In this concluding chapter, we summarize the story of the National Cancer Act of 1971 in a more analytical manner and point out some of the current and future issues raised by the Act for cancer research and the entire biomedical research enterprise. First we analyze the work of Mary Lasker and the Panel of Consultants in setting the agenda for the 1971 cancer legislation and consider the extent to which pre-legislative activity determined the legislative outcome. Then we analyze the debate over the cancer legislation from the viewpoint of the parties to the debate and draw attention to the oft-overlooked, but frequently important, legislative "end game" for the outcome of legislation. Third, we indicate some of the developments and problems raised in the implementation of the National Cancer Program and discuss several of the more important impacts of the cancer initiative on the NIH. Finally, we conclude with reference to the nature of the policy debate on biomedical research in mid-1976 and to what this suggests for "the drift of biomedical science" in the years ahead.

SETTING THE AGENDA OF LEGISLATIVE ACTION

The enactment of the National Cancer Act of 1971 represented a reformulation of federal government policy toward cancer. Cancer had long been a matter of public concern and had been on the formal governmental agenda continuously since 1937. Its persistence as an agenda item is attributable to several facts: a definitive answer or "cure" to cancer has not been found, the problem of cancer remains of very intense interest to many people, and the response to the problem has been institutionalized in the National Cancer Institute and the National Institutes of Health.

The source of the initiative leading to the cancer legislation was not the Congress, the executive branch, the relevant professional community of medical scientists, or an organized interest group, but was instead a unique constellation of private citizens grouped around Mary Lasker. In this respect, the cancer initiative was rather unlike most agenda-setting efforts in public policy.

Also unlike many prominent policy initiatives, the cancer initiative had no single dramatic triggering event to which it could be attributed. Rather, there were a number of converging events that, when taken together, constituted a "slow trigger" to the cancer crusade. Several general factors were at work. The most obvious factor was the continuing substratum of pain and suffering associated with cancer, and the deep anxiety and fear that cancer inspires in many people. The role of research was a second general factor, simultaneously holding out hope and promise for answers to the disease problems of man as well as creating frustration at the lengthy time from laboratory to bedside. The question asked with persistence since the mid-1960s had been "What are we getting for our biomedical research dollar?" So another factor was the increasing suspicion felt by Mrs. Lasker and some of her associates that the NIH and the greater portion of the scientific community were much less interested in solving categorical disease problems than in attending to the broad range of biomedical research. Finally, management developments within the National Cancer Institute reinforced the thinking of some cancer advocates that there was certainly a different way, and perhaps a better way, to manage the research effort.

There were more specific factors at work also that propelled Mrs. Lasker into action. One such factor was her concern for the displacement of leadership that had occurred in biomedical research. The loss of old allies, the dominance of NIH over NCI, and her own remoteness from the Nixon White House compelled her to search for a way to offset these developments. Another factor was Mrs. Lasker's continuing concern about money for research, especially cancer research. This concern was aided symbolically by the broader efforts in the Congress to reallocate resources from the Vietnam war to domestic problems, the effort to reorder national priorities. The disappointment with the Regional Medical Program's failure to bridge the gap between research and patient care was a third factor giving impetus to the cancer crusade. The search for effective models of action was symbolically reinforced here by the success of the U.S. space program in landing a man on the moon. Finally, after a political lifetime of substantial accomplishment, Mrs. Mary Lasker had still not witnessed the triumph of research over the major killer diseases and time was running out. A major new initiative was sorely needed.

There are four discernible stages through which the cancer initiative progressed in its transformation from an inchoate belief that something should be done to a specific piece of legislation that was the subject of Senate hearings. The first stage centered on Mrs. Lasker and her search for a strategy to enter cancer on the formal agenda. This stage included the consensus among her, Sidney Farber, Solomon Garb, and others that an initiative was needed. It also included the complicated, year-long discussions between Mrs. Lasker and Senator Yarborough in 1969, which involved securing the Senator's commitment to action and searching for the appropriate vehicle of action. This stage concluded with the establishment of the Panel of Consultants by the Senate in the spring of 1970.

The second stage, of course, involved the selection of the panel members and the chairman, and the organization and conduct of its work. The Panel of Consultants and its chairman Benno Schmidt identified the issues, defined the problem, and wrote the report that dominated the legislative debate on the cancer act. In addition to establishing the bases for legislative action in their recommendations, the panel and especially the chairman also made a substantial investment of time in personally preparing the ground for the reception of the report.

The third stage, logging the cancer initiative onto the formal agenda of the Congress, began the very day the Panel of Consultants presented its report, December 4, 1970, with Senator Yarborough's introduction of legislation. This was followed in January 1971 by the more authoritative introduction of the same legislation—S. 34—by Senators Kennedy and Javits and the announcement by them of scheduled hearings on the bill. Accompanying actions in the House of Representatives occurred both in late 1970 and in early 1971 to extend the logging-in requirement to that body.

The proposed reformulation of federal government policy toward cancer was securely entered on the formal agenda with the public expression by President Nixon of his interest in the matter. Thus, the fourth stage included the processes by which he and his key White House advisers were persuaded of the seriousness of the panel's demands and the manner in which he dealt with them in the State of the Union message and in the special budgetary request for an additional $100 million for cancer research.

THE POLITICAL RESOURCES
OF THE CANCER CRUSADERS

Mrs. Lasker and the individuals involved in the Panel of Consultants, it is worth recapitulating, constituted an elite group that brought substantial resources to bear on both the agenda-setting and legislative phases of their activity. They had clear ideas about how to use their resources and were remarkably persistent in seeking to employ them to maximum advantage. Foremost among their resources were wealth, status, and the ability to confer status. The wealth was remarkable. In addition to Mrs. Lasker, at least eight members of the Panel of Consultants were million-aires—Rockefeller, Bobst, Funston, Schmidt, Wasserman, Lawrence, Foote, and Parten—a number of them many times over! These individuals also had high status in the worlds of finance, industry, philanthropy, public relations, entertainment, and politics, to say nothing of the personal and institutional status represented by the physician-scientists. The ability to confer status was another resource of this group, exercised directly and indirectly through influence over employment, research awards, honors and rewards, participation in blue-ribbon panels of ex-perts, and by the simple fact of association.

Equally impressive as a resource, however, was the fact that the Panel of Consultants spanned and resolved within itself sev-eral traditional lines of social and political cleavage in this coun-try. The panel members were bipartisan in politics, ranging from loyal supporters of Senator Kennedy to close friends and ardent defenders of President Nixon. Labor and management (or capital) were both represented. Indeed, it is hard to imagine any other major health policy issue on which the representatives of organ-ized labor would be as willing to cede to those of capital the re-sponsibility for formulating a position as they were in the case of the cancer crusade. The panel also spanned elite-mass divisions, not within its own highly elite membership, but in its knowledge of the underlying public support for its activity and its ability to mobilize that support at critical junctures. Finally, the division be-tween the lay community and the medical-scientific communities was effectively bridged within the Panel of Consultants.

The phenomenon of wealthy individuals spearheading a major legislative initiative in health, with full proxy support from the representatives of organized labor and the latent support from a broad stratum of public opinion, is sufficiently rare to deserve further comment. Is there reason to expect altruism as a motive in the case of cancer where it is so conspicuously absent in more mundane areas of health care, like health insurance for the eld-erly, the indigent, the disabled, those with catastrophic illness, or the general public? Altruism need not be a requirement in the choice between biomedical research and patient-care financing. Research can be regarded as a prudent, and rather limited, search for scientific solutions to vexing medical problems, while the patient-care financing often appears as an open-ended drain on public resources. But why should cancer be the focus of support and why not heart disease? We may assume that cancer is the most feared disease for the wealthy as it is for the rest of the public. Furthermore, cancer is a disease for which the best of care— always obtainable by the wealthy—is none too good. One is left with the speculation, then, that cancer represents the major health problem for which private wealth cannot purchase solutions. Thus, the wealthy are prepared to advocate the expenditure of large sums of public resources in the pursuit of solutions to cancer, not out of altruism, but out of a clear assessment of their own self interest. That their concern is congruent with the rest of the public makes its political appeal all but irresistible.

Politically speaking, access was one of the greatest resources possessed by the Panel of Consultants. They had access to everyone regardless of their views on the cancer initiative. This included access to loyal supporters, to attentive audiences who were not loyalists, and to their opponents based upon the grudging recognition of their power. Access to the medical-scientific community ran primarily, but not exclusively, to those with a clinical orientation. It also reached to political leadership in both houses of Congress, into the inner circle of White House advisers and to the president himself. This access extended to the populace through the effective publicizing of the cancer cause and the tapping of the latent, but strong, public support for government action.

One of the most substantial resources possessed by Mrs. Lasker and the Panel of Consultants was strategic and tactical knowledge of how to use power and influence for securing their desired ends. One root of this knowledge, obviously, stemmed from the fact that these were efficacious people—they knew how to get things done. They also had detailed knowledge of how to use the blue-ribbon panel of experts. Schmidt, for instance, understood the internal management of such a group completely. The importance of a chairman in charge of the staff and leading the members was central to his thinking. Also important was the value of participation, the importance of listening to all points of view, the need to avoid forcing divisive issues to a vote, and the significance of securing unanimous agreement on final recommendations.

The panel, and Schmidt in particular, also understood the importance of the report. It had to be brief with a few clearly stated, easily understood recommendations. Schmidt also knew the relative importance of the nontechnical and technical portions of the report. He was aware that analysis was not the key to the report's reception, but that the appearance of analysis was necessary. Finally, the recommendations of the report had to be ones that could be easily translated into legislation.

The Panel of Consultants also knew that substantial preparation was required if their report was to be well received. Even the high face-validity of a report from a blue-ribbon group, written in brief and understandable language, was not enough to guarantee its favorable reception. Groundwork with the president's advisers, the Senate, and key members of the House was essential.

Davies has argued that neither congressional hearings on an agenda item nor that item's inclusion in a president's legislative program can be considered as sure signs of the government's intention to act. He writes, "there are no unequivocal indicators of such commitment because the long path from initial attention to actual changes that do something about the problem can be subverted or sidetracked at any stage."[1] No group understood this argument better than Mrs. Lasker and the Panel of Consultants; one critical resource they brought to bear upon events was their attentiveness to the entire legislative process. They were involved in the tasks of advocacy, counterattack, and compromise both in public and in private, at every step of the way. Their time was allocated as any scarce resource might be. No detail was so small that it could safely be ignored. Because of the nearly full-time involvement of Schmidt, the panel chairman on several critical occasions came to play a broker's role in practically everyone's behalf.

AGENDA SETTING AS A POLICY DETERMINANT

Schattschneider has written, "Political conflict is not like an intercollegiate debate in which the opponents agree in advance on a definition of the issues. As a matter of fact, *the definition of the alternatives is the supreme instrument of power*; the antagonists can rarely agree on what the issues are because power is involved in the definition. He who determines what politics is about runs the country, because the choice of alternatives is the choice of conflicts, and the choice of conflicts allocates power"[2] (Emphasis in original). The Panel of Consultants, in retrospect, achieved its greatest influence through the definition of the alternatives. Their dominance of the legislative debate by means of the prelegislative activity was so extensive that it deserves consideration.

First, the efforts of the panel established two major assumptions that went unquestioned in the entire debate. The panel established the proposition that new legislation was needed for cancer, though it was not at all clear that this was the case. Support for the position of the administration in opposing new legislation steadily eroded, as the precariously placed Senate Republicans argued in May of 1971. Paul Rogers benefited from this with his recognition that "you can't fight something with nothing." It should be noted in passing, however, that Rogers stood to benefit from new legislation because it extended the authority of his subcommittee over a major portion of the National Institutes of Health, authority that he could not exercise as long as the 1937 National Cancer Act remained unchanged.

Also unquestioned was the proposition that cancer needed a new and higher priority than it had received in the past. Acceptance of the need for new legislation meant new priority by virtue of a new mandate. But a new priority was implicit in the president's $100 million budget request. It was also implicit in the widespread assumption that substantial new amounts of money would be forthcoming for cancer, an assumption that only Shannon and Marston criticized, and then weakly. But the question of priority was explicit in the initial legislation submitted by Paul Rogers; this bill included the provision for the NCI budgetary by-pass of NIH and HEW and direct submission to OMB, a provision adopted against the explicit recommendation of the AAMC. The scientific community was never well placed to counter this assumption. Initially, some thought if cancer got better then everyone would get better; they were thus trapped by their own expectations and preferences. Then they were left to argue that money would not buy a solution to cancer, a hard argument to make and one that raised the question of whether money would buy any progress through medical research. Finally, they could argue the opportunity costs of allocating funds for cancer with respect to the other biomedical sciences; but arguments that cancer was getting more than its share left those proposing them vulnerable to charges of self-interest. The need for priority for cancer argued by the Panel of Consultants went largely unquestioned in the legislative debate.

There were several secondary issues, involving no specific provisions in bills under consideration, that were weakly debated. The Panel of Consultants propounded the view that there were major opportunities confronting cancer research in 1971. This view was never seriously challenged, though it was easy to find medical scientists of the view that no major clinical or scientific advances had occurred to warrant the cancer initiative. Internal to the administration, quite obviously, a general decision was reached not to attack the scientific-opportunities proposition. Indeed, both Marston and David embraced it, thus reducing the possibilities of a careful critique. While it was undoubtedly true that the scientific and clinical situation in 1971 was better than it had ever been, that fact provided no clear policy guidance. The same could have been said in 1966 or 1976. The real question was whether the situation was that much better to warrant an expanded cancer program. The scientific community did respond on the point of the limits of scientific knowledge, but this was anecdotal, illustrative, and impressionistic, not analytical in character. It remains unclear exactly why no detailed critical response was forthcoming to the scientific portion of the panel's report. The answer may be that the scientific community, as a matter of deeply ingrained style, is simply not analytical about itself. On the other hand, there is no effective counter to the scientific-opportunities argument. Nor was there any genuine response to the "experts know best" argument put forward by the Panel of Consultants; the charge of expert bias was potentially a two-edged sword. The argument of scientific opportunities was only weakly joined.

Another issue to which no legislative provision was addressed was that of the philosophy and strategy of research management. This issue generated substantial defense of the peer-review process of NIH and some criticism of research contracts. Research management, however, had been primarily a concern of the staff to the Panel of Consultants and the NCI staff. Schmidt, who committed himself on the matter only late in the work of the panel, played the management question in a rather low key and never got out in front on this issue. The debate did not involve a detailed examination of either targeted research support by the contract mechanism or of the research planning advocated by NCI. Rather, it was at the level of general philosophy and had as much to do with contending expectations about resource allocation as anything.

The primary issue of the legislative debate dealt with the organizational status and autonomy of the National Cancer Institute, both as an indicator and guarantor of priority for cancer research. The panel's recommendation that a separate cancer agency be established independent of the NIH and the commitment of the Senate committee to this view gave the issue its special character. The scientific community mobilized its resources against this view, stressing the consequences such a severance would have for tearing the fabric of biomedical research. Everyone would lose if cancer were removed from the rest of biomedical research: both cancer research and the broader biomedical research effort would be impoverished. Furthermore, a separate agency would encourage greater fragmentation of NIH, a prospect both real and greatly feared. On the separate agency proposal, the scientific community prevailed.

The scientific community never responded forcefully on the alleged six or seven layers of bureaucracy under which NCI was presumed to suffer. Nor did they fully explicate the need for access by the NIH director, or the NCI director, to higher levels of authority. Shannon dealt with the need for access to the secretary of HEW but few other voices were raised on this point and none adequately dealt with the limited attention that the White House could give to any agency.

But though separate agency status was not secured, substantial autonomy for NCI was obtained. The new law gave NCI a renewed and expanded mandate, raised the formal status of its director, provided the expectation for vastly increased funds, and in the budgetary by-pass mechanism established a procedure for asserting autonomy. The resolution of the legislative debate left the cancer crusade advocates with much of what they wanted, but gave the opponents the symbolic and material accomplishment of defeating the proposed separate agency recommendation.

THE NATURE OF THE POLITICAL DEBATE

The Panel of Consultants derived substantial political advantage in the legislative debate from the momentum it gained in writing its report, preparing for its reception, and securing the commitment to action of the key senators. This momentum carried the panel through a good portion of the debate in the Senate. When momentum slowed in late March and early April of 1971, the cancer crusade advocates recaptured it dramatically through Ann Landers' column and firmly secured its position through direct intervention at the White House. They employed their substantial political resources skillfully in a manner consistent with their established style.

The cancer crusaders, however, were also aided by the response of the Nixon administration and by that of the scientific community. The only consistent thread in the position of the executive branch was to deal with the politics of the cancer issue without a clear policy on the matter. The initial position in January 1971 was to recommend more money without legislation. From February to early April, through several administration spokesmen, the position was elaborated to include internal NIH and NCI reorganization. In May, the White House yielded to pressure from the public, the panel, and Republican senators and proposed its own legislation. By June, that legislation had been made consistent with the preferences of the Panel of Consultants and the Democrats controlling the Senate Subcommittee on

Health. Small wonder, then, that it had lost credence among House Republicans in the autumn and could muster no support for its position in the House Commerce Committee. The administration's vacillation became the subject of ridicule by Representative Staggers and its current position the subject of controversy by Representative Tiernan on the floor of the House. The ambivalence of President Nixon's early December response to the conference committee capped this behavior of politics without policy.

The behavior of the administration reflected its general internal confusion on health policy, the absence of orderly procedures for generating policy, and the administrative weakness of those who might be presumed responsible for developing such policy. One immediate consequence of this behavior for the legislative debate was that NIH, which might have been the source of a consistent policy view on cancer, found its degrees of freedom steadily reduced as the White House became increasingly involved with the issue. Another consequence was that the Panel of Consultants and its supporters found themselves greatly helped in the contrastingly clear-minded pursuit of their objectives.

The Panel of Consultants was also aided in securing its objectives by the behavior of the scientific community. The response of the scientific community was characterized by several factors. First, it basically accepted the definition of the issues as provided by the Panel of Consultants. Second, it accepted several key propositions made by the panel without challenge. Third, the absence of a publicly active NIH spearheading the opposition meant that the leadership of the scientific community had to be exercised through the Association of American Medical Colleges and the Federation of American Societies of Experimental Biology. The result of this leadership was a counterattack from a large number of sources, both individual and institutional, addressing a very significant proportion of the members of Congress on a wide range of debatable points. But this decentralized counterattack was diffuse and noncumulative. The same points were reiterated in each instance, by each witness in a hearing or by each medical scientist speaking to his representative, and did not build to an effective case except on the issue of a separate organization for cancer.

A fourth characteristic of the response by the scientific community was its anecdotal and illustrative rather than analytical character. It may be, as suggested above, that medical scientists have a deeply ingrained style of argument that is nonanalytical. Perhaps they see critical decisions as based upon the successful exercise of influence with key officials in private session. Whatever, the scientific community demonstrated very limited analytical capability in their response to the cancer advocates.

It is unclear that an analytical approach would have affected the outcome of the debate. But Comroe and Dripps have recently argued the need for something more substantial than the "informal let-me-give-you-an-example arguments by concerned scientists" in the formulation of the nation's biomedical research policy.[3] Comroe has further elaborated this argument in a report prepared for the President's Biomedical Research Panel,[4] and a Rand Corporation report to that same body argued some of the possibilities for policy research on biomedical science.[5] More substantively, it should be recognized that neither the cancer crusaders nor their opposition from the larger scientific community differed on objectives. Both groups desired to realize scientific and clinical advance in the war against cancer. The differences were on means, not ends. Though analysis is hardly the way by which public objectives are chosen, it is where ends are not in dispute that analysis has the most to contribute to the choice of alternative means. It is conceivable, at any rate, that the response of the scientific community to the cancer initiative could have been much more analytical and with potential effect upon the outcome of the legislative debate.

The position of the various senators and representatives who participated in the debate deserves mention in relation to these comments about policy analysis. The members of the Senate Subcommittee on Health were committed quite early to endorsing the views of the Panel of Consultants. Some of them, however, might have been responsive to counterarguments that involved more than choosing among contending anecdotes. The House, by contrast, was clearly receptive to any ammunition it might have been supplied that might have helped it deal with the Senate. The nature of the scientific opportunities in cancer, the administrative situation and requirements of NIH and NCI, the significance of various forms of research management, the need for more money for cancer research—all these issues might have been dealt with analytically in communication with the House. Such analysis was not forthcoming.

The role of the press and media in the debate was relatively limited, owing in large measure to the complexity of the issues and the speed with which they became visible and were resolved. There were three types of publicizing institutions active in this debate. First were the house organs or "dedicated" information systems of the various contending parties. These included the several outlets of the American Cancer Society and the cancer establishment, as well as the newsletters and communications networks of the AAMC and FASEB. Next were the specialized journals read by the attentive public in the health and medical-science communities. *Science, Nature, "The Blue Sheet,"* and others were included here. Finally, the national press was the third type of publicizing institution—the *New York Times*, the *Washington Post*, the *Wall Street Journal*, and the national wire services.

The messages transmitted through these publicizing institutions were threefold in nature. Information about events was communicated to the attentive and general publics. Efforts to shape preferences among readers was a second type of message. Finally, some communications were directed to eliciting a response of political support for a given position by the reader. In relation to shaping preferences, which often existed in latent form, and to eliciting a supportive response, both the cancer advocates and the scientific community had significant resources in the debate and used these resources extensively.

The cancer crusaders had a comparative advantage in the publicizing of their cause, however, derived not from access to these differing types of media so much as from their exploitation of various symbols.[6] The arguments for a reordering of national priorities that arose in 1969 and 1970 had their origins in a Democratic Congress increasingly restive with the Vietnam war, no longer constrained by a Democrat in the White House, and desirous of directing resources to domestic social objectives. The Panel of Consultants was able to skillfully set the cancer initiative in this context. They were strongly aided also by the success of the U.S. space program in landing a man on the moon, and the moon-shot analogy—no matter how inappropriate to cancer—provided them a powerful aid.

THE LEGISLATIVE "END GAME"

It is clear that the agenda-setting activities of Mary Lasker and the Panel of Consultants had great effect on the outcome of the legislative debate leading to the National Cancer Act. It is also clear that the outcome was not entirely anticipated by the panel and that several key provisions of the act emerged in the late stages of the legislative process. Two provisions, in particular, were "end game" provisions—the cancer control program and the President's Cancer Panel provisions.

Cancer control, established in the 1937 legislation, had been transferred out of NCI into the Public Health Service in the mid-1950s. There it had languished and was on the threshold of expiration in 1971. Representative Rogers, however, distressed at the prospect of its demise and concerned with the apparent contradiction between its projected fate and the program advocated by the Panel of Consultants, challenged Secretary Richardson on the point. The logic of making the best diagnostic and therapeutic capability widely available was highly attractive and was the basis for Rogers' insistence that the program be revived within an expanded National Cancer Program. On this issue, as with so many in the cancer debate, there was relatively little examination of what was implied, a fact that has contributed to the complicated implementation history of the control program.

The second "end game" provision was that which established the President's Cancer Panel. This arose from Representative Ancher Nelsen's search for a means to accommodate the president's desire for access to the cancer program with the desires of the Panel of Consultants. It was accepted in the House Subcommittee mark-up session by Paul Rogers, in large measure to secure unanimity within the subcommittee. The formulation by Nelsen was contrary to the desires of the panel for access by the NCI director to the president, and was resisted on those grounds. The provision was not the subject of any hearings and was poorly understood at the time. This latter point is underscored by the fact that Benno Schmidt, who is, at this writing, in his second three-year term as chairman of the President's Cancer Panel, was one of the strongest opponents in the final stages of the legislative process.

The National Cancer Act raises some questions about "end game" provisions. How do they arise? It would appear that such provisions are suggested as a consequence of the individual legislator's search for a "solution" to some problem identified in the hearings, a problem that proposed legislation has neither identified nor dealt with. Why are such provisions accepted? The answer to this question would seem to consist of several parts: these provisions have a high face validity, they involve the allocation of power among participants in important situational ways, and they permit further differentiation among individual legislators and between legislative bodies in the quest for authorship of key provisions. What do such provisions mean? The answer here is that meaning is often unclear at the time of enactment, that the ambiguity of meaning may affect successive efforts at implementation, and that an "end game" provision may turn out to be important in retrospect in ways quite unanticipated at the time. If nothing else, the National Cancer Act of 1971 demonstrates both the importance of the pre-legislative stages of an Act of Congress and the final stages of legislative history. In these ways, it enriches our understanding of the legislative process.

It is appropriate to turn now from agenda-setting and legislative history to a brief indication of the major outcomes that are discernible to date. Most of these have to do with the National Cancer Program established by the new legislation. Many of the important outcomes, however, have also affected the National Institutes of Health and the biomedical scientific community.

THE NATIONAL CANCER PROGRAM

One of the more interesting outcomes of the 1971 legislation has to do with the leadership of the National Cancer Institute. Previously, the NCI director was subordinate to the NIH director within the overall framework of the government's effort in biomedical research. With respect to the NCI itself, however, the director exercised great control over the National Advisory Cancer Council and had substantial influence on resource allocation and program management, as we have seen, through the several directed research programs of chemotherapy, virology, and carcinogenesis. These arrangements were overturned as a consequence of the new law.

The National Cancer Advisory Board (NCAB) has far broader authorities and responsibilities than its predecessor, the NACC. Members of the NCAB are presidential appointees, as is its chairman. By law, the NCAB has authority to investigate all the activities of NCI it deems desirable. It has exercised this authority with respect to contract research, annual review of the proposed budget, and all matters related to the National Cancer Program—behavior that represents a sharp departure from that of the NACC. It is appropriate to think of the NCAB as the "board of directors" of the National Cancer Program.

The President's Cancer Panel, through 1976, at least, has proved to be the major locus of policy leadership for the new cancer program. Benno Schmidt, though rebuffed in early 1972 in an effort to be named chairman of both the panel and the NCAB, has nevertheless managed to establish himself as the most powerful figure in the leadership of the cancer program. He was instrumental in the selection of the initial NCAB members, and acted as their agent in firing Carl Baker as NCI director and replacing him with Dr. Frank J. Rauscher by early May 1972. The three-member President's Cancer Panel, though conceived as an oversight group to assist the president, has in reality become the "executive committee" of the "board of directors." Schmidt is the undisputed chairman of the "executive committee" and is in fact the "chairman of the board" of the National Cancer Program, though formally the chairman of the NCAB is Dr. Jonathan Rhoads.

The "chief executive officer" responsible for the day-to-day operations of NCI had been Dr. Frank Rauscher, until his resignation on November 1, 1976. Rauscher, a Ph.D. in virology from Rutgers University, came to NCI in 1959. He was serving as the

Institute's scientific director for etiology at the time of his appointment as director. Internal management of NCI has been exercised through an executive committee of senior NCI officials, and Rauscher is credited with having strong working relationships with Schmidt and with the NCAB.

But, as a former NCAB member said in 1974, "the head of the program is Benno Schmidt and Rauscher is executive director." Rauscher does not disagree. In a farewell interview he said, "The program's going to be OK. As long as Benno Schmidt is at the helm, the program will be OK."[7] Schmidt dominates the President's Cancer Panel and the NCAB to a great degree. The panel has involved itself in every major aspect of the National Cancer Program, especially budgetary matters. Panel transcripts read like a continuing seminar in cancer research and treatment and in the management of a half-billion dollar program. Schmidt's constant questioning requires NCI staff to explain and defend their programs to a greater degree and with a greater frequency than they have ever done before.

The Schmidt-Rauscher team has been seen as a "fortuitous" combination by NIH leadership. The NCI director, one former NIH official said, "could cause a convulsion in NIH if he tried. A director who wished to exercise the full range of authority could beg a hell of a large degree of autonomy from NIH. But Rauscher has played it very carefully. Also, Benno Schmidt has grown tremendously in stature and understanding. He started as a narrow partisan and has become a broad generalist. He has been a restraining hand at NCI."

The priority for the cancer program established by the new law is revealed most dramatically in the appropriations history for NCI. Table 1 reveals that the appropriated funds to NCI have almost quadrupled from fiscal 1971 to 1977, indicating how conspicuously successful Mary Lasker and her associates have been.

The original recommendation of the Panel of Consultants was that the NCI budget reach a level of $800 million to $1 billion by fiscal 1976. Congressional authorizations, moreover, for fiscal years 1972 through 1977, respectively, are $420 million, $530 million, $640 million, $804 million, $899 million, and $1,074 million. The NCI, in its strategic plan for the National Cancer

TABLE 1

National Cancer Institute Appropriations History,
Fiscal Years 1971-1977
($ in Millions)

| | | Increase over Prior Year: | |
Fiscal Year	Appropriation	Amount	Percentage
1971	$233	$ 53	29%
1972	379	146	62
1973	492	113	30
1974	527	35	7
1975	692	165	31
1976	762	70	10
197T*	154	—	—
1977	815	53	7

*The three-month transition period from 1 July 1976 to 30 September 1976 to the new fiscal year for the federal government.
Source: National Institutes of Health.

Program, suggested an institute program of $852 million for fiscal 1978 out of a total national cancer effort in excess of $1.2 billion. While recommendations, authorizations, and planning figures do not represent appropriated funds, the appropriations history does suggest widespread consensus on a nearly billion dollar a year program.

Questions of research management and planning have engaged the attention of those responsible for the National Cancer Program. The 1970 report of the Panel of Consultants called for "a comprehensive national plan" for cancer. The need for such a plan echoed throughout the 1971 legislative hearings. The actual language of the Cancer Act directed NCI to establish a five-year plan for an expanded cancer program. In anticipation of this legislative outcome, Carl Baker initiated an NCI-directed effort that must rank as the largest, most extensive planning effort ever undertaken within biomedical research.

NCI staff initially stated the program goal to be the development of means "to reduce the incidence, morbidity, and mortality of cancer in humans." Seven objectives, couched in nonscientific terms, were developed as targets for the realization of this goal. The hierarchy of goals and objectives was then extended to include major approaches, approach elements, and project areas. For this the NCI involved one panel of scientists for each objective, developed thirty-five approaches, and formulated the initial statement of approach elements. Following this, approximately 250 scientists, working in forty panels, developed project area statements for each element in a series of three-day conferences stretching over a four-month period. The result was seven volumes of documentation, one for each objective. A single volume digest of scientific recommendations was then prepared on the basis of the panel's work.[8]

The NCI staff then developed a two-part strategic plan.[9] The scientific recommendations became the program strategy, while the implementation strategy consisted mainly of general guidelines for managing the five-year cancer program plan. The detailed relationships between the strategic plan and the actual conduct of the cancer program were addressed in an operational plan issued in August 1974.[10]

The plan, though the object of much concern and criticism, has proved useful in explaining the cancer program to the Congress and the public and in providing general directions for NCI. It has not been used to any significant degree in governing the actual day-to-day management of various NCI programs. Neither Schmidt nor Rauscher have displayed more than mild support for the management importance of the plan itself.

The contract-supported, directed-research programs of NCI have also been of management concern. Attention has focused on the Virus Cancer Program (VCP), the strongest contract program. External criticism from the scientific community and internal NCI discussions led the NCAB to establish an *ad hoc* committee in March 1973 to review the scientific and management aspects of the virology program.[11] Known as the Zinder Committee, after its chairman, Dr. Norman Zinder of Rockefeller University, this *ad hoc* group submitted a critical report that was received by the NCAB at its March 1974 meeting.[12]

The Zinder Committee accepted the scientific rationale of research on viral causes of human cancer, but took issue with the objective of developing antiviral vaccines. The ignorance of the basic disease mechanism was so great that the analog to poliomyelitis and infectious diseases was deemed inappropriate. The dominance by the VCP of viral oncology raised the question of whether the contract mechanism was the right way to support this area of science. The committee's answer was a resounding "No." The problems were that the contract proposal review process was dominated by VCP program officials (or segment chairman), included potential and actual contractors in the review of proposals, lacked scientific rigor, and was inaccessible to the larger virology community. The contract-research program, moreover, represented an extension of the intra-mural research work of some VCP scientists. "We are quite sure," the committee wrote, "that much more would have been accomplished if equal support had been provided on a competitive basis to many more laboratories with greater capability and experience in particular areas."

The Zinder Committee recommended "opening up" the VCP program, specifically through the establishment of an NCAB oversight committee, a reconstitution of the contract review groups, and the elimination of contract work that was an extension of VCP scientists' intramural research. The NCAB did establish a subcommittee, chaired by Dr. Harold Amos of Harvard Medical School, to monitor the program's response to the Zinder Committee's recommendations. The VCP, on its part, has established an advisory committee of nonprogram scientists to provide advice on broad directions of resource allocation, promising lines of scientific inquiry, and means of application of research findings. The contract review process was also modified to increase the rigor of review of individual contract proposals. The Amos subcommittee, in its report to the NCAB of June 1975, indicated its general approval of the program changes.

Partly as a consequence of the VCP review, the NCI has moved to tighten the review of all contract-research programs.[13] In addition, NCI has sought to emphasize to the scientific community its concern for the balance between contract research and that supported by the traditional grant mechanism. Rauscher, in a 1974 article in *Science*, for instance, pointed out that the ratio of research grants to contracts (both support and research) had shifted from 50-50 to 55-45 in the three-year period from 1972 through 1974. He emphasized a desire to be responsive to the "concerns and advice from the scientific communities of this country," and pointed to the need for selecting the best funding mechanism to perform the job after priorities had been determined.[14]

Beyond the issues of research planning and management, three major programs of NCI have been the focus of much discussion and debate in the implementation of the National Cancer Program. These programs are the comprehensive centers program, the cancer control program, and the carcinogenesis program. In the early 1960s, a program initiated within NCI established a number of centers engaged in clinical research on cancer. Though much basic research was also done in these centers, the emphasis was on clinical research, usually on one aspect of the cancer problem. Section 408 of the National Cancer Act committed NCI to move from these specialized centers to more-comprehensive centers, and to the creation of a network of cancer care institutions reasonably distributed relative to geography, population, and medical-scientific talent. The 1971 legislation authorized the establishment of fifteen such centers, but the reauthorization of the act in 1974 removed the statutory limit on the number.

The purposes of the comprehensive cancer centers are threefold. First, they represent an effort to create interdisciplinary research teams more suited to the cancer problem than discipline-oriented groups. Secondly, they are to demonstrate the most advanced techniques of diagnosis and treatment of cancer patients. Finally, the centers are to establish community outreach efforts in order to improve the quality of cancer care provided by community health institutions.[15]

The centers program continues to have a number of problems. A report of the General Accounting Office issued in early 1976 pointed out that the congressional concern for geographic distribution of centers was not being achieved in the first 17 centers established.[16] Geographic distribution was needed to insure that the greater proportion of the population lived within 120 miles of such a center of specialized care. The particular distribution realized was a function of the requirement that designated centers be located at institutions having established scientific excellence. How to balance geographic and scientific criteria remains a difficult problem.

The second problem identified by the GAO report was that comprehensive centers have not become focal points for the demonstration of the latest findings in cancer prevention, detection, diagnosis, treatment, and rehabilitation. The absence of clear NCI guidelines for centers, the competition among institutions in a given area, the multi-institutional nature of many "centers," and the lack of a designated geographic area of responsibility have limited the demonstration role of comprehensive centers.

Not mentioned by the GAO, but of great concern in those institutions that have been designated as comprehensive centers, is the impact of such centers upon the total life of the institution. Comprehensive cancer centers, by virtue of their financial support from NCI and as a result of the authority given to the center director, have become more than just another department of a medical school. The drawing together of the resources related to oncology has often resulted in the center coming to exercise as much control over resources as the parent organization. While serious disruptive effects are alleged, the extent of institutional impact of the comprehensive cancer centers remains to be fully evaluated.

The reestablishment of the cancer control program, provided by Section 409 of the new act, authorized NCI to undertake programs designed to improve the diagnosis, prevention, and treatment of cancer. The basic purpose of the program was to insure the rapid translation of new knowledge from the research laboratory into the treatment of the cancer patient, with the ultimate objective being to decrease the mortality and morbidity caused by cancer.[17] In this regard, the test of success would be the premise often asserted by the Panel of Consultants that the five-year survival rate among cancer patients could be increased from one-in-three to one-in-two if the best techniques of diagnosis and treatment were made widely available.

The program has dealt with a number of organizational problems internal to NCI. It established the boundary between its activity and research programs by deciding to support only rehabilitation research, and by emphasizing that its mission was to encourage the application of existing knowledge. It established the boundary between control and the delivery of patient care by restricting its role to demonstration, dissemination, and field testing but not to include direct support of cancer patients. The organization of the control program evolved from an activity within the NCI director's office to division status within NCI, with branches for prevention, detection, treatment, rehabilitation, education and training, and program support. The initial budget in fiscal 1973 was $5 million, but this jumped substantially to $34 million in the next year. Criteria for project selection were developed that included program need, project relevance, feasibility of implementation, number of beneficiaries, types of cancer, intervention impact, acceptability potential, cost-benefit ratio, potential for success, and ease of evaluation. Program planning has involved outside scientific and medical personnel working with NCI program staff. Most projects are contract supported and a contract-review process has been developed that involves an internal program review for need and relevance, technical evaluation by one of three committees composed primarily of outsiders, and final review by the program's advisory committee. This review process seeks to avoid some of the problems encountered by the VCP program.

One external problem has been coordination with the centers program. The GAO report which indicated that centers had not become the focal point of cancer demonstration efforts in the communities where they were located is an indication that coordination between the two programs has not been particularly successful. Another external problem has been to establish relations between the control programs and the practicing medical community in the grey area between demonstration and delivery of care. The control program and the outreach program of the comprehensive centers have already created difficulties in relations with the American Cancer Society and its traditional education and demonstration programs.

The basic problem of the control program, however, is that even the best of existing treatment and diagnosis techniques are the subject of controversy within the medical-scientific community.[18] Some critics are skeptical of the expected impact of the program and openly question the wisdom of allocating resources to control in too generous a manner.

In no area of activity, however, has the National Cancer Institute encountered greater difficulties in mounting a program than in that of chemical carcinogenesis. Though an NCI program was initiated in the 1960s by Dr. Endicott, little attention was paid to it until 1974. The report of the Panel of Consultants mentioned it, but practically no discussion occurred in the legislative debate in 1971. The 1972 plan for the National Cancer Program discussed carcinogenesis, but no special emphasis was placed upon it.

The basic challenge to NCI, with its cellular- and molecular-oriented research program, is that the preponderance of cancer is caused by insults received by the individual from chemicals in the environment.[19] This charge is supported by NCI data demonstrating greater than expected incidence of certain forms of cancer in areas where industrial chemicals are produced and in the industrial "hot spots" of the country. Those making the challenge argue that NCI ought to devote itself more vigorously to establishing a scientific basis for withdrawing carcinogenic chemicals from production and use in the economy by government regulation. Thus, they believe, one could prevent a good deal of cancer, even if knowledge of underlying mechanisms did not exist.

Technical problems confront the carcinogenesis program. The criteria for determining whether a given chemical is carcinogenic are very difficult to establish. In vivo tests using animals are very expensive and require several years of testing for each compound. In vitro tests (bioassays) are promising in both speed and cost but require further development. Internal to NCI there have also been problems, some reflected in the early May 1976 resignation of Dr. Umberto Safiotti, the head of the carcinogenesis program since 1968.[20] Safiotti has indicated extreme dissatisfaction with inadequate staffing for his program and with increasingly bureaucratic decision-making in NCI. Rauscher and Dr. James Peters, Safiotti's immediate superior, criticized the scientist for failure to reassign personnel within his program, with not pushing the development of bioassays rapidly enough, and with bottlenecks in the publication of in vivo test results. The validity of charge and counter-charge is not possible to establish, but it is clear that internal problems exist.

Carcinogenesis also creates several new external problems for NCI. There is only a small scientific community involved in cancer epidemiology and carcinogenesis research, thus there is no strong research clientele group pressing for action. The pressure for action has emerged as much from publicity and public opinion about egregious instances of chemically-induced cancer as from normal decision-making channels. But this area involves the NCI to a far greater degree than before in supporting the regulation of a major American industry. This role is far afield from its traditional research orientation, and carries it into potential conflict with powerful chemical firms and a powerful industry. NCI hardly relishes its posture in this David-and-Goliath situation.

ACCOUNTABILITY

What should the nation expect from the national cancer program? The average citizen expects a reduction in the incidence of mortality and morbidity due to cancer. Is it a fair expectation?

In our judgment, it is entirely fair. For one thing, the drive to separate the Cancer Institute from NIH was partially justified on the grounds that NIH was more concerned with medical research than with finding cures to the major killer diseases. NCI, having received substantial autonomy from NIH, can now reasonably be asked how its enlarged authorities and responsibilities have contributed to the conquest of cancer. Furthermore, since the Panel of Consultants and its supporters claimed in 1970 and 1971 that cancer was on the threshold of major scientific and clinical advances, we are entitled to know the progress made in crossing that threshold. Finally, since funds appropriated to NCI have increased from approximately $200 million in fiscal 1971 to over $800 million in fiscal 1977, an accounting in the use of these funds is not only a reasonable but a prudent expectation.

But answering the question about progress in the war against cancer is not easy. There are immense technical difficulties—the complexity of cancer and its numerous manifestations, the vast ignorance about its etiology and underlying mechanisms, the empirical nature of much knowledge about treatment—that limit the possibility of understanding by even informed citizens. There are administrative barriers—the sheer size and rapid growth rate of the national cancer program, the internal complexity of the NCI—that constrain the acquisition of reliable knowledge. There are problems arising from the way promising research results are reported to the public.

The means for informing the public about progress against cancer are currently inadequate. The internal reports of NCI, available to the interested public upon request, are too detailed and voluminous to be of general help. The annual reports required by the Act that are submitted to the president and the Congress by the NCI director, the NCAB, and the President's Cancer Panel are superficial.[21] The hearings and reports by the relevant congressional committees, especially those related to reauthorization of the legislation, have been very cursory and uncritical.[22] And press releases about research results are often too optimistic.

Several improvements might be considered. For example, all grantees and contractors of NCI could be required to follow a specified format when issuing press releases about advances in cancer research. This format could require that the research being reported on include information on the status of the research (laboratory research, animal studies, clinical research), the stages of work through which the research must pass before it could be utilized for therapeutic, diagnostic, or preventive purposes, an estimate of the time required for work to progress through these stages, the type of cancer to which the work reported is related, the incidence of that type, and the likely effect if the most optimistic expectations of the research were realized.

Another means for improving the reporting on progress would be the development of an annual report *by the Congress*, independently of NCI, on end-results data pertaining to changes in the five-year survival statistics for the various forms of cancer.[23] It is important that such data include the incidence of the particular form of cancer, so the public may know whether progress is being made against the rare or common forms of cancer. Such a report might be issued at an annual joint congressional hearing between the Senate subcommittee on health and the House subcommittee on public health and environment. The end-results data could be presented publicly at this time, providing Congress an opportunity to question the statistical information generated by the experts. It would be essential, however, that any end-results data be presented in a form that is readily understandable to the public. To this end, the Congress might seek the direct assistance of expert consultants—an epidemiologist, a science writer, and a graphics designer—to assist in designing the reporting format for presentation of the data.

The National Cancer Act of 1971 was reauthorized in 1974, extending authority from fiscal 1975 through 1977. It will be reauthorized again in 1977, either for one year or for three years. Regardless of which course is followed, it would be appropriate now for Congress to begin planning for a detailed, critical review of the national cancer program at the end of its first decade. Such a review, given appropriate preparation, would certainly be warranted; no one could argue that not enough time had passed for results to be valid. The feasibility of an earlier review would be highly contingent upon crowded legislative calendars. This decade review of the program, along with the two other measures suggested above, could provide the public with a greater understanding of how well the national cancer program is doing in the war against cancer.

PERTURBATIONS WITHIN NIH

There have been a number of perturbations in the National Institutes of Health since the 1971 cancer legislation. Though caused by several factors, these disturbances have been attributed in no small measure to the effects of the cancer initiative on NIH.

The National Cancer Act, in upgrading the status of the NCI director to that of a presidential appointee, had to similarly upgrade the NIH director if the Cancer Institute was to remain within the larger organization. But consequently, when President Nixon requested the resignations of nearly 2,000 senior government officials immediately upon his overwhelming reelection victory in November 1972, Robert Q. Marston's was among those received. On December 15, it was announced by the White House that the

NIH director was among several HEW officials who had indicated a "desire to return to private life." In fact, Marston had indicated no such thing. He had been fired.[24]

It was not until May 1973, six months later, that Robert S. Stone, M.D., dean of the school of medicine and vice-president for health sciences at the University of New Mexico, was appointed as Marston's successor.[25] Stone, relatively unknown in biomedical research and from an obscure medical school, had a reported orientation toward management. He was, however, unable to retain the other top NIH leadership.

In the wake of Marston's dismissal, Robert Berliner, the NIH deputy director for science, and a twenty-three-year veteran of the Bethesda campus, announced his resignation in July to accept the position of dean of the Yale Medical School. "I would be less than candid," he said, "if I were to claim that the administration's attitude toward the conduct and support of biomedical research played no part in my decision."[26] Six months later, in January 1974, John Sherman, the NIH deputy director, and another twenty-year veteran of NIH, resigned with a blast at the Nixon administration.[27] He was "discouraged and disillusioned," by the administration's attitude toward science. Things at NIH, Sherman said, "aren't as well as they used to be or should be." The situation was still "salvageable," he said, "but it's going to take a different atmosphere and tangible support" from HEW and OMB.

Then, in January 1975, less than two years after his appointment, Robert Stone was fired as NIH director. The reasons given for this action were that he had become an advocate of medical research rather than an emissary of the HEW secretary's office, he had failed to relate the federal government's health research effort to the developing health services activities, and he had failed to give strong direction to NIH. His successor, Donald S. Frederickson, M.D., was appointed in May 1975, after another six-month hiatus.[28]

The leadership situation in 1975 and 1976 was remarkably stable compared with the turbulence of 1973 and 1974. The assistant secretary of health since February 1975 has been Theodore Cooper, M.D., a former director of the National Heart and Lung Institute. Frederickson, the NIH director, had been president of the Institute of Medicine (of the National Academy of Sciences), and also a long-time veteran of the Heart Institute. A cooperative relationship between the NIH director and the assistant secretary has been accompanied by cooperative relations between Frederickson, on the NIH side, and Rauscher and Schmidt, on the cancer side. To the extent this stability in leadership prevails, the prospect for stability in the entire biomedical research enterprise increases.

Throughout the 1971 legislative hearings, it was argued that funds for cancer should not come at the expense of funds for the rest of medical research. The experience since the rapid growth in funds for NCI began, however, suggests that precisely the opposite has occurred, as Table 2 indicates. While the NCI budget has climbed dramatically, only the National Heart and Lung Institute, among the other NIH institutes, has managed to maintain its relative share of resources. Funds for the rest of NIH, while up, have declined from about one-half of the total to exactly two-fifths. The appropriations for the National Institute of General Medical Sciences (NIGMS), moreover, the institute most closely identified with basic biomedical research, increased by only 14 percent from 1971 to 1976. This represents an actual loss in research support when inflation is taken into account.

The budget difficulties faced by NIH provide no satisfaction to supporters of an expanded cancer program. In fact, these difficulties have been the source of continuing embarrassment to them. It was never their intention, they argue, that medical research in general should suffer as cancer research benefited, and they say it cannot be conclusively demonstrated that money for cancer has come from the rest of medical research; in any event, they do not feel responsible for the situation which has developed. In fact, both Benno Schmidt and Frank Rauscher have argued to the OMB, the White House, and the Congress that funds for the rest of NIH should grow in reasonable relation to those for the NCI.

Whatever the cancer program supporters have thought about the rest of the medical research enterprise, they have had little effect on checking the declining fortunes of the other NIH institutes. This is true for two essential reasons. First, the budgetary by-pass provision of the National Cancer Act permits NCI to submit its

TABLE 2

National Institutes of Health, Research Institutes and
Divisions Budget Authority, Fiscal Years 1971-1976
($ in Millions)

Fiscal Year	NCI amt.	NCI pct.	NHLI amt.	NHLI pct.	NIGMS amt.	NIGMS pct.	Other Research amt.	Other Research pct.	Total Research amt.	Total Research pct.
1971	$233	19.7	$195	16.5	$160	13.5	$595	50.3	$1,183	100
1972	379	25.8	233	15.9	173	11.8	682	46.5	1,467	100
1973	492	28.7	300	17.5	183	10.7	738	43.1	1,713	100
1974	527	30.2	290	16.6	168	9.6	760	43.6	1,745	100
1975	692	33.8	325	15.9	187	9.1	840	41.1	2,044	100
1976	762	34.6	370	16.8	187	8.5	882	40.1	2,201	100

Source: National Institutes of Health.

annual budget request directly to OMB; NIH and HEW may comment on but not change the NCI submission. This means that the NCI submission is developed without reference to anything but its own plans and aspirations, and that there is no intermediate level between the Cancer Institute and OMB capable of making a balanced allocation of resources to the entire medical research enterprise. A motion did reach the floor of the Senate in the spring of 1976 to amend the cancer legislation by eliminating the by-pass provision. This provision was explicitly rejected, however, indicating for the moment at least little congressional interest in modifying the priority-maintaining mechanism of the by-pass.

Second, the Congress had appropriated funds for medical research including cancer research, in such a way as to indicate strong support for expanding the cancer program and weaker support for offsetting the effects upon the other institutions of NIH. In 1976, however, in relation to the fiscal 1977 appropriations, there were indications that Congress was acting to constrain the growth of NCI to a more modest rate than has been the case, though with limited benefit to the rest of NIH. Though HEW and OMB have attempted to exercise budgetary discipline over medical research, including constraining the growth of funds for cancer, they have been ineffective without congressional support. It may now be the case that a new period of more balanced growth is in prospect.

A third effect of the National Cancer Act on the NIH has been to reinforce the pressures toward categorical research. The 1971 law became a model for heart research legislation in 1972. A bevy of new laws has been adopted that has not created new institutes within NIH, but has articulated the full spectrum of responsibilities, from research through demonstration and control to professional and public education.[29]

This new legislation is forcing a piecemeal divergence of the contemporary NIH from its traditional policy, practice, and organization. In the past, as John Sherman put it, NIH as a research and development (R&D) organization has represented "a large R and a small d." The contemporary pressures are to force NIH in the direction of greater attention to development, that is, exploitation of basic research ideas, and beyond R&D entirely to a concern for the modification of medical practice.

These developments leave many at NIH uneasy about the future. Two basic questions are raised. Should NIH move in this direction, or what is its role? Second, can NIH move in this direction, or what is it competent to do? Two basic organizational options exist in response to those questions. First, the NIH could undergo a more thorough internal differentiation as an organization, developing a more rationalized division of labor among basic research, clinical research, and demonstration-and-control activities. Or, alternatively, NIH could remain an institution committed mainly to the support of fundamental biomedical research, with reasonable attention to applied and clinical research, but stopping short of the responsibility for carrying research results into medical practice. Under this second option, it would seem, HEW would then have to articulate another organization whose principal mission would be to exploit the results of medical research and seek to translate them into modifications in medical practice. Organizational differentiation must occur, however, either within NIH or within the HEW health establishment, given the substantial pressures emanating from the public, Congress, and the executive branch to utilize the results of the research being supported.

There are several unresolved problems in this situation confronting NIH. One is the fear, widespread among senior NIH officials and prominent medical scientists, that in a more highly differentiated biomedical R&D establishment the basic research might be swallowed by the applied research and development efforts. This fear is based in part on self-interest, in part on the honest view of many medical scientists that it is premature to expect technical solutions to emerge from the results of medical science and useless to attempt to force such results.

Another problem is the relationship of biomedical research, largely federal-government financed, to the health care system, which also is increasingly financed by the government. It is now widely assumed that national health insurance is not a question of whether but when. One effect of government financing of medical treatment could be sharply increased pressure to allocate the biomedical R&D dollar toward those diseases that constitute the greater proportion of health care financing costs to the government. This would mean substantially more differentiation of NIH and much greater articulation of the relations between biomedical research and the provision of health services. This large policy issue, however, is not likely to be seriously addressed until passage of national health insurance legislation and not likely to be resolved sooner than five to ten years after such legislation.

TRANSITION TO A NEW ERA?

A number of these policy questions were being prominently discussed in 1973 and 1974. HEW Secretary Caspar Weinberger was considering the establishment of a special panel to review the entire NIH. In an "end game" addition to the 1974 reauthorization of the cancer act, Senator Kennedy proposed the creation of a permanent presidential panel on biomedical research analogous to the cancer panel. Because Representative Rogers opposed a permanent body, a compromise was found to the satisfaction of the White House, HEW, Kennedy, and Rogers. There was created, by Title II of Public Law 93-352, the President's Biomedical Research Panel.[30]

This panel was to consist of seven members, including the

chairman of the President's Cancer Panel. It was to function for fifteen months after its appointment and to report "simultaneously" to the president and the Congress on its findings. Its mission was to review, assess, and recommend on policy issues dealing with "the subject and content" and "the organization and operation" of biomedical and behavioral research supported by NIH and the National Institutes of Mental Health.

The panel, though authorized in July 1974, was not appointed until December of that year, and began its work with meetings on January 29 and 30, 1975. It submitted its report on April 30, 1976. In many ways, the work of the panel was very impressive. In addition to the main report, it also issued four supporting volumes of appendices, and another four volumes of supplementary materials, including written and verbal statements of witnesses heard by the panel.[31] One important appendix included reviews of eleven interdisciplinary "clusters" of scientific activity and an overview report of these reviews. Two appendices included supporting studies done for the panel under contract, while the fourth consisted of selected staff papers. The volume and richness of the material released by the panel was substantial.

The initial reception of the report by Senator Kennedy's Senate health subcommittee in mid-June 1976 made clear that there were some major issues of deep concern to the Congress that the panel had failed to resolve.[32] In particular, the mission of the NIH was the focus of sharp differences of opinion. The panel's view on this, as stated in its report, was, "The primary mission of the NIH, as constituted today, is fostering, supporting, and conducting laboratory and clinical research to increase our understanding of life processes and the etiology, treatment, and prevention of diseases."[33] The panel's report acknowledged several difficulties in fulfilling this role. One was the increasing competition for resources between the research activity of NIH and that devoted to the application and dissemination of knowledge. Another was an increasing and changing public demand for allocating resources "according to public perceptions of important health goals, rather than on the basis of scientific opportunities."

Senator Kennedy challenged this view of the NIH mission by reference to the language of the Public Health Service Act, which includes concern for the "prevention, diagnosis, and treatment" of disease in addition to "understanding."[34] The hearing produced further criticism about the allocation of resources between basic and applied research, with Kennedy and his colleagues favoring greater investment in the latter.

This opening to a planned year-long series of related hearings on biomedical research was not an auspicious indication that the biomedical research community and the elected representatives of the public were approaching a consensus on the role and mission of NIH. Rather, it indicated that the concerns that motivated Mrs. Mary Lasker and her associates in 1970 and 1971 in regard to cancer continue to trouble the entire biomedical research enterprise five years later. Indeed, Mrs. Lasker was among those supplying some of Senator Kennedy's colleagues with questions for challenging the presentation of the panel.

It would be quite inappropriate, however, to overestimate the extent to which the conflict can be personalized. Mrs. Lasker, though in her late 70s, is still very active. But with time, the Lasker forces are passing from the scene. In March 1972, barely three months after passage of the legislation in which he had such an important role, Luke Quinn died of cancer at the NIH Clinical Center. On March 30, 1973, Sidney Farber, long the doyen of the cancer research community in the United States, died of a heart attack at the age of 69.

The point of view reflected by Mrs. Lasker, which emphasizes and reemphasizes that the public funds are invested in biomedical research for the purpose of improving the health of the American people, is, if anything, more ascendant now than before. The "drift of the decade" should reveal whether medical science can deliver clinical answers on demand to the problem of cancer. If so, then the cancer crusaders will be entitled to public praise and we shall all be forever in their debt. If not, it will not be for lack of effort. We shall then be obliged to review our progress and to choose prudently the most appropriate future course.

Afterword

In mid-1977, Congress enacted and the president signed legislation that provided a simple one-year reauthorization of the National Cancer Program for fiscal year 1978. This means that reauthorization legislation will have to be enacted by May 15, 1978, for fiscal years 1979 and beyond.

It is possible that the 1978 reauthorization legislation will involve thorough hearings on the National Cancer Program. It is also possible that serious attempts will be made to amend key aspects of the law. Senators Nelson, Cranston, and Eagleton, and Representatives L. H. Fountain (D., N.C.) and David Obey (D., Wisc.) have emerged as critics of the cancer crusade on several issues. Efforts to eliminate the budgetary bypass authority, to change the function of the National Cancer Board, and even to eliminate the President's Cancer Panel could come from these sources.

On the other hand, referring to the prospective 1978 reauthorization, a senior aide to Representative Rogers observed: "Cancer is carved in stone. Nothing is going to happen this year. And there are a lot of people working to see that nothing does happen." The most likely occurrence in 1978, therefore, is that the National Cancer Program will be reauthorized for three years—fiscal years 1979, 1980, and 1981. It is also likely that no major changes will be made in the law and that hearings will be rather perfunctory.

The next reauthorization, then, will have to be enacted by May 15, 1981, for fiscal years 1982 and beyond. At that time, there will be a golden opportunity for the Congress to review the first full decade of the National Cancer Program. In the judgment of this observer, the Congress should seize this opportunity and conduct a thorough, full-scale review of the entire program and its accomplishments.

There are several reasons why such a review should be undertaken. First, in the period covering fiscal years 1972 through 1981, 7 to 8 *billion* dollars will have been appropriated to NCI for the National Cancer Program. This amount will exceed by more than three times the funds appropriated in the prior thirty-five years from fiscal year 1938 through 1971! On this basis alone the public will deserve a thorough accounting of performance.

Second, the National Cancer Program has been subjected to a constant drumfire of criticism since its inception. Scientific critics have continued to insist that the domain of ignorance is too great to warrant a program of such magnitude, that clinical advances are being promoted prematurely, and that the rest of biomedical research is suffering unfairly. Congressional critics have warned against slackening the commitment to transmitting cancer research into medical practice and against redirecting resources to basic research. Environmentalists and others have claimed that the cancer effort is misdirected toward cures rather than a prevention strategy that seeks the withdrawal of chemical carcinogens from food, the workplace, the environment, and consumer products. The public deserves to have the confusion raised by these claims and counterclaims dealt with in the most authoritative manner possible.

Finally, the rhetoric of the cancer crusaders has undergone a dramatic shift from that of 1971. The language of the cancer crusade was that of exaggeration, of hyperbole. The Citizen's Committee for the Conquest of Cancer, in 1969, petitioned the President: "MR. NIXON: YOU CAN CURE CANCER. . . . There is not a doubt in the minds of our top cancer researchers that the final answer to cancer can be found." Senator Yarborough justified the proposed Panel of Consultants on the grounds that it would be asked to recommend how to achieve cures for the major forms of cancer by 1976—the bicentennial of the Republic. House Concurrent Resolution 526 boldly declared: ". . . it is the sense of Congress that the conquest of cancer is a national crusade to be accomplished by 1976 as an appropriate commendation of the two hundredth anniversary of the independence of our country." The report of the Panel of Consultants argued that "a national program for the conquest of cancer is now essential if we are to exploit effectively the great opportunities which are presented as a result of recent advances in our knowledge." Dr. Wendell Scott, a Panel member, said there was no question that an expanded cancer effort would lead to "vast inroads . . . [on cancer] in a relatively short period of time." Dr. Sidney Farber expressed the hope that the cancer effort be organized "so that the mission can clearly be defined as the conquest of cancer . . . and the progress of that effort measured against that standard."

The claims of 1977, however, are put forward in a far more modest and subdued manner. This is nowhere more evident than in Benno Schmidt's fifth report to the president on the National Cancer Program.[1] Schmidt, who orchestrated the 1971 chorus, writes in a very restrained way as chairman of the President's Cancer Panel.

> There is no question that there has been during this period an enormous extension of our science base and our knowledge as a result of the vast amount of highly excellent fundamental basic research that has been supported. But this extension of our knowledge only underlines how vast are the areas of ignorance which remain. Just as the past five years have brought a greatly enlarged science base, they have also brought important improvements in the clinic in dealing with cancer, but here again our progress only serves to emphasize how far we have to go.

Schmidt lays great stress on NCI's support of basic research, arguing that "we cannot afford not to support basic research" for without it "there will be no payoffs." What is needed if cancer is to be eliminated as a major human disease, he writes, "is an understanding of the underlying processes of cancer at a far more profound and sophisticated level than we possess today." He continues, in words that remind one of the opponents of the 1971 Act: "For we are, in truth, profoundly ignorant about the real nature of cancer. We do not really understand what happens."

To be sure, the President's Cancer Panel chairman indicates, "attractive clues" for explaining cancer are now to be found "all over the place." Though there is a search for new ground, he writes, "we have come far enough along in the biological revolution to know, for an absolute certainty, that the problem of cancer is an approachable and soluble biological problem, even though none of us can predict when or at what cost."

At the clinical level, Schmidt finds that advances have been made even though no spectacular breakthroughs have occurred. He points to the "new era" in which the more common tumors are now treated by combination therapies. This era represents a "drastic alteration" in the practice of medicine, involving the cooperation of the several medical specialties involved in "a joint effort [that] has not been and is not today common medical practice." The cancer patient, consequently, "has a better chance today *in the hands of good cancer doctors* than has been true at any time in the past" (emphasis added).

· On the management issues, Schmidt clearly acknowledges the limits of targeted research, points to the diminishing proportion of contracts-supported research relative to grants, all but dismisses the National Cancer Plan as having any value for program administration, and underlines the importance of scientific initiative and peer review as the primary determinants of the cancer program.

The conclusion of Schmidt's report emphasizes that "the cancer program is a vast undertaking which will require long-term support and great patience."

> We are still far away from being able to put either a date or a price tag on the ultimate conquest of cancer. We are making progress in our understanding of this disease, and there is no question that the benefits of our research are increasingly available to the American people in the form of better treatment as time goes by. But it is a long road that will require patience and constancy on the part of the Congress, the Administration, and the public. In fact, at this stage of our progress, it is true in a very real sense that "the goal is the course we travel together, and the end is only the beginning."

The sober words of Schmidt contrast so sharply with the hyperbole of 1970 and 1971 that one must ask why. Were the cancer crusaders engaged in polite deception of the Congress and the public about the prospects for advances against cancer? Or were the self-appointed experts ignorant of the true difficulties involved? Whatever the answer, the hard sell of 1971 must be followed by the hard facts in 1981. The public deserves no less than a full accounting of the accomplishments of the National Cancer Program.

What should a review of the cancer crusade's first decade address? It should focus on progress in the prevention, diagnosis, and treatment of the cancer patient, and it should examine the following areas: quantitative measures of progress in the war on cancer; the cancer control program; general advances in cancer treatment; the detection and treatment of breast cancer; major developments in clinical research; progress in development of the cancer-related science base; and the carcinogenesis program.

Sorely needed for evaluation of the cancer crusade are quantitative measures of progress. One frequently used measure is the five-year survival rate, which gives the probability of an individual's being alive five years after being diagnosed as having cancer. A controversy arose in 1975 about whether these five-year survival data revealed any progress in the treatment of cancer. Greenberg, using NCI data, developed the argument that there had been little actual improvement in survival rates since the 1950s.[2] The NCI response acknowledged the correctness of the data, but argued that this was only part of the story "concerning what has happened to cancer mortality since the 1950s."[3] Recently, Enstrom and Austin have pointed out the deficiencies of survival rates as measures of progress, arguing instead that they should not be the sole or primary measures of progress, and suggesting that cancer incidence and mortality rates are far more sensitive indicators of progress.[4]

The Congressional review should aim to clarify the conceptual basis for measuring progress in the cancer crusade. A background document is needed that reviews existing cancer data over time, indicating the sources of this data, its representativeness, the means of acquisition, the biases in the reporting system (e.g., the counting of "undiagnosed lesions" as cancerous before malignancy is confirmed), and the effects of revisions of classification procedures on these data. The document should also include a discussion of the advantages and disadvantages of the several available performance measures, and of how these measures might be used together to portray accurately the true state of affairs. Then, *for each classified form of cancer*, longitudinal data should be presented on incidence, mortality, and survival rates, broken down by age, sex, race, and geographic and occupational distribution of the cancer. An interpretive statement should also accompany the data to elaborate the dynamics of change for that particular form of cancer. Throughout the document, expert controversy should be identified and elaborated, not ignored or suppressed. Finally, every effort should be made to see that the document, or a special version of it, is readable by a broad, intelligent lay audience. In this way, the public can perhaps be provided with a quantitative statement of progress in the cancer crusade.

The cancer control program should be evaluated next for its contribution to improved performance in the management of the cancer patient. Major treatment modalities should also be examined for indications of their contributions over the decade. Progress in treatment by surgery, radiotherapy, and chemotherapy should be carefully reviewed, as should progress in the use of combination therapies. Quantitative measures of performance should be developed for all means of treatment, and again expert controversy should be identified and elaborated.

Breast cancer, which accounts for nearly a quarter of all cancers in women and is the leading cause of mortality for women, should receive special attention. The controversy over the use of mammography for early detection of breast cancer should be thoroughly analyzed and the appropriate policy lessons thoughtfully developed.[5] The controversy over the appropriate mode of therapy for breast cancer—whether the radical Halstead mastectomy or modified forms of this surgical procedure—should be reviewed in light of the results of the NCI controlled clinical trial on this subject. Senator Kennedy, through hearings on breast cancer in 1976, for example, has made an important contribution to this end.[6]

Obviously not all progress in the cancer crusade will be manifest in improved patient management, even after a decade. Much will be found in the changing knowledge base of clinical research and fundamental science. Major developments in the important subfields of research should be documented for the past decade and future prospects set forth in a manner similar, perhaps, to the "cluster analyses" performed for the President's Biomedical Research Panel.[7] Management and organizational issues, including the performance of the comprehensive cancer centers, should be reviewed for their contributions to clinical and scientific progress.

A full-scale review of the carcinogenesis program is warranted.[8] This review should indicate, on a cancer-by-cancer basis, those forms of malignancy for which the evidence strongly suggests an environmental carcinogen. A distinction should be made between naturally occurring and man-made carcinogens, and, in the latter category, between those that are self-administered, like tobacco, and those that are introduced by industrial processes into our food, the workplace, the environment, or consumer products. Epidemiological data should be presented that indicate what has been learned about carcinogenesis in the past decade. Progress in, the status of, and prospects for improved testing for carcinogenicity should be analyzed, including the identification and elaboration of the key issues of expert controversy. The level, scope, and nature of NCI's carcinogenesis program should be examined in relation to the cancers of greatest threat. NCI's relations to related research agencies, like the National Institute of Environmental Health Sciences and the National Institute of Occupational Safety and Health, should be reviewed, as should its relations with the appropriate regulatory agencies—Food and Drug Administration, Environmental Protection Agency, Occupational Safety and Health Administration, and the Consumer Products Safety Commission. Beyond this, a ten-year agenda of research should be developed for the coming decade, or alternatively an explicit strategy for identifying and responding to research needs. In the last analysis, it may be that the greatest potential for the cancer crusade lies in dealing with the *chemical* nature of our industrial society in the years ahead, a potential scarcely appreciated at the passage of the National Cancer Act.

The purposes of this proposed review are not to generate new scientific and clinical information, but to synthesize existing data for understanding by the public and for policymaking by the Congress. The public would be well served by a document—or a series of documents—that explained progress in and prospects for the war against cancer. The Congress would be served by having an evidentiary basis for judging the need for more, less, or stable funding for the National Cancer Program and the need for internal reallocation of resources within the program.

The time and personnel required to undertake a review of the cancer program would not be trivial, though the costs would be trivial as a percent of the cancer program. To be effective, such a review should probably begin in mid-1978 and conclude in late 1980. It should be conducted by one or several of the appropriate Congressional committees aided by a special staff established for the review. The staff should have sufficient latitude to generate extensive analyses of the cancer program, which could then be published as staff reports. The members of Congress should use the staff work as the basis for periodic, focused hearings on particular topics. This series of reports and hearings could provide Congress the basis for the reauthorization of the cancer program as it begins its second decade. The final product made available to the public, in addition to the public record established by the review, should be a readable report that summarizes the full dimensions of the war against cancer. What is required, in short, is a thorough, balanced, analytical review and report, not an advocacy-oriented report like that of the Panel of Consultants in 1970.

It is extremely unclear whether the Congress would find any incentive to conduct such a review. There are political benefits in favoring the cancer crusade and perceived costs in criticizing it. Perhaps more importantly, cancer is so pervasive that many members of Congress have either had it themselves or have a close family member who has. Representatives Daniel Flood and Tim Lee Carter, and Senators Philip Hart (who died of cancer in 1976), Hubert Humphrey, Warren Magnuson, Birch Bayh, and Edward Kennedy are in this set. Also having direct experience are former President Gerald R. Ford and former Vice-President Nelson A. Rockefeller (through the breast cancer mastectomies of Betty Ford and Happy Rockefeller), and President Jimmy Carter. Were the members of the cancer crusade coalition to oppose an end-of-decade review, certainly they would call upon many of the above to sidetrack such an effort. The institutional stake of the American Cancer Society, for instance, appears to be in maintaining an image that "we're-always-making-progress," rather than in supporting a critical review of the National Cancer Program.

On the other hand, supporters of the type of review proposed here can perhaps draw encouragement from cancer patients themselves. It is more frequently the case, unlike the situation several decades ago, that cancer patients are told of their condition. Patients increasingly want to understand the truth about cancer for themselves. Moreover, a small but impressive literature increasingly reveals that individuals are capable of great personal strength, courage, and wisdom as they face near-certain death from this disease.[9] We can all learn from them on a personal level and perhaps draw an important public lesson from their experience as well. To wit, the American people increasingly want to understand the full meaning of the cancer problem and to have a realistic assessment of progress across the range of efforts to combat it. Furthermore, it is an affirmation of confidence in democratic government to assert that the people are fully capable of responding with collective courage and wisdom to the situation that will confront us in 1981. Although the review proposed herein will not guarantee that wise choices will be made, it should at least increase the likelihood that they will and reduce the occurrence of egregious ones. It should also confer the additional benefit that full discussion and debate provide the polity on issues of great public moment. Such benefits are to be fervently desired, for our personal and collective interests in the conduct and progress of the cancer crusade are profound.

Notes

• PREFACE •

1. There are relatively few items that pertain centrally to agenda-setting. The most prominent include: E. E. Schattschneider, *The Semi-Sovereign People*, New York, Holt, Rinehart and Winston, 1961; Charles O. Jones, *An Introduction to the Study of Public Policy*, Belmont, Calif., Wadsworth Publishing Company, Inc., 1970; Roger W. Cobb and Charles D. Elder, *Participation in American Politics: The Dynamics of Agenda-Building*, Boston, Allyn and Bacon, Inc., 1972; Jack L. Walker, "The Diffusion of Knowledge and Policy Change: Toward a Theory of Agenda Setting," Paper presented at the 1974 Annual Meeting of the American Political Science Association, Chicago, Illinois, August 29-September 2, 1974; Jack L. Walker, "Performance Gaps, Policy Research, and Political Entrepreneurs: Toward a Theory of Agenda Setting," *Policy Studies Journal*, vol. 3, No. 1 (Autumn 1974); and Roger Cobb, Jennie-Keith Ross, and Marc Howard Ross, "Agenda Building as a Comparative Political Process," *American Political Science Review*, vol. 70 (March 1976). In addition, I have had the benefit of reading an unpublished manuscript by J. Clarence Davies, III, "Setting the National Agenda," Washington, D. C., 1976.

2. See in fn. 1 above: Cobb and Elder; Cobb, Ross, and Ross; and Davies.

3. See Jones, op. cit., p. 11.

4. Cobb, Ross, and Ross, op. cit., p. 126.

5. Stephen P. Strickland, *Politics, Science and Dread Disease*, Cambridge, Massachusetts, Harvard University Press, 1972.

• 1 •

1. The report of the Panel of Consultants was initially printed in two parts in November 1970, reprinted as a single report on December 4, 1970, and released again with an additional two hundred pages of material on April 14, 1971. See U.S. Congress, Senate, Committee on Labor and Public Welfare, *National Program for the Conquest of Cancer*, Report of the National Panel of Consultants on the Conquest of Cancer, parts I and II, 91st Cong., 2nd sess., November 1970, Senate Report No. 91-1402, 91st Cong., 2nd sess., December 4, 1970, and Senate Document No. 92-9, 92nd Cong., 1st sess., April 14, 1971. Hereafter, reference will be to the Panel Report.

2. Panel Report, p. 3.

3. Ibid., p. 4.

4. Ibid., p. 5.

5. Ibid., p. 7.

6. "The War on Cancer: Progress Report," *Newsweek*, February 22, 1971.

7. See John Cairns, "The Cancer Problem," *Scientific American*, November 1975, pp. 64-78 for a useful background article on cancer. See also the "Scientific Report" of the Panel of Consultants, in Panel Report, pp. 11-150.

8. See American Cancer Society pamphlet, *'72 Cancer Facts and Figures*, for these and subsequent figures quoted here.

9. It is seldom appropriate to speak of a "cure" for cancer. The typical standard of successful treatment is the patient-survival rate for a specified number of years, usually five.

10. Panel Report, p. 49.

11. Aaron Wildavsky's *The Politics of the Budgetary Process*, Boston: Little, Brown and Company, 1964, though now somewhat dated, is still the best general introduction to the politics of the federal budget process.

12. The best single treatment of these events is found in Stephen P. Strickland, *Politics, Science and Dread Disease*, Cambridge, Massachusetts, Harvard University Press, 1972. See also chapter 2, *supra*.

13. Don K. Price, *The Scientific Estate*, Cambridge, Harvard University Press, 1965, p. 12.

14. Robert C. Wood, "Scientists and Politics: The Rise of an Apolitical Elite," pp. 41-72 in Robert Gilpin and Christopher Wright, eds., *Scientists and National Policy-Making*, New York, Columbia University Press, 1964.

15. Wallace S. Sayre, "Scientists and American Science Policy," pp. 97-112 in op. cit., Gilpin and Wright.

16. "Report of the Medical Advisory Committee," p. 54, in Vannevar Bush, *Science: The Endless Frontier, A Report to the President on a Program for Postwar Scientific Research, July 1945*, Reprinted Washington, D. C., National Science Foundation, July 1960.

17. Ibid., pp. 55-56.

18. U.S., Congress, House, *Department of Labor and Federal Security Agency Appropriations for 1953*, Hearings, 82nd Cong., 2nd sess., February 20, 1952, p. 197.

19. U.S., Congress, House, Committee on Appropriations, *Departments of Labor and Health, Education, and Welfare Appropriations for 1960*, Hearings, 86th Cong., 1st sess., April 15, 1959, p. 212.

▪ 2 ▪

1. See John Gunther, *Taken At the Flood: The Story of Albert D. Lasker*, New York, Harper & Brothers, 1960, pp. 235-243 and p. 257ff. for background on Mary Lasker.

2. Gunther's biography is the principal source of information on Albert Lasker and thoroughly describes his remarkable advertising career as well as his later years with Mary.

3. The most extensive account of Mary Lasker's involvement in medical research is in Stephen P. Strickland, *Politics, Science, and Dread Disease*, Cambridge, Mass., Harvard University Press, 1972. See also "Who Sets U.S. Research Policy?" *Medical World News*, February 3, 1961, pp. 34-36; Elizabeth Brenner Drew, "The Health Syndicate: Washington's Noble Conspirators," *The Atlantic Monthly*, December 1967, pp. 75-82; and Joseph D. Cooper, "The Medical Research Lobby: Influence for NIH Declines," *Medical Tribune*, December 23, 1968.

4. Background on the foundation is in Gunther, op. cit., pp. 325-328.

5. The account of the takeover of the American Cancer Society is described in ibid., pp. 321-325; Richard Carter, *The Gentle Legions*, Garden City, New York, Doubleday & Co., 1961, pp. 139-172; and Elmer Bobst, *Bobst: The Autobiography of a Pharmaceutical Pioneer*, New York, David McKay Co., Inc., 1973, pp. 223-232.

6. Bobst, op. cit., pp. 226-228.

7. Carter, op. cit., p. 158.

8. See Strickland, op. cit., pp. 51-53, for the account of the heart institute legislation. In 1948, with the passage of this legislation the name of the parent organization was changed to the National Institutes of Health—plural.

9. Mrs. Lasker subsequently served on a number of NIH advisory councils, as did her medical and lay friends and associates. From these positions they often sought to exercise substantial influence upon the course of medical research.

10. For Kennedy ties, see: Bess Furman, "Antiques Sought for White House," *New York Times*, February 24, 1961, and "President Names Job Rights Panel," *New York Times*, April 4, 1961. For Johnson relationship, see: Nan Robertson, "First Lady Here For a 2-Day Visit," *New York Times*, January 21, 1964; "Mrs. Johnson and Lynda Attend Dinner-Dance Given by Mrs. Lasker," *New York Times*, June 3, 1966; Stephen Conn, "Antique Cars Recreate Chaplin Era," *New York Times*, March 16, 1967; early in 1973, Mrs. Johnson's Europe travel plans included a stop at Mrs. Lasker's villa on the French Riviera; and "Lasker Research Awards Presented by Mrs. Lyndon B. Johnson," *New York Times*, November 15, 1962. For Mary Lasker's efforts at beautification, see John Gunther, op. cit., p. 298, involving the donation of thousands of chrysanthemums, tulips, and daffodils, and numerous flowering trees to New York City; Nan Robertson, "Capital to Bloom with Gift Plants," *New York Times*, March 28, 1965; "First Lady Leads Beautification Tour Through the Capital," *New York Times*, April 14, 1966; Douglas Robinson, "New Yorkers Are Asked to Help Beautify Capital," *New York Times*, November 25, 1965; and Enid Nemy, "Mary Lasker: Still Deter-

mined to Beautify the City and Nation," *New York Times*, April 28, 1974. For Lasker award to Johnson, see "Lasker Foundation Special Award to President Johnson for His Contributions to American Health," *New York Times*, April 8, 1966.

11. Walter Rugaber, "Study Find 15 Gave Over $50,000 Each to Political Parties," *New York Times*, January 31, 1971. A Common Cause study of contributors to the 1972 campaign lists Mrs. Lasker as having given $39,000 to 29 Democrats: William Claiborne, "Common Cause Lists Big '72 Donors," *Washington Post*, February 17, 1974. Senator Daniel Inouye (D., Hawaii) recalls that in his 1959 campaign for the House, he received a $5,000 check and a one-line letter wishing him "all the best" from Mary Lasker, whom he had never met: "Sen. Daniel Inouye: Watergate Changed His Life," *Parade*, November 11, 1973.

12. Mrs. Lasker was quoted in a 1967 interview as saying: "I would love to have someone write about me when I have finally given up trying to get things done. Then nothing could any longer do much harm. My life might even make quite a good musical": "The Great Persuader," *Vogue*, May 1967, pp. 194-197. Now in her seventies, Mrs. Lasker has hardly retired.

13. See Strickland, op. cit., pp. 91-108.

14. U.S. Department of Health, Education, and Welfare, *The National Institutes of Health Almanac, 1972*, Bethesda, Maryland, Office of Information, NIH, 1972, p. 98.

15. See Strickland, op. cit., pp. 109-133, Milton Viorst, "The Political Good Fortune of Medical Research," *Science*, April 17, 1964, pp. 267-270, and Drew, op. cit.

16. Mike Gorman, quoted in Strickland, op. cit., p. 145.

17. See the "Nomination Blank for the 1970 Albert Lasker Medical Research Awards" distributed by the Albert and Mary Lasker Foundation.

18. The account of the "Bo" Jones Committee is found in Strickland, op. cit., pp. 160-162.

19. U.S. Congress, Senate, Committee on Appropriations, *Labor-Health, Education, and Welfare Appropriations for 1961*, Hearings, 86th Cong., 2nd sess., May 20, 1960, p. 1463.

20. See Thomas J. Kennedy, Jr., Ronald Lamont-Havers, and John F. Sherman, "Factors Contributing to Current Distress in the Academic Community," *Science*, February 11, 1972, pp. 599-607 for a detailed analysis of how this financial crunch affected the academic biomedical community.

21. The controversial John H. Knowles, M.D., then general administrator of the Massachusetts General Hospital, was HEW Secretary Robert Finch's choice to fill the position, but this appointment was blocked by extensive AMA lobbying and the influence of Senator Everett Dirksen (R., Ill.) on the White House.

22. *The Budget of the United States Government, Fiscal Year 1970*, Washington, D. C., GPO, January 1969, was prepared by the outgoing Johnson administration. The pertinent portions are found on pp. 289-293. See U.S., Congress, House, *Reductions in 1970 Appropriations Requests: Communication from the President of the United States*, House Document No. 91-100, 91st Cong., 1st sess., April 15, 1969, and U.S., Congress, House, *Amendments to Requests for Appropriations in 1970 Budget: Communication from the President of the United States*, House Document No. 91-113, 91st Cong., 1st sess., May 5, 1969, for the overall reductions suggested by the incoming Nixon administration. See also John Walsh, "NIH: Another Tight Budget, Fewer Friends in High Places," *Science*, April 11, 1969, pp. 165-167.

23. Sandra Blakeslee, "Cancer Experts Ask Research Funds," *New York Times*, March 29, 1969.

24. See U.S., Congress, House, Committee on Appropriations, *Departments of Labor, and Health, Education, and Welfare, and Related Agencies Appropriation Bill, 1970*, Hearings, pt. 7, 91st Cong., 1st sess., May 21, 1969, pp. 549-555, 560-564.

25. U.S., Congress, House, Committee on Appropriations, *Departments of Labor, and Health, Education, and Welfare, and Related Agencies Appropriation Bill, 1970*, House Report No. 91-391, 91st Cong., 1st sess., pp. 16-17, 51.

26. See John Walsh, "NIH: Agency and Clients react to Retrenchment," *Science*, September 26, 1969, pp. 1332-1334.

27. Harold M. Schmeck, Jr., "Nineteen Clinical Units Facing Shutdown," *New York Times*, September 9, 1969.

28. Schmeck, "Cancer Institute to Kill Monkeys," *New York Times*, September 15, 1969. This was a page-one story.

29. See, for example, Schmeck, "Inflation and Budget Cuts Cause Alarm Among Scientists Seeking Research Funds," *New York Times*, October 5, 1969; and Victor Cohn, "U.S. Scientists Fear Further Cuts in Aid," *Washington Post*, October 13, 1969.

30. See Schmeck, "Scientists Warn of Health Crisis," *New York Times*, November 12, 1969. The news conference was sponsored by the Ad Hoc Committee on the Nation's Health Crisis, a short-lived organization supported by the American Heart Association, American Cancer Society, National Committee Against Mental Illness, Association of American Medical Colleges, American Dental Association, and American Nurse Association.

31. "Magnuson to Fight Research Cutback." *New York Times*, November 14, 1969.

32. Personal interview with Edwin A. Mirand, M.D., AACI secretary-treasurer.

33. U.S., Congress, Senate, Committee on Appropriations, *Departments of Labor, and Health, Education, and Welfare, and Related Agen-*

cies Appropriations Bill, 1970, Report No. 91-610, 91st Cong., 1st sess., December 16, 1970, pp. 32-33, 101.

34. U.S., Congress, House, Committee of Conference, *Appropriations for the Departments of Labor, and Health, Education, and Welfare, and Related Agencies, 1970: Conference Report*, Report No. 91-781, 91st Cong., 1st sess., December 20, 1969, p. 9.

35. Nineteen sixty-nine witnessed an unusual, serious, and sustained congressional attack upon the Department of Defense budget. Inefficiency in defense procurement constituted one issue. Equally compelling to many, however, was the oft-voiced refrain of the need to reorder national priorities. See U.S., Congress, Joint Economic Committee, *The Military Budget and National Economic Priorities*, Hearings, pts. 1, 2, and 3, 91st Cong., 1st sess., 1969.

36. See U.S., Congress, House, *Veto Warning on Labor-HEW-OEO Appropriations; Communication from the President of the United States*, Document No. 91-206, 91st Cong., 1st sess., December 19, 1969.

37. Republican senators and congressmen received a letter from Bryce N. Harlow, counselor to the president, on January 6, 1970, reiterating the president's veto intention. See "HEW-Labor-OEO Appropriations Bill," *Weekly Compilation of Presidential Documents*, vol. 6, No. 2 (January 12, 1970), pp. 33-34.

38. Nixon, "HEW-Labor-OEO Appropriations Bill: The President's Veto Message to the House of Representatives," January 26, 1970, in *Weekly Compilation of Presidential Documents*, vol. 6, No. 5 (February 2, 1970), pp. 78-84.

39. See Public Law 91-204, March 5, 1970, in U.S. *Statutes-at-Large*, 84, pt. 1, 23.

40. Nixon, op. cit., pp. 82-83.

41. Thomas J. Kennedy, Jr., Ronald Lamont-Havers, and John F. Sherman, "Factors Contributing to Current Distress in the Academic Community," *Science*, February 11, 1972, pp. 599-607.

42. *The Budget of the United States Government, Fiscal Year 1971*, Washington, D. C., GPO, January 1970, p. 317.

43. U.S., Congress, House, Committee on Appropriations, *Departments of Labor, and Health, Education, and Welfare, and Related Agencies Appropriation Bill, 1971*, Report No. 91-1310, 91st Cong., 2nd sess., July 16, 1970, pp. 17-18, 53. Luke Quinn, working closely with his old friend, Rep. Dan Flood, was mainly responsible for the $25 million addition by the House.

44. U.S., Congress, Senate, Committee on Appropriations, *Departments of Labor, and Health, Education, and Welfare, and Related Agencies Appropriation Bill, 1971*, Report No. 91-1335, 91st Cong., 2nd sess., October 13, 1970, pp. 33-35, 89.

45. See Public Law 91-667, January 11, 1971, in U.S. *Statutes-at-Large*, 84, pt. 2, 2001.

46. See Strickland, op. cit., pp. 55-74.

47. Ibid., p. 74. See American Medical Association, *Report of the Commission on Research*, Chicago, Illinois, February 1967, for a generally favorable assessment of federally sponsored medical research.

48. The origins of the commission lie in the national committee on heart disease and cancer appointed by President Kennedy in 1961. The report of this early group was reportedly delivered to the White House in April 1961 on the day when U.S.-supported Cuban exiles attempted their ill-fated invasion of Cuba. This coincidence, plus the poor quality of the report, led people to refer to it as "The Bay of Pigs Report." The document was apparently lost and the entire effort remains shrouded in obscurity. When the president's father, Joseph Kennedy, suffered a disabling stroke, the idea for the commission was revived and extended to include that disease. See Elinor Langer, "Presidential Medicine: Johnson Panel, Lay and Medical, To Study Heart Disease, Cancer, and Strokes," *Science*, March 20, 1964, pp. 1308-1309. See David Price, *Who Makes the Laws*, Cambridge, Mass., Schenkman, 1972, for a discussion of the legislative history of the "Heart Disease, Cancer, and Stroke Amendments of 1965," especially pp. 216-227.

49. Lyndon B. Johnson, "Special Message to the Congress on the Nation's Health," February 10, 1964, U. S. President, *Public Papers of the President: Lyndon B. Johnson, 1963-64*, vol. I, Washington, D. C., GPO, 1965, p. 282.

50. See the President's Commission on Heart Disease, Cancer and Stroke, *Report to the President: A National Program to Conquer Heart Disease, Cancer and Stroke*, vol. I, Washington, D. C., GPO, December 1964, pp. iii, 84-86.

51. Ibid., pp. 28-34. Also recommended were grants for the development of "medical complexes" to enable medical schools, hospitals, and research agencies "to work in concert." Grants were recommended to increase the number of "excellent" medical schools. Other recommendations dealt with the development of new scientific knowledge and its application in the community, health manpower, and health facilities.

52. Ibid., p. xi.

53. R. Lee Clark, M.D., "M. D. Anderson's Legacy to Texas," *Texas Medicine*, vol. 64 (October 1968), p. 85. See *The First Twenty Years of the University of Texas M. D. Anderson Hospital and Tumor Institute*, Houston, Texas, The University of Texas, M. D. Anderson Hospital and Tumor Institute, 1964, for a detailed account of the development of this impressive institution.

54. The President's Commission, op. cit., vol. II, February 1965, p. 107. See pp. 105-117 for the cancer subcommittee report.

55. Clark, op. cit., p. 85. Memorial Sloan-Kettering Cancer Center is also an autonomous corporate entity, though it has complex legal and institutional relations with the Cornell University Medical School. Roswell

Park Memorial Institute is part of the New York State Department of Health, and thus has organizational autonomy from any medical school.

56. See The President's Commission, op. cit., vol. I; Lyndon B. Johnson, "Remarks Upon Receiving Report of the President's Commission on Heart Disease, Cancer, and Stroke," December 9, 1964, *Public Papers*, op. cit., vol. II, pp. 1650-1651; and Elinor Langer, "U.S. Medicine: LBJ Commission on Heart Disease, Cancer and Stroke Offers Sweeping Recommendations," *Science*, December 25, 1964, pp. 1662-1664.

57. Lyndon B. Johnson, "Special Message to the Congress: 'Advancing the Nation's Health,' " January 7, 1965, in U.S. President, *Public Papers of the President: Lyndon B. Johnson, 1965*, vol. I, Washington, D. C., GPO, 1966, pp. 16-17. Drew, "The Health Syndicate,"pp. 79-80, quotes a person who worked on the drafting of the legislation as saying, "in all my experience I never saw a piece of legislation leave the White House on which there was less clarity on what the federal government was going to do."

58. Elinor Langer, "Heart, Cancer, and Stroke: Bill Based on Presidential Commission Calls for Regional Medical Centers," *Science*, May 14, 1965, pp. 930-933.

59. U.S., Congress, Senate, Committee on Labor and Public Welfare, *Combating Heart Disease, Cancer, Stroke, and Other Major Diseases*, Hearings, 89th Cong., 1st sess., February 9 and 10, 1965.

60. Langer, "Heart, Cancer, Stroke: Rising Opposition From Doctors May Slow Passage of Johnson Program," *Science*, August 20, 1965, pp. 843-845.

61. Langer in *Science*, May 14, 1965, p. 932.

62. *Langer*, "New Health Act: AMA Criticism Reflected in Adoption of Bill on Heart, Cancer, and Stroke," *Science*, October 15, 1965, pp. 323-324. The bill was passed as Public Law 89-239, "Heart Disease, Cancer, and Stroke Amendments of 1965," 89th Cong., October 6, 1965; see U.S., *Statutes-at-Large*, 79, 926-931.

63. In Houston, for example, the local RMP funds were controlled by DeBakey, which meant that Baylor University School of Medicine, of which he was president, benefited substantially. M. D. Anderson Hospital, a next-door neighbor in the Texas Medical Center, received very little benefit from the program.

▪ 3 ▪

1. See "Bureau of Health," Message from the President of the United States, in U.S., Congress, House, Committee on Interstate and Foreign Commerce, *Health Activities of the General Government*, Hearings, pt. I, 61st Cong., 2nd sess., June 2, 1910, pp. 3-4.

2. Press release, Roswell Park Memorial Institute, Buffalo, New York, 1972.

3. The memorandum can be found in U.S., Congress, House, *Cancer in Fishes*, Message from the President of the United States, House Document No. 848, 61st Cong., 2nd sess., April 9, 1910, pp. 3-4.

4. Ibid., pp. 1-3.

5. Roswell Park Memorial Institute, op. cit.

6. U.S., Congress, House, Committee on Interstate and Foreign Commerce, *Health Activities of the General Government*, Hearings, pts. I-XI, 61st Cong., 2nd sess., June 2, 3, 4, 6, and 7, 1910, and 62nd Cong., 1st sess., January 18 and 19, 1911.

7. See Michael B. Shimkin, "Research Activities of the National Cancer Institute," *Journal of the National Cancer Institute* (hereafter *JNCI*), vol. 5, (October 1944), pp. 77-80, and H. B. Andervont, "J. W. Schereschewsky: An Appreciation," *JNCI*, vol. 19, No. 2 (August 1957), pp. 331-333.

8. *Congressional Record*, 69th Cong., 2nd sess., 1927, 68, pt. 3, 2922. For a more complete account of this episode, see Strickland, *Politics, Science and Dread Disease*, pp. 1-8.

9. *Congressional Record*, 70th Cong., 1st sess., 1929, 69, pt. 8, 9050.

10. See Strickland, op. cit., pp. 11-14; see also, J. R. Heller, "The National Cancer Institute: A Twenty-Year Retrospect," *JNCI*, vol. 19 (August 1957), pp. 147-151.

11. See Irwin Stewart, *Organizing Scientific Research for War*, Boston, Little, Brown and Company, 1948, pp. 313-319, for the account of this transfer.

12. *Supra*, chapter 2, p. 23.

13. See the following for historical background: Carl Voegtlin and R. R. Spencer, "The Federal Cancer Control Program," *JNCI*, vol. 1 (August 1940), pp. 1-9; Ora Marashino, "Administration of the National Cancer Institute Act, August 5, 1937, to June 30, 1943," *JNCI*, vol. 4 (April 1944), pp. 429-443; Shimkin, op. cit., pp. 77-88; and Heller, op. cit., pp. 147-190.

14. The original six council members were: Dr. James Ewing, Memorial Hospital for Cancer and Allied Diseases, New York City; Dr. Frances Carter Wood, Columbia University; Dr. C. C. Little, American Society for the Control of Cancer; Dr. Ludvig Hektoen, McCormick Institute for the Study of Infectious Diseases, Chicago; Dr. James B. Conant, Harvard University; and Dr. Arthur H. Compton, University of Chicago. Hektoen served as the first executive secretary. See Marashino, op. cit., pp. 429-430, and Heller, op. cit., pp. 151-152, 163.

15. See "Fundamental Cancer Research," *Public Health Reports*, vol. 53 (December 2, 1938), pp. 2121-2130, reprinted in *JNCI*, vol. 19 (August 1957), pp. 317-325.

16. Personal interview with G. Burroughs Mider.

17. Shimkin, op. cit., p. 80.

18. Heller, op. cit., pp. 178, 181.
19. Mider, "Research at the National Cancer Institute," *JNCI*, vol. 19 (August 1957), p. 192.
20. See Marashino, op. cit., p. 441, Shimkin, op. cit., p. 88, and Heller, op. cit., p. 157.
21. Heller, op. cit., pp. 167, 178, 181.
22. Ibid., pp. 153, 154. See also Marashino, op. cit., pp. 431-432.
23. Heller, op. cit., pp. 170-171, 174.
24. Ralph G. Meader, O. Malcom Ray, and Donald T. Chalkey, "The Research Grants Branch of the National Cancer Institute," *JNCI*, vol. 19 (August 1957), pp. 241-243.
25. See 42 U.S.C. § 281-290 (1970 ed.). The NCI received the largest annual appropriation of any NIH institute until fiscal 1964, when it was exceeded by the National Institute of Mental Health (NIMH). NIMH was made a separate bureau of the Public Health Service in 1967, and NCI resumed first place in fiscal 1968. See U.S. DHEW, *NIH Almanac 1972*, Bethesda, Maryland, 1972, pp. 99-100.
26. Heller, op. cit., pp. 157-158, 161-162. See also James W. Hawkins, "Organizational Structure and Activities of State Cancer Programs," *JNCI*, vol. 4 (February 1944), pp. 347-350, and Raymond F. Kaiser and Rosalie I. Peterson, "Activities of the Field Investigations and Demonstrations Branch," *JNCI*, vol. 19 (August 1957), pp. 259-260.
27. See "Cancer Facilities and Services: A Report from the National Advisory Cancer Council," *JNCI*, vol. 6 (April 1946), pp. 239-302.
28. Heller, op. cit., pp. 169-174; Kaiser and Peterson, op. cit., p. 259.
29. Personal interview with James A. Shannon.
30. This account is based upon interviews with several former NIH officials.
31. See C. J. Van Slyke, "New Horizons in Medical Research," *Science*, December 13, 1946, pp. 559-567; Meader et al., op. cit., p. 229.
32. Personal interview.
33. NACC, "Cancer in the Medical School Curriculum," *JNCI*, vol. 8 (August 1947), pp. 1-6. This report was stimulated by the 1946 NACC report, "Cancer Facilities and Services," especially pp. 239-242.
34. "Cancer in the Medical School Curriculum," p. 6, and Heller, op. cit., pp. 165, 169.
35. Personal interview with G. Burroughs Mider.
36. See, for instance, Donald R. Green and Leonard W. Towner, "The Relation of Medical School Teaching Methods to Student Performance on a Test of Cancer Knowledge," *JNCI*, vol. 28 (October 1959), pp. 605-616.
37. See E. C. Andrus et al., *Advances in Military Medicine*, vol. II, Boston, Little, Brown and Company, 1948, pp. 717-745, for an account

of the development of penicillin, and pp. 665-716, for the antimalaria story.

38. See C. G. Zubrod, S. Schepartz, "The Chemotherapy Program of the National Cancer Institute: History, Analysis, and Plans," *Cancer Chemotherapy Reports*, vol. 50 (October 1966), pp. 349-352.

39. Ibid., p. 353.

40. U.S., Congress, Senate, Committee on Appropriations, *Labor—Health, Education, and Welfare–Appropriations for 1954*, Hearings, 83rd Cong., 1st sess., June 3, 1953, p. 1387.

41. U.S. Senate, Committee on Appropriations, *Departments of Labor and Health, Education, and Welfare, and Related Independent Agencies Appropriation Bill, 1954*, Report No. 478, 83rd Cong., 1st sess., June 29, 1953, p. 15. The Senate added $1 million to the fiscal 1954 budget for leukemia research.

42. See Andrus et al., op. cit., pp. 698-716, for Shannon's account of this experience.

43. Zubrod and Schepartz, op. cit., p. 354; Kenneth M. Endicott, "The Chemotherapy Program," *JNCI*, vol. 19 (August 1957), pp. 275-276.

44. See U.S., Congress, Senate, Committee on Appropriations, *Departments of Labor and Health, Education, and Welfare, and Related Independent Agencies Appropriation Bill, 1955*, Report No. 1623, 83rd Cong., 2nd sess., June 22, 1954, p. 8. The report read: "It would appear that ultimate success in this endeavor will require active collaboration among the essential disciplines in the overall effort and [the committee] instructs the Institute to maintain mechanisms which will facilitate such collaboration as well as free interchange of information among the cooperating scientists."

45. These figures do include funds for chemotherapy contracts and grants and were obtained from the table on p. 369 of Zubrod and Schepartz, op. cit.

46. See ibid., pp. 353-356, and Endicott, op. cit., pp. 275-276. See also U.S., Congress, Senate, Committee on Appropriations, *Departments of Labor and Health, Education, and Welfare, and Related Agencies Appropriation Bill, 1956*, Report No. 410, 84th Cong., 1st sess., June 2, 1955, p. 19. Senate "pressure" reflected Sidney Farber's influence.

47. Zubrod and Schepartz, op. cit., pp. 356-360, and Endicott, op. cit., pp. 276-277.

48. Contract authority, though initially provided in fiscal 1955 appropriations, represented a new problem for NCI and NIH, and it took some time for the Department of Health, Education, and Welfare to work out a reasonable contract administration policy. Cost reimbursement was one thorny problem, since the original contract authority provided only for fixed-price contracts. Several years elapsed before cost reimbursement authority was provided to NCI. Patent policy was another difficult

problem that took time to resolve. See Endicott, op. cit., p. 277, and Zubrod and Schepartz, op. cit., pp. 357, 360-61, 367.

49. See, for example, the discussion by Endicott, op. cit., pp. 278-285. The chemotherapy program also included a related effort to understand the role of endocrine factors in cancer and to identify endocrine agents useful in treating cancer.

50. See ibid., pp. 279-282, for a discussion of this problem. See also Stuart M. Sessoms, "Review of the Cancer Chemotherapy National Service Center Program," *Cancer Chemotherapy Reports*, No. 7 (May, 1960).

51. A. Gellhorn and E. Hirschberg, eds., "Investigation of Diverse Systems for Cancer Chemotherapy Screening," *Cancer Research*, Supplement No. 3 (1955), p. 13.

52. Substantial criticism from the scientific community preceded this modification of the screening effort. See especially the severe critique by Alfred Gellhorn before the NACC on June 15, 1959: "Invited Remarks on the Current Status of Research in Clinical Cancer Chemotherapy," *Cancer Chemotherapy Reports*, No. 5 (1959), pp. 1-12. A response by I. S. Ravdin, a supporter of the program and a close Lasker ally, was made before the NACC on November 23, 1959 (see ibid., pp. 13-17). Ravdin, a surgeon, was one of the leaders in the development of adjuvant chemotherapy. Zubrod and Schepartz, op. cit., pp. 371-374, contains the best account of this controversy. The official NCI response is found in Sessoms, op. cit.

53. Sessoms, op. cit., p. 33.

54. Telephone interview with Howard E. Skipper.

55. Kenneth M. Endicott, "Progress Report," Cancer Chemotherapy National Service Center, December 1, 1957, p. 10.

56. Zubrod and Schepartz, op. cit., p. 371.

57. U.S., Congress, Senate, Committee on Appropriations, *Departments of Labor and Health, Education, and Welfare, and Related Agencies Appropriation Bill, 1961*, Report No. 1576, 86th Cong., 2nd sess., June 14, 1960, p. 24.

58. Telephone interviews of Gellhorn and Skipper.

59. Some NCI staff were quite critical of the directed chemotherapy program, partly because one of earliest and most successful developments of chemotherapy was pioneered by an intramural NCI scientist working prior to and independent of the CCNSC. Roy Hertz and his colleagues demonstrated the dramatic curative effects from using methotrexate in cases of disseminated choriocarcinoma. The initial results were reported in 1956. See the letter from Michael B. Shimkin, *Science*, July 12, 1974, p. 99. I am indebted to G. Burroughs Mider for drawing this to my attention.

60. *Supra*, this chapter, note 64.

61. See especially p. 12 of U.S. DHEW, *Report of the Secretary's*

Advisory Committee on the Management of National Institutes of Health Research Contracts and Grants, Washington, D. C., March 1966 (hereafter referred to as the Ruina report).

62. Ibid., p. 20.

63. Ibid.

64. Ibid., pp. 20-21.

65. The source of the complaints was Mary Lasker and Sidney Farber, each of whom had been reappointed to the NACC in 1962. Former NCI staff, however, remember supplying Mrs. Lasker with copious amounts of information. Farber had previously served on the council from 1953-1957, Lasker from 1954-1958.

66. *Congressional Record*, 88th Congress, 2nd session, 1964, 110, part 16, 21575.

67. The Ruina report, p. 20.

68. See *Biomedical Science and its Administration: A Study of the National Institutes of Health*, The White House, Washington, D. C., February 1965, pp. 37, 39, 40.

69. Quoted in the Ruina report, pp. 22, 23.

70. See U.S., Congress, Senate, Committee on Appropriations, *Supplemental Appropriation Bill, 1966*, Report No. 912, 89th Cong., 1st sess., October 19, 1965, pp. 32-33.

71. Quoted in the Ruina report, p. 1.

72. See U.S., Congress, House, *Supplemental Appropriations, 1966*, Conference Report, Report No. 1198, 89th Cong., 1st sess., October 21, 1965, pp. 7-8.

73. See Ruina report, op. cit., pp. 3-5.

74. Personal interview.

75. Ibid.

76. Ibid.

77. See U.S. DHEW, *Special Virus Cancer Program, Progress Report #6*, National Cancer Institute, July 1969, for background on the program.

78. U.S. DHEW, *National Cancer Institute: 1972 Fact Book*, January 1972, p. 30.

79. See Ruina report, pp. 5, 21.

80. Personal interview.

81. See Louis M. Carrese and Carl G. Baker, "The Convergence Technique: A Method for the Planning and Programming of Research Efforts," *Management Science*, vol. 13 (April 1967), pp. B-420-438.

82. See *Special Virus Cancer Program, Progress Report #6*, op. cit., pp. 2-3.

83. Zubrod and Schepartz, op. cit., pp. 480-539.

84. This account is based upon interviews with several former NIH and NCI officials.

85. The Ruina report, pp. 26-30, gave substantial attention to the

problem of recruiting competent program managers for programs like the
NCI's directed-research efforts, but practically no attention to the incentive system where dual responsibility existed for conduct of intramural
research and management of contract programs.

▪ 4 ▪

1. Garb, *Cure for Cancer*, New York, Springer Publishing Company,
Inc., p. 3.
2. Ibid., p. 303; see pp. 6-18 for Garb's discussion of the space program.
3. Strickland, op. cit., pp. 152-153. Quinn's other clients included the
National Cystic Fibrosis Research Foundation, the National Committee
for Research in Neurological Disorders, Research to Prevent Blindness,
the National Multiple Sclerosis Society, the United Cerebral Palsy Association, and the Neuro Research Foundation.
4. *Times*, p. 61. Sympathetic readers noted an invitation: "This advertisement may be reproduced in whole or in part without credit or permis-
sion. Run it in your newspaper." The advertisement indicated Mary
Lasker's resourcefulness in promoting cancer research, but it also revealed that she no longer had direct access to the White House—a serious
political liability.
5. Morse, in fact, wrote a letter (see *Science*, October 11, 1968, p.
214) in which he pledged to the scientific community that after reelection
"I shall do my best to obtain the necessary funds for research."
6. Gorman became executive director of the National Committee
Against Mental Illness in 1949, an organization established by Mrs.
Lasker and Mrs. Florence Mahoney; see Drew, op. cit., and Strickland,
op. cit., pp. 137-138. Gorman and Quinn constituted Mary Lasker's
full-time Washington representatives, even though it was widely known
in Washington that the two men despised each other. In both cases, the
scope of their responsibilities often exceeded the areas of cancer and
mental health research. For instance, as a consultant Gorman had drafted
large portions of the report of the President's Commission on Heart Disease, Cancer, and Stroke. Quinn, more recently, had been instrumental
in creating the National Eye Institute within NIH.
7. The account that follows is based upon personal interviews and
telephone conversations with former Senator Ralph Yarborough, Mrs.
Mary Lasker, and several former Senate staff members.
8. Yarborough, "Senate Resolution 376—Submission of a Resolution
Relating to Cancer Research," *Congressional Record*, 91st Cong., 2nd
sess., 1970, 116, part F, 9260-9262.
9. See the press releases issued by the office of Senator Yarborough of
April 15, 23, and 27, 1970.
10. U.S., Congress, House, 91st Cong., 2nd sess., 1970, H. Con.

Res. 526. Rooney introduced a substitute resolution in July (H.Con. Res. 675) that deleted "massive" from the earlier one.

11. Rooney had been successfully operated on for lung cancer in 1966, which had much to do with his personal interest in the resolution. See *Congressional Record*, 92nd Cong., 1st sess., 1971, 117, pt. 31, 41153.

12. Yarborough, "Cancer Cure Must Be a National Goal," *Congressional Record*, 91st Cong., 2nd sess., 1970, 116, pt. 22, 30307-30308. See also Senator Alan Cranston, "Conquest of Cancer—A National Goal," *Congressional Record*, 91st Cong., 2nd sess., 1970, 116, pt. 23, 30656-30657.

13. The director of Roswell Park, Dr. James T. Grace, had been in an automobile accident on March 8, 1970, which killed his wife and left him in a coma until his death on August 13, 1971. Were it not for this tragedy, Grace would have been a member of the panel.

14. Lederberg had testified in early 1970 before the House appropriations committee on behalf of the Cystic Fibrosis Foundation, an organization represented by Luke Quinn, and was so effective that an additional $10 million above the administration's budget request was added to genetics research. See U.S., Congress, House, Committee on Appropriations, *Departments of Labor, and Health, Education, and Welfare, and Related Agencies Appropriation Bill, 1971*, H.R. No. 91-1310, 91st Cong., 2nd sess., July 16, 1970, pp. 26-27.

15. Abel indicated in advance that he would not attend panel meetings but would do anything the senator asked of him.

16. Mrs. Lawrence, with a salary in excess of $400,000 a year, is reported to be one of the highest paid women in the world. See Judy Klemsrud, "On Madison Avenue, Women Take Stand in Middle of the Road," *New York Times*, July 3, 1973; and Marilyn Bender, "America's Corporate Sweethearts," *New York Times*, January 20, 1974.

17. O'Neill, a reporter for *Medical World News* a decade earlier, was familiar with many of the issues confronting medical research.

18. See Bobst, op. cit., pp. 267-277, 317-318, 323-331 for the autobiographical account of Bobst's friendship with Nixon. See also Dom Bonafede, "President's Inner Circle of Friends Serves as Influential 'Kitchen Cabinet,' " *National Journal*, January 22, 1972, pp. 126-135. Bobst was instrumental in bringing Nixon to the New York law firm of Mudge, Rose, Guthrie, and Alexander after the Nixon's defeat in the California gubernatorial election of 1962. Bobst also established the $25,000 trust fund for Tricia Nixon, which was later used by the president to purchase real estate in Key Biscayne. The president and his wife hosted Bobst's 85th birthday party at the White House on December 16, 1969.

19. The following account is based largely upon two interviews with Benno C. Schmidt.

20. Memorial Sloan-Kettering Cancer Center is the parent corporation for two other corporations, Memorial Hospital for Cancer and Allied Diseases—the treatment arm of the center—and Sloan-Kettering Cancer Research Institute—the research arm. Each organization has its own board of trustees and while overlapping memberships are frequent, the three boards are not identical. In addition, the Sloan-Kettering Cancer Research Institute is one of the graduate schools of the Cornell University School of Medicine and staff at the hospital and the institute are all on the Cornell faculty.

21. See T. A. Wise, "Jock Whitney: Unclassified Capitalist," *Fortune*, October 1964, pp. 116-119ff. for background on J. H. Whitney & Co.

22. Schmidt has also been a director of Marine Colloids, Inc., Memorex Corporation, Global Marine, Inc., Transcontinental Gas Pipe Line Corporation, and Freeport Minerals.

23. Personal interview with Benno C. Schmidt.

24. Kennedy was responsible for the establishment of two nonprofit corporations—the Bedford-Stuyvesant Development and Services Corporation (D&S), and the Bedford-Stuyvesant Reconstruction Corporation (Reconstruction). The D&S board is drawn from New York City's business community and exists to attract private financial investment to the Bedford-Stuyvesant area and to provide technical assistance to Reconstruction. Reconstruction, governed by a Black board of trustees, exists to formulate projects for the community.

25. See Kathleen Teltsch, "Ford Fund to Give City $5-Million," *New York Times*, May 8, 1968; and "Fund for the City," editorial, *New York Times*, May 13, 1968.

26. See Seth S. King, "Mayor Discloses Welfare Is. Plan," *New York Times*, February 13, 1969; and Ada Louise Huxtable, "A Plan for Welfare Island is Unveiled," *New York Times*, October 10, 1969.

27. Telephone conversation with Dr. Solomon Garb.

28. Much of the account that follows is based upon an extended personal interview with Sweek and on materials from his files.

29. Yarborough, in announcing Sweek's appointment, drew attention to his "impressive record of success in managing and directing large scale, multi-disciplined research and development efforts in defense, space, and nuclear programs, both in government and in industry, and is ideally suited to undertake this demanding assignment." The Senator referred to Sweek's deputy and long-time personal friend, Carl Fixman, as having "an imposing record of achievement in both government and industry in large scale research and development programs." See "Yarborough Appoints Advisory Committee on Cancer Study," press release from the office of Senator Yarborough, June 12, 1970. These staff appointments of men with no prior biomedical research experience reinforced the fear of many observers that the panel would recommend a

crash program for cancer analogous to the Manhattan Project or the Apollo Project, an anxiety that was never fully laid to rest.

30. Personal interview.

31. Schmidt had met with Farber the previous Friday to iron out details on the agenda.

32. Subpanel 1, "Where do we stand?" was to be chaired by Dr. Jonathan Rhoads. Dr. Joseph Burchenal was named chairman of subpanel 2, "Where is good work going on and who is doing it?" Farber was to be chairman of subpanel 3, "Delineation of areas of greatest promise." Subpanel 4, "Mechanics for coordination, and promulgation of information," was to be headed by Dr. Lee Clark.

33. The Rhoads subpanel met on at least two occasions after the first meeting, once in late July and again in late August. The group working on "areas of greatest promise" met in mid-July and early August, Dr. Clark's subpanel met twice in mid-July and then again in early September.

34. He also asked Dr. James Holland to take charge of subpanel 3, with Dr. Mathilde Krim as co-chairman.

35. Quinn, with more than two decades of experience on Capitol Hill, was skeptical of Schmidt's ability to deal with the Congress. This luncheon, however, impressed even the veteran lobbyist.

36. See Jerry E. Bishop, "National Goal: Curing Cancer by 1976," *Wall Street Journal*, August 26, 1970, and Jonathan Spivak, "Cancer Panel to Seek More U.S. Spending and New Agency to Spur Search for Cure," *Wall Street Journal*, November 23, 1970.

37. Independent agencies are to be distinguished from cabinet-level departments: they are headed by individuals of lesser official status than cabinet secretaries, and they tend to have narrower purposes and more homogeneous functions than do departments. A combination of political and administrative reasons are usually responsible for their autonomous status. Beyond the agency leadership, policies and programs are subject to review only by the White House, the Domestic Council, and the Office of Management and Budget (OMB), and budgets are submitted directly to OMB for review. In short, there is substantially more administrative freedom in such an agency than there is for subordinate units of a cabinet-level department.

38. The report recommended that the Authority be headed by an administrator, appointed by the president with the advice and consent of the Senate, who would "report directly to the President and present his budgets and programs to the Congress." U.S., Congress, Senate, Committee on Labor and Public Welfare, *National Program for the Conquest of Cancer*, Report of the National Panel of Consultants on the Conquest of Cancer, Report No. 91-1402, 91st Cong., 2nd sess., December 4, 1970, p. 4.

39. *Cure for Cancer*, pp. 2 and 3.

40. Telephone conversation with Yarborough.

41. Yarborough, "The Need for an Investigation of the Causes and Cures of Cancer," *Congressional Record*, 91st Cong., 2nd sess., 1970, 116, pt. 7, 9260-9262.

42. The account that follows is based upon extensive interviews and examination of personal files of a number of participants.

43. In late August, Sweek brought to the chairman's attention the difficulty that celebrity Danny Thomas, a member of the National Advisory Cancer Council, was having in getting a film about cancer approved by a minor clerk in the HEW visual aids office.

44. Reports were circulating by mid-August, however, that making cancer a national goal might also mean "splitting the Cancer Institute away from NIH." See *The Blue Sheet*, August 19, 1970.

45. The five were Lee Clark and Solomon Garb, panel members; Bob Sweek and Carl Fixman, staff; and Luke Quinn. Garb wryly observed that the only ones favoring the idea were those who had had experience running large-scale research-and-development efforts.

46. *The Blue Sheet*, December 9, 1970. *The Blue Sheet*, a publication of *Drug Research Reports*, is a widely read weekly newsletter in health and medicine that covers the Washington scene.

47. Personal interview with Carl G. Baker.

48. *The Blue Sheet*, July 15, 1970, p. 3, and August 19, 1970, p. 3.

49. Pp. 13, 14. The article stated: "The future of all biomedical research—a subject worthy of a major natl. [sic] debate, at least among all scientists—is perhaps being decided, or at least prejudiced, by a congressional commission that is operating in a framework of 'closed-door' meetings, 'secret hearings' such as those held August 24-25, and 'confidential' questionnaires circulated among well-meaning people who submit answers without a full realization of all that might be at stake" (p. 14).

50. Robert J. Bazell, "Cancer Research: Senate Consultants Likely to Push for Planned Assault," *Science*, October 16, 1970, p. 304. *Science*, a weekly scientific journal with news coverage of science policy issues, has a readership of over 100,000 and thus reaches a broader cross-section of scientists and engineers than any other scientific journal in the country.

51. The proposed "National Advisory Cancer Board" would be to the "National Cancer Authority" what the existing NACC was to the NCI.

52. Yarborough, "Introductory remarks upon the presentation of the report of the National Panel of Consultants on the Conquest of Cancer to the Labor and Public Welfare Committee," December 4, 1970, press release.

53. Memorandum from Alan C. Davis, American Cancer Society, to Lane W. Adams, executive vice-president, ACS., December 7, 1970, "Report on the Yarborough Commission Hearings Before the Senate

Committee on Labor and Public Welfare, Subcommittee on Health, December 4, 1970."

54. Richard Witkin, "Senate Rejects SST Fund in 52-41 Vote After Drive by Environmental Lobby," *New York Times*, December 4, 1970.

55. The report of the Panel of Consultants was initially printed in two parts in November 1970, reprinted as a single report on December 4, 1970, and released again with an additional two hundred pages of material on April 14, 1971. See the Panel Report (cf. n. 38), and Senate Document No. 92-9, 92nd Cong., 1st sess., April 14, 1971.

56. Panel Report, p. 3.

57. Ibid., p. 4.

58. Ibid.

59. Ibid.

60. Ibid.

61. The creation of the deputy director for science position resulted from the assignment to NIH of the Bureau of Health Manpower, an assignment that included responsibility for the training of nonmedical allied health professionals. The reorganization was designed to allay the anxieties of two clientele groups. Allied health professionals had traditionally distrusted NIH with its medical-school orientation. The organizational change was designed to enable Marston and Sherman to give more attention to the sensitivities of this new clientele. Simultaneously, the biomedical research community, which thought research would be eclipsed by NIH's new health manpower responsibilities, was reassured when Berliner, with his impeccable scientific credentials, was made responsible for "research NIH" (interview with John F. Sherman).

62. Schmidt, in a personal interview, described the panel's problem as how "to get rid of the NIH director's problem of how to keep the other institutes moving along in relation to NCI. We did not want to try to ram down the throat of the NIH director a priority which, by definition, he could not respond to." Burchinal, in a personal interview, described the problem in a similar manner: "We were tryng to get NCI out from under the complete authority of NIH. We felt we were going to ask for a big chunk of money. We knew that doubling the budget of NCI would make it the envy of everyone else. The NIH director would try to siphon off funds to the other institutes. The idea was to get away from this."

63. This and following extracts are from Panel Report, pp. 4-7.

64. Personal interview with Michael J. O'Neill.

65. Panel Report, p. 7.

66. Ibid.

67. Ibid., p. 8.

68. Davis, op. cit.

69. Copy of letter from Mathilde Krim, Ph.D., July 2, 1970, to numerous cancer scientists.

70. Personal interview.
71. Panel Report, pp. 11-12, 38, 41, 44. The differences in 5-year survival rates between localized and disseminated cancers were uniformly dramatic for all types of cancer and attested to the value of early detection.
72. Ibid., p. 47.
73. See ibid., pp. 13-14, 50-56.
74. Ibid., p. 61.
75. See ibid., pp. 16-29, 69-148.
76. Ibid., p. 67, 70, 73.
77. Ibid., pp. 102-118.
78. Ibid., pp. 22-23, 129-144.
79. See ibid., pp. 86, 91, 98, 117-118, 123-124, 143-144, and 146; see 24-29 for the panel's overall summary.
80. Yarborough, "S. 4564—Introduction of a Bill to be Cited as the 'Conquest of Cancer Act,' " *Congressional Record*, vol. 116, pt. 30, December 4, 1970, p. 39945.
81. See Victor Cohn, "Multi-Billion Cancer War By a New Agency Urged," *Washington Post*, December 5, 1970, p. 1, and Richard D. Lyons, "U.S. Crusade on Cancer Urged," *New York Times*, December 5, 1970, p. 22.
82. Nixon, "The National Cancer Act of 1971," *Weekly Compilation of Presidential Documents*, vol. 7, No. 52, December 27, 1971, p. 1710.

· 5 ·

1. Cohn, "Multi-Billion Cancer War," December 5, 1970.
2. Marston, "Report of the Panel of Consultants on the Conquest of Cancer to the Senate Committee on Labor and Public Welfare," Memorandum to Assistant Secretary for Health and Scientific Affairs, DHEW, December 8, 1970, from which Marston's following quotes derive.
3. See Burton Hersch, "The Thousand Days of Edward M. Kennedy," *Esquire*, February 1972, for an appraisal of the impact of Chappaquiddick on Kennedy's 1972 presidential prospects.
4. See John W. Finney, "The Session Opens: Coup by Byrd Dismays the Liberals—Scott Keeps His Post," *New York Times*, January 22, 1971.
5. "Russell's Proxy Kept Byrd in the Race," *New York Times*, January 22, 1971.
6. "Richard B. Russell Dead at 73; Georgian Was Dean of Senate," *New York Times*, January 22, 1971.
7. Mary Lasker had backed John Kennedy's campaign for president and had been involved with Jacqueline Kennedy's efforts at restoration of

the White House. She had supported Robert Kennedy's successful 1964 bid for the Senate seat in the State of New York, and had publicly declared her support for him as the 1968 Democratic presidential nominee immediately after President Johnson announced he would not be seeking re-election. Eunice Shriver, one of the Kennedy sisters, had been involved with mental retardation activities, a long-time concern of Mary's. The relations between Mrs. Lasker and the Kennedys were extensive and close over many years.

8. Senator Edward M. Kennedy, "Address at the Testimonial Dinner for Dr. Sidney Farber," Boston, Massachusetts, September 18, 1969, press release from the office of Senator Kennedy.

9. Cohn, op. cit.

10. See U.S., Congress, Senate, Committee on Labor and Public Welfare, *Conquest of Cancer Act, 1971*, Hearings, 92nd Cong., 1st sess., 1971, pp. 3-19.

11. *Supra*, pp. 92-93.

12. Personal interview with Benno C. Schmidt.

13. Ibid.

14. See Dom Bonafede, "President's Inner Circle of Friends Serves as Influential 'Kitchen Cabinet,' " *National Journal*, January 22, 1972, p. 130.

15. His bias against medical research was so strong, in fact, that in the fiscal 1972 budget review the $1.8 million request for medical research did not emerge as an element in its own right but was buried in other health categories. Lewis Butler, the HEW assistant secretary for planning and evaluation, was of the same persuasion. Butler, moreover, had more influence in health matters with HEW Secretary Robert Finch than did Dr. Roger Egeberg, the assistant secretary for health and scientific affairs.

16. These concerns led in 1971 to an extensive government-wide study headed by William McGruder, former director of the SST development project of the Department of Transportation. The study resulted in President Nixon's message to the Congress, "Science and Technology," March 16, 1972, *Weekly Compilation of Presidential Documents*, vol. 8, No. 12 (March 20, 1972), pp. 581-590, and two modest size programs in the National Science Foundation and the National Bureau of Standards.

17. Personal interview with Office of Management and Budget official.

18. Personal interview with Robert Q. Marston.

19. See "Milk Price Decision," and "ITT Antitrust Decision," *Weekly Compilation of Presidential Documents*, vol. 10, No. 2 (January 14, 1974), pp. 20-28 and 28-32 respectively, for a fascinating glimpse at the inside workings of the Nixon White House during this period and for the critical roles played by Shultz and Ehrlichman.

20. Richard M. Nixon, "The State of the Union," January 22, 1971, *Weekly Compilation of Presidential Documents*, vol. 7, No. 4 (January 25, 1971), pp. 91-92.

21. Nixon, "Budget Message of the President, January 29, 1971," p. 21 in *The Budget of the United States Government: Fiscal Year 1972*, Washington, D. C., GPO, 1971. See also pp. 306 and 309 of the *Budget*.

22. Nixon, "The President's Remarks to the 20th Annual Scientific Session of the American College of Cardiology," February 4, 1971, *Weekly Compilation of Presidential Documents*, vol. 7, No. 6 (February 8, 1971), p. 161.

23. Dr. Edward E. David, Jr., "Address by Dr. Edward E. David, Jr., Science Adviser to the President, Before the Assembly of the Association of American Medical Colleges," Palmer House Hotel, Chicago, February 13, 1971, p. 1, mimeograph.

24. Ibid.

25. Ibid., p. 6.

26. Ibid., pp. 1, 5, 6.

27. Ibid., pp. 4, 5, 6.

28. Ibid., p. 4.

29. Ibid., pp. 2, 4.

30. Ibid., pp. 6-7.

31. This and following excerpts are from Nixon, "National Health Strategy," February 18, 1971, *Weekly Compilation of Presidential Documents*, vol. 7, No. 8 (February 22, 1971), pp. 253-254.

· 6 ·

1. Luke C. Quinn, Jr., "S. 34 For The Conquest of Cancer," Letter memorandum, February 15, 1971.

2. U.S., Congress, Senate, Committee on Labor and Public Welfare, *Conquest of Cancer Act, 1971*, Hearings, 92nd Cong., 1st sess., 1971, pp. 3-19. Hereafter, reference will be to Senate cancer hearings. The objectives of the new agency were to continue the research being conducted by NCI, use existing resources for an "accelerated exploration . . . in areas of special promise," and conduct "an expanded, intensified, and coordinated cancer research program." Industrial research was to be encouraged where capacity for such research existed. Established cancer centers were to be strengthened and new ones established for multidisciplinary clinical research and teaching and for the development and demonstration of new means of cancer treatment. Other functions were to include the collection, analysis, and dissemination of data useful to the prevention, diagnosis, and treatment of cancer, and the support of specialized facilities for the large-scale production of biological ma-

terials—viruses, cell cultures, and animals. The international emphasis included the support of foreign scientists, encouragement of U.S. and foreign scientist collaboration, and the training of U.S. scientists abroad and foreign scientists in the United States.

3. Ibid., p. 16.

4. Ibid., pp. 14, 19.

5. Robert J. Bazell, "Cancer Research Proposals: New Money, Old Conflicts," *Science*, March 5, 1971, pp. 878-879.

6. Senate cancer hearings, pp. 1-2.

7. Ibid., p. 25.

8. Ibid., p. 27.

9. Ibid., pp. 29, 30.

10. Ibid., pp. 30-31.

11. Ibid., p. 32.

12. Ibid., p. 33.

13. The exchange between Kennedy and Baker is found at ibid., pp. 33-35.

14. Ibid., p. 40.

15. Ibid., p. 41.

16. Ibid., pp. 60-61.

17. Ibid., p. 62.

18. Ibid., p. 45; see pp. 46-51 for copies of these letters.

19. Cooper, in his representative capacity, spoke for the 103 medical schools in the country, 401 of the major teaching hospitals, and 47 academic medical societies representing medical school faculties. Ibid., pp. 88, 92.

20. On April 9, 1970, Gallagher (D., N. J.) introduced a resolution in the House calling for a national commitment to cure and control cancer within a decade ("Cure and Control of Cancer," *Congressional Record* 91st Cong., 2nd sess., 1970, 116, pt. 8: 11098-11099). The resolution was reintroduced two months later with additional House co-sponsors and with a list of 600 "prominent physicians, scientists, and medical educators" who allegedly endorsed the resolution (see "Legislation Aimed at Creating a Firm National Commitment to Cure and Control Cancer in This Decade," *Congressional Record*, 91st Cong., 2nd sess., 1970, 116, pt. 15: 20268-20273). After the presentation of the Panel of Consultants report to the Senate, the New Jersey Democrat introduced a new resolution into the House, identical to the earlier one, except it also called for a NASA-like cancer authority to administer the program ("Gallagher Introduces Expanded Resolution to Cure and Control Cancer," *Congressional Record*, 92nd Cong., 1st sess., 1971, 117, pt. 1: 245-251). The list of academicians endorsing the earlier Gallagher resolution was appended such that it appeared as if they now endorsed a separate cancer agency, though they had never been asked to comment on one. Messages to this effect soon began to filter back to the AAMC. Cooper,

in his testimony, sought to clarify the circumstances around this episode (Senate cancer hearings, p. 108).

21. Senate cancer hearings, pp. 89-90, 95.
22. Ibid., pp. 95-96, 97, 102.
23. Ibid., p. 103.
24. Ibid., p. 90, 101.
25. The following Kennedy-Cooper exchange is in ibid., pp. 105-106.
26. Ibid., p. 110.
27. Ibid., p. 112.
28. Ibid., p. 114.
29. Ibid., p. 116.
30. Ibid., pp. 117, 118.
31. Ibid., pp. 119-120, 124.
32. Ibid., pp. 126, 127.
33. Ibid., p. 129.
34. See ibid., p. 120, for a copy of the Hitch letter.
35. Shannon's letter appears at ibid., pp. 121-123. The quotes which follow are excerpted from it.
36. See Byron Scott, "The Saga of S. 34," *Medical Opinion*, April 1971, p. 37.
37. Ibid., p. 36.
38. Senate cancer hearings, pp. 173-189.
39. Ibid., p. 190.
40. Ibid., p. 191.
41. Ibid., pp. 191-194.
42. Ibid., pp. 193-194.
43. Ibid., p. 194.
44. Ibid., p. 195.
45. Ibid., p. 196.
46. Ibid., pp. 197, 198.
47. Scott, op. cit., p. 37.
48. Senate cancer hearings, p. 200.
49. Ibid., pp. 200-201.
50. Ibid., p. 202.
51. Ibid., p. 203.
52. Ibid.
53. Ibid., p. 207.
54. Ibid., pp. 214-220.
55. Ibid., p. 220-221.
56. Ibid., pp. 222-230.
57. Ibid., pp. 230-231.
58. Ibid., p. 231.
59. Ibid.
60. Ibid., p. 232.

• 7 •

1. John K. Iglehart, "Proposals for Cancer Research: A New Agency, or an Expanded NIH Program?" *National Journal*, March 27, 1971, p. 680.

2. See U.S., Congress, Senate, Committee on Labor and Public Welfare, *Conquest of Cancer Act, 1971*, Hearings, 92nd Cong., 1st sess., 1971, pp. 269-270, for a copy of Lederberg's letter. (Hereafter referred to as Senate cancer hearings.)

3. Ibid., p. 270.

4. R. J. B., "Lederberg Opposes Cancer Authority," *Science* March 26, 1971, p. 1220.

5. Comptroller General of the United States, *Administration of Contracts and Grants for Cancer Research*, Report to the Committee on Labor and Public Welfare, Washington, D. C., March 5, 1971. See also Iglehart, op. cit., p. 680.

6. See *Administration of Contracts . . .*, pp. 40-43, for a copy of Richardson's letter.

7. Ibid., p. 29.

8. Iglehart, op. cit., pp. 675-676. See also *The Blue Sheet*, March 10, 1971, p. 1.

9. Iglehart, p. 676, and personal interview with Robert Q. Marston.

10. Personal interviews with Benno C. Schmidt and Robert Q. Marston.

11. Iglehart, op. cit., pp. 676, 681.

12. Ibid., pp. 674, 676.

13. Ibid., p. 676.

14. Ibid., pp. 675, 676.

15. Senate cancer hearings, p. 64.

16. Iglehart, op. cit., p. 674.

17. See Senate cancer hearings, op. cit., pp. 64-68, for Richardson's letter.

18. See ibid., pp. 65-68, for the plan of organization.

19. See ibid., pp. 247-248, for a copy of Dominick's letter.

20. See ibid., pp. 265-266, for a copy of Richardson's letter.

21. See Iglehart, op. cit., pp. 680-681; "Cancer Authority Backers Aim at Senate Health Committee Members," *The Blue Sheet*, April 14, 1971, p. RN-5; and "National Cancer Authority Bill Doesn't Have the Votes in Senate Subcommittee," *The Blue Sheet*, April 28, 1971, pp. 6-8, for several accounts of the situation within the health subcommittee.

22. See Senate cancer hearings, p. 50, for a copy of the letter to President Nixon from Walter C. Bornemeier, M.D., AMA president.

23. Ibid., pp. 235-237, for a copy of the letter from Ernest B. Howard, M.D., executive vice-president of the AMA, to Senator Kennedy.

24. The institutional members of AAMC are organized in three

councils—the Council of Academic Societies, the Council of Deans, and the Council of Teaching Hospitals; see the annual *AAMC Directory of American Medical Education* for details about AAMC organization.

25. The six societies are: American Physiological Society, American Society for Biological Chemists, American Society for Pharmacology and Experimental Therapeutics, American Society for Experimental Pathology, American Institute of Nutrition, and American Association of Immunologists. See annual *Directory of Members: Federation of American Societies for Experimental Biology*, for details of FASEB's organization.

26. The FASEB public affairs committee had been organized in 1968 by Mr. Robert Grant, who became the full-time director of the newly created office of public affairs upon his retirement from nearly 20 years of administrative service in NIH. The committee consists of one member from each of the six scientific societies. Its purpose is to consider legislative and administrative matters affecting science at the federal level. Committee representatives routinely testify on a number of bills before Congress and also appear annually before the House and Senate appropriations committees on behalf of the NIH budget. The entire FASEB membership receives public affairs information through the *FASEB Newsletter*. In addition, approximately four hundred "national correspondents" are kept informed by Grant of fast-breaking developments as they occur and may be asked to provide timely letters to Congress in response.

27. For example, Thomas D. Kinney, M.D., secretary-treasurer of the Intersociety Committee for Research Potential in Pathology, sent identical letters to Senators Kennedy, Dominick, and Packwood. So did Robert E. Stowell, M.D., president of the American Association of Pathologists and Bacteriologists. Similarly, the secretary of the Association of Pathology Chairman, Inc., Rolla B. Hill, Jr., M.D., sent identical letters to Dominick and Packwood. Each writer urged that the cancer effort be administered through the existing peer review mechanisms and within the NIH. All three individuals had AAMC affiliation, Kinney as Duke University's medical school dean on the Council of Deans, and Stowell and Hill as representatives of member organizations of the Council of Academic Societies. Furthermore, all were members of the American Society for Experimental Pathology, a FASEB constituent, while Kinney and Stowell were among the four hundred "national correspondents." In a related instance, Clifford A. Barger, M.D., president of the American Physiological Society, a FASEB constituent and member organization of the Council of Academic Societies, wrote Kennedy opposing a separate cancer agency. See Senate cancer hearings, pp. 238, 250, and 268.

28. Ibid., p. 51.
29. Ibid., p. 258.
30. Ibid., p. 239.

31. Ibid., p. 250. Aldrich, in the early 1960s, had been the first director of the National Institute of Child Health and Human Development, an institute created in large measure because of influence from the Kennedy family.

32. Ibid., p. 239.

33. Ibid., pp. 275-276.

34. Ibid., pp. 240-242. Handler, in 1964, had been a member of the President's Commission on Heart Disease, Cancer, and Stroke, and had been at Duke University Medical School before becoming president of the National Academy of Sciences.

35. Ibid., pp. 255-256.

36. Ibid., p. 257.

37. Ibid., p. 238. Kety, also a professor at Harvard Medical School, had been the scientific director of the National Institute of Mental Health during the years when that same position had been held by James Shannon in the National Heart Institute and by G. Burroughs Mider in the National Cancer Institute.

38. Philip H. Abelson, "Federal Support of Cancer Research," *Science*, April 2, 1971, p. 15.

39. American Cancer Society Interorganization Memorandum, "Questions and Answers on the National Program to Conquer Cancer," March 18, 1971.

40. Jane E. Brody, "Cancer Society for Kennedy Bill," *New York Times*, April 3, 1971.

41. Telephone conversation with Solomon Garb.

42. "Letters to the Editor," *Ithaca Journal*, April 20, 1971. Garb failed to identify himself as a member of the Panel of Consultants or with the Citizens Committee for the Conquest of Cancer, only as the director of a hospital that "has cared for cancer patients from your area."

43. By a Parent, "Cancer Research—Only $1 Per Person," *Wall Street Journal*, April 14, 1971. The anonymous parent, who wished to protect his son from the knowledge that he had leukemia, was a member of a Washington, D. C. organization, the Candlelighters, a group of parents of children who have cancer or have died of it.

44. See, for example, "The War on Cancer: Progress Report," *Newsweek*, February 22, 1971, a lengthy cover story on the subject; "The Search for a Cancer Cure," *Time*, April 19, 1971, p. 44; and Jill Hirschy, "Cancer Breakthrough: A New Test Can Detect One Form of the Disease in Its First Stages," *Life*, April 23, 1971, pp. 28-31.

45. The column appeared under the familiar "Ask Ann Landers," on April 20, 1971. Ann Landers' column appears in nearly 750 newspapers across the country, with an estimated daily readership of fifty-four million people. All the material quoted below is directly from this column.

46. See, for example, "National Cancer Authority Bill Doesn't Have the Votes in Senate Subcommittee," *The Blue Sheet*, April 28, 1971, p.

7; R. J. B., "To Cure Cancer," *Science*, May 7, 1971, p. 545; and Iglehart, "House Subcommittee Questions Wisdom of Establishing Independent Cancer Agency," *National Journal*, July 31, 1971, p. 1615.

47. U.S., Congress, Senate, Committee on Appropriations, *Supplemental Appropriations for Fiscal 1972*, Hearings, 92nd Cong., 1st sess., November 2, 1971, p. 878.

48. See the *Cleveland Plain Dealer*, May 2, 1971, p. 3-AA, for a copy of the advertisement.

49. See Dom Bonafede, "President's Inner Circle of Friends Serves as Influential 'Kitchen Cabinet,' " *National Journal*, January 22, 1972, p. 130.

50. *Bobst: The Autobiography of a Pharmaceutical Pioneer*, New York, David McKay, 1973, p. 341.

51. A reorganization plan takes effect within 60 days after submission by the president, unless either house of Congress adopts a resolution of disapproval by a majority vote.

52. Personal interview by Stephen P. Strickland of Kenneth Cole, as told to author by Strickland.

53. Nixon, "Cancer-Cure Program," *Weekly Compilation of Presidential Documents*, vol. 7, No. 20, May 17, 1971, pp. 751-754. See also Harold M. Schmeck, Jr., "President Vows To Lead a Drive Against Cancer," *New York Times*, May 12, 1971; and see "Nixon Unveils 'Cancer Cure' Program; Kennedy Subcommittee Vote Delayed; Two Approaches Are Similar In Concept, Nixon's Nominally Within NIH," *The Blue Sheet*, May 12, 1971, pp. 17-19.

54. Nixon, op. cit., p. 751.

55. Ibid.

56. Ibid., pp. 753, 754.

57. Ibid., p. 751.

58. Office of the White House Press Secretary, "The White House Press Conference of Hon. Elliot Lee Richardson, Secretary, Department of Health, Education, and Welfare; Dr. Robert Q. Marston, Director, National Institutes of Health; and Dr. Edward E. David, Jr., Science Adviser to the President," May 11, 1971, p. 4.

59. Editorial, "Playing Politics with Cancer Research," *Washington Post*, May 13, 1971.

60. Senator Harrison Williams, "Amendments Nos. 69, 70, 71, 72, and 73," *Congressional Record*, 92nd Cong., 1st sess., 1971, 117, pt. 11, pp. 13844-13845.

61. See Amendment No. 69, ibid., p. 13845.

62. See Amendment No. 72, ibid., p. 13845.

63. Schmeck, "Medical Teachers Oppose Separate U.S. Cancer Unit," *New York Times*, May 3, 1971.

64. See Senate cancer hearings, op. cit., p. 408, for a copy of the resolution adopted by the board of directors of the American Association for

Cancer Research. The AACR did present two special citations, one to Sidney Farber and one to Benno Schmidt, for their respective outstanding contributions to the conquest of cancer, at this same meeting. See also the letter by Emil Frei, *Science*, June 25, 1971, p. 1293, for a copy of the resolution.

65. Nixon, "Cancer-Cure Research," *Weekly Compilation of Presidential Documents*, vol. 7, No. 22, May 31, 1971, p. 822.

66. Quinn, "HEW Request for Cancer Funds Far Short of Needs," letter memorandum, April 23, 1971.

67. Schmeck, "President Vows to Lead a Drive . . .," May 12, 1971. 12, 1971.

68. *The Blue Sheet*, op. cit., May 12, 1971, p. 17.

69. Letter from Secretary of Health, Education, and Welfare Elliot L. Richardson, to Honorable Carl Albert, Speaker of the House of Representatives, May 11, 1971, transmitting the draft bill, "Act to Conquer Cancer."

70. See Senate cancer hearings, pp. 298-304, for a copy of the legislation.

71. Letter from Quinn to Schmidt, May 12, 1971.

72. Letter from Quinn to Bobst, May 14, 1971.

73. Ibid.

74. See Senate cancer hearings, pp. 487-490, for a copy of Bobst's letter to Adams.

75. "Nixon Cancer Bill," *The Blue Sheet*, May 19, 1971, p. 3. The American Cancer Society, after some handwringing about whether it should issue a public statement critical of the divergence between the president's statement and the administration bill, adopted a rather bland resolution at the May 22, 1971, meeting of its board of directors. The statement applauded the president "for placing the full support and influence of the White House behind what we believe is a direct, and effective means of mounting a meaningful assault on this dread disease." The ACS affirmed support for administrative and budgetary independence for cancer research, but clearly refrained from anything more critical. The ACS board acted after hearing a report from Quinn that "letters of clarification and strong assurances" had been received that indicated a willingness by the administration to bring the legislation in line with the president's May 11 statement (American Cancer Society, "Report of the Legislative Committee to the Board of Directors," May 22, 1971).

76. Personal interviews with Schmidt, Lee Goldman, and Jay Cutler.

77. See Senate cancer hearings, pp. 297-413, for the transcript of the June 10 hearing.

78. Senate cancer hearings, p. 305.

79. Ibid., p. 313.

80. See ibid., pp. 314-319, for Richardson's statement. The HEW secretary also indicated that the cancer program director would have au-

thority to review and comment on cancer-related research of other NIH institutes. The program would have construction authority. The program director and the NIH director would be expected to work out mutually agreeable arrangements for the use of NIH central facilities and services by the program. The National Advisory Cancer Council would be retained as the working arm of the new Cancer Cure Advisory Committee, and the existing NCI would become the "working nucleus" for the new program and the base for future effort.

81. See ibid., pp. 334-335, for this exchange.

82. Ibid., p. 336.

83. Iglehart, "Proposals for Cancer Research," p. 676.

84. The Wisconsin senator had concluded that it was more important to establish the NIH as a separate agency outside of HEW than it was to break cancer research out of NIH. On May 21, Nelson and Senators Cranston and Schweiker announced their intention to propose an amendment to S. 34 which would do just that. Reaction to this proposal varied. Early supporters of S. 34 continued to prefer that legislation. Others were enthusiastic about the proposed amendment, since it maintained the integrity of NIH while removing medical research from the HEW bureaucracy. A third group expressed a guarded preference for the Nelson proposal over S. 34. Finally, there were those who would take the amendment over S. 34 if forced to choose, but who had serious misgivings about considering priorities for medical research independently of other federal health programs. Secretary Richardson, however, dismissed the proposal from further consideration as "impossible to support." The amendment drew no serious attention within the subcommittee. See Senate cancer hearings, pp. 351-371, for a copy of the amendment, and pp. 372-376 and pp. 422-496, for reaction to it.

85. Ibid., p. 337.

86. See ibid., pp. 338-339, for this exchange which is rare, if not unique, in the annals of legislation.

87. See, for example, Iglehart, "House Subcommittee Questions Wisdom . . ."

88. Lee Goldman, staff director of the health subcommittee, and Jay Cutler, minority counsel to the Labor and Public Welfare Committee, it was reported, "addressed themselves to making the President's bill consistent with his statement by using a scissors and a copy of S. 34" (Iglehart, "House Subcommittee Questions Wisdom . . . ," p. 1613).

89. Stuart Auerbach, "Accord Reached on Cancer Program," *Washington Post*, June 11, 1971.

90. Senate cancer hearings, p. 337.

91. Ibid., pp. 396-397.

92. Ibid., p. 397.

93. Ibid., pp. 397-398.

94. Ibid., p. 402. Nelson and his staff were informed at one point that the division within the panel on the separate agency question had been

about 60-40. They translated these percentages into numbers and came up with the 16-to-10 count.

95. Ibid., p. 378.

96. Ibid., p. 379. Senator Nelson was pleased that this point had been made. He quoted Bobst's letter to Lane Adams: "I think anyone with any knowledge of the respective fields is aware of the fact that in the development of atomic power and in the outer space program, there was a wonderful control in a most highly businesslike type of approach. It is my thinking that we must strive to break businesslike methods into the fight against cancer." Nelson had had the "disturbing thought" he said, that Bobst might have been thinking of Lockheed or the Edsel (Ibid., pp. 385-386).

97. Cooper responded that the composition of the proposed advisory council was weighted more heavily to lay representation than the AAMC and the scientific community would prefer (ibid., p. 382). Nelson, Richardson, and Marston had had a lengthy discussion on this point earlier that morning with Marston expressing the preference for the traditional NIH ratio of 2:1 scientific to lay members on councils (ibid., pp. 343-346).

98. See ibid., pp. 382-383.

99. Richardson's statement was: "We think . . . that this legislation should, while mandating the requirements of peer review, nonetheless leave open the possibility that the director of the new effort might, in mutual consultation with the director of the National Institutes of Health, and on the advice of his advisory body, decide to do it some other way." See ibid., pp. 341-343, for the Nelson-Richardson exchange.

100. Ibid., pp. 384, 385.

101. Ibid., pp. 386-387.

102. Ibid., p. 402.

103. U.S., Congress, Senate, Committee on Labor and Public Welfare, *Conquest of Cancer Act*, Report No. 92-247, 92nd Cong., 1st sess., June 29, 1971, pp. 25-26.

104. Personal interview with Lee Goldman.

105. See "Conquest of Cancer Act," *Congressional Record*, 92nd Cong., 1st sess., 1971, 117, pt. 18, pp. 23762-23799, for the account of action on the Senate floor.

106. Ibid., pp. S 10614-10619 and 10621-10622, for Nelson's full statement.

107. Ibid., pp. S 10622-10624.

• 8 •

1. Marston, "The Management of National Research Programs and Implications of the Cancer Initiative," Remarks to the Office of Science

and Technology Seminar on "The Role of the Executive Branch in Biological Science," July 31, 1971, Woods Hole, Massachusetts, mimeo.

2. See "Search for Cancer Head," *Nature*, July 2, 1971, p. 8; John K. Iglehart, "House Subcommittee Questions Wisdom of Establishing Independent Cancer Agency," *National Journal*, July 31, 1971, p. 1612, and "Proposed Top Cancer Post's Vast Program Power Draws Intense Interest to Leadership Selection," *The Blue Sheet*, September 1, 1971, p. 14. The White House search team was reportedly looking for a director with managerial skills, not necessarily a doctor. *Nature* speculated on the prospects of Clark, Schmidt, Burchenal, Lederberg, Kaplan, Holland, Rusch (all panel members), and Albert Sabin, Emmanuel Farber, Emil Frei, and Maclyn McCarty.

3. "Cancer Conquest or Setback?" editorial, *New York Times*, July 11, 1971. The *Washington Post*, taking a similar view, asked "Can 79 Senators Be Wrong?" in an editorial on Sunday, July 18, and hoped with Senator Nelson that the House would give the cancer agency proposal "a deliberate, careful look."

4. Javits, "Coordinated Attack on Cancer," *New York Times*, July 24, 1971.

5. Pollard, "Upgrading the Crusade Against Cancer," *New York Times*, August 1, 1971.

6. Robert G. Martin et al., "Cancer Bill Assessed by Scientists," *New York Times*, August 9, 1971. The five Nobel winners were Julius Alexrod, Arthur Kornberg, Marshall Nirenberg, Severo Ochoa, and Edward L. Tatum.

7. See U.S., Congress, House, Committee on Interstate and Foreign Commerce, *National Cancer Attack Act of 1971*, Hearings, pts. 1 and 2, 92nd Cong., 1st sess., 1972, pp. 66-82. Reference hereafter will be made to House cancer hearings.

8. "Rep. Rogers (D-Fla.) May Chair House Health Subcommittee," *The Blue Sheet*, February 3, 1971, p. RN-1. See also "Health's New Strong Man in Congress," *Medical World News*, April 28, 1972, pp. 30-38, for background on Rogers.

9. U.S., Congress, House, Committee on Interstate and Foreign Commerce, *Investigation of HEW*, Hearings, 89th Cong., 2nd sess., April-May-June, 1966; and *Investigation of HEW*, Report, 89th Cong., 2nd sess., October 13, 1966.

10. "Fourth of House Membership Co-Sponsors Cancer Bill; But One Key Legislator Balks," *The Blue Sheet*, February 17, 1971, p. 18.

11. Iglehart, "Proposals for Cancer Research: A New Agency, Or an Expanded NIH Program?" *National Journal*, March 27, 1971, p. 681.

12. The unidentified lobbyist was widely known to be Mike Gorman; see "House Health Subcommittee Members Join Rogers in Opposition to

Separate Agency; Lobbyists Feel Floor Approval Threat Will Force Change," *The Blue Sheet*, July 21, 1971, p. 1.

13. "House Action on Cancer Agency Not Imminent," *The Blue Sheet*, July 14, 1971, p. 1.

14. "House Health Subcommittee Members Join Rogers," loc. cit.

15. "Rep. Rogers Schedules Cancer Hearings," *The Blue Sheet*, August 11, 1971, p. 1.

16. "Proposed Top Cancer Post's Vast Program Power," p. 1.

17. Letter from Mr. Lane W. Adams to Mr. Alfred H. LeDoux, August 16, 1971.

18. Personal interview with Schmidt.

19. See House cancer hearings, pp. 59-65.

20. Ibid., pp. 117-128.

21. Ibid., p. 118.

22. Ibid., p. 121.

23. Ibid., pp. 119-120.

24. Personal interview with Spencer C. Johnson.

25. See "Remarks of Congressman Paul G. Rogers (D-Fla.) . . . announcing introduction of 'The National Cancer Attack Amendments of 1971,' " September 14, 1971, press release.

26. Personal interview with Robert Maher.

27. House cancer hearings, p. 2.

28. See ibid., pp. 147-154 for Richardson's formal statement.

29. Ibid., p. 149.

30. Ibid., p. 161.

31. Ibid., p. 156.

32. Ibid., p. 183.

33. Ibid., p. 174.

34. Ibid., pp. 158-159.

35. Ibid., p. 173.

36. Letter from Robert S. Adelstein et al., "Javits Disputed on Cancer Agency Role," *Washington Post*, September 5, 1971. The signers were essentially the same individuals whose letter to the *New York Times* was published on August 9, 1971. Supra, n. 6.

37. House cancer hearings, pp. 177, 178.

38. Ibid., p. 184. No one in HEW, NIH, or NCI anticipated that questions might arise regarding the defunct cancer control programs.

39. Ibid., p. 3.

40. Ibid., p. 196.

41. Ibid., pp. 197-198.

42. Ibid., p. 199.

43. Ibid., p. 200.

44. Ibid., p. 217.

45. See ibid., pp. 219-222.

46. See ibid., pp. 222-230.

47. See ibid., pp. 234-247, for Ingelfinger's testimony and responses to questions, and p. 245 for a copy of the AMA letter to Rogers.

48. Ibid., pp. 261, 263.

49. Ibid., p. 265.

50. Ibid., p. 268.

51. Ibid., p. 269.

52. Ibid., p. 270.

53. Ibid., pp. 270-271.

54. Ibid., pp. 294-295; see also pp. 186-189 for Rogers' questioning of Elliot Richardson on this same point.

55. Ibid., p. 299.

56. Ibid., p. 296.

57. See ibid., pp. 296-298 for the discussion on this point among Rogers, Schmidt, and Scott; see also note 47 above.

58. Ibid., p. 299.

59. Ibid., pp. 303, 304.

60. Ibid., p. 301.

61. Ibid., pp. 301, 302.

62. Ibid., p. 302.

63. Ibid., p. 307.

64. Ibid., p. 306.

65. Ibid., see pp. 310-320 for the full discussion on this critical point.

66. Ibid., p. 165.

67. Ibid., p. 310; this comment, not surprisingly, was widely reported in the press.

· 9 ·

1. These congressmen included 20 Democrats and 11 Republicans from 14 states.

2. Some congressmen, however, were spokesmen for constituent interests. Rep. Goodloe E. Byron (D., Md.) urged the use of Fort Detrick for cancer research, while Rep. Dan Kuykendall (D., Tenn.) made a plea for the financial situation of St. Jude Children's Research Hospital in Memphis.

3. See House cancer hearings, pp. 426-427. Rep. John H. Buchanan (R., Ala.) indicated that he had co-sponsored the Gallagher resolution and the Pepper legislation. The director of cancer research and training at the University of Alabama Medical Center had expressed reservations to him about an expanded national effort, and the congressman, as a result, took the view that the need for an independent cancer agency was not obvious (ibid., p. 430).

4. Ibid., p. 554.

5. The HEW secretary testified on September 15, the first day of hearings, the Panel of Consultants on September 20, and the American Cancer Society officials on September 21. There was a counterpoint to even this, however, since Senator Nelson and Dr. Ingelfinger testified on September 16.

6. House cancer hearings, p. 364. Pollard urged Rogers to use as a "fine filter" in evaluating advice on cancer the distinction between those directly involved in the field, and those only indirectly or peripherally involved or even uninvolved (ibid., p. 364).

7. Ibid., p. 378. The ACS later approached Rep. Carter to see if this section could be "sanitized" out of the transcript.

8. Ibid., pp. 383-385. Rhoads did comment at the hearing that the writer in question built up his readership "by his attacks on existing institutions in medicine."

9. Ibid., p. 392. See pp. 389-393 for Kyros' aggressive questioning of the ACS witnesses.

10. Ibid., pp. 397-398.

11. Ibid., p. 400. This entire discussion, including a table, "Estimated Cancer Deaths by Sex and Site—1970," may be found on pp. 398-401. Rogers was referring to the testimony of Howard H. Hiatt, M.D.; see pp. 329-330, 332.

12. Ibid., p. 417.

13. Ibid., pp. 420, 421.

14. Ibid., p. 497. St. Jude was founded by Danny Thomas, the entertainer.

15. Ibid., pp. 502-503. He later wrote the chairman that the bill was inadequate because it did not provide sufficient access to the president and the Congress for the cancer program. Also its scope was too broad since it included all degenerative disease. He went on to attack the basic scientists who had opposed S. 1828. They were "far removed from the sick," had "little appreciation" of what was involved in turning scientific findings into "effective prevention and treatment of disease," and "have not and never will take part in direct application of scientific research to the health of the American people." The choice before the subcommittee, he concluded, was whether it wished to "represent American science or the American people." The animus was clear.

16. Ibid., pp. 548-549.

17. Ibid., p. 563.

18. See ibid., pp. 563-564.

19. Ibid., pp. 588, 600.

20. Ibid., p. 329.

21. Ibid., p. 337.

22. Temin and Baltimore received the 1975 Nobel Prize for Physiology and Medicine for this work. They shared the prize with Dr. Renato Dulbecco, a pioneer in tumor virology.

23. House cancer hearings, pp. 339-340.
24. Ibid., p. 341.
25. Ibid., pp. 432-433.
26. Ibid., p. 447.
27. Ibid., p. 463.
28. Ibid., pp. 479, 480.
29. Ibid., p. 486.
30. Ibid., p. 488.
31. Ibid., p. 491.
32. Ibid., p. 585.
33. See ibid., p. 652, for a copy of "A Resolution Adopted by the Executive Council of the Association of American Medical Colleges on the National Debate Over the Conquest of Cancer—September 17, 1971."
34. Ibid., p. 678.
35. The others, again, were Furchgott, Handschumacher, and Philips on behalf of the American Society for Pharmacology and Experimental Therapeutics, McCarty for the American Society of Immunologists, and Schneider for the FASEB public affairs committee, who was a member of the American Institute of Nutrition.
36. Ibid., p. 357.
37. Ibid., p. 633. For Kaplan's following testimony, see ibid., pp. 634-635.
38. Ibid., p. 735.
39. Ibid., pp. 735, 736.
40. Ibid., p. 740.
41. Ibid., p. 743.
42. Ibid., p. 746.
43. Ibid., pp. 747, 750.
44. Ibid., p. 755.
45. Ibid., p. 766.
46. Ibid., pp. 744, 778.
47. Ibid., pp. 786, 787.

▪ 10 ▪

1. Personal interview with Robert Maher.
2. Personal interview with Metzenbaum. See also the press release issued by Metzenbaum in early September, and the brochure for the annual luncheon meeting of the Cuyahoga Unit of the American Cancer Society.
3. Personal interview with Alan C. Davis, and files of the American Cancer Society.
4. The advertisements were run in the following newspapers, arranged by congressional district of the subcommittee members:

Paul G. Rogers (D., Fla.)	*Palm Beach Post Times* *Miami Herald*
David E. Satterfield (D., Va.)	*Richmond Times-Dispatch*
Peter N. Kyros (D., Me.)	*Portland Telegram* *Augusta-Kennebec Journal*
Richardson Preyer (D., N. C.)	*Greensboro News*
James W. Symington (D., Mo.)	*St. Louis Post-Dispatch* *Kansas City Star*
William R. Roy (D., Kans.)	*Topeka Capitol Journal*
Ancher Nelsen (R., Minn.)	*Minneapolis Tribune* *Owatonna Peoples Press* *Mankato Free Press*
Tim Lee Carter (R., Ky.)	*Middleboro News* *Louisville Courier-Journal* *Lexington Herald-Leader*
James F. Hastings (R., N. Y.)	*Buffalo Courier-Express* *Jamestown Post-Journal* *Corning Leader*
John G. Schmitz (R., Cal.)	*Los Angeles Times* *Orange County Metro Papers* *San Diego Union*

Constituents of both Symington and Roy would have seen the advertisement in the *Kansas City Star*.

5. A "clean bill" results from rewriting an existing bill, in this case H.R. 10681, which has been the subject of extensive hearings. The new text gives no indication of changes from the original, it is submitted to the legislative hopper in the appropriate chamber, and it is given a new number. Hence, the adjective "clean."

6. See the text of H.R. 11302, sec. 2. (a) (5), in *Congressional Record*, 92nd Cong., 1st sess., 1971, 117, pt. 31, pp. 41143-41145, from which the following extracts are derived.

7. "Health's New Strong Man in Congress," *Medical World News*, April 28, 1972, p. 36.

8. See House cancer hearings, p. 209.

9. Ibid., pp. 631-632.

10. Ibid., p. 673.

11. Ibid., p. 676.

12. Harold M. Schmeck, Jr., "House Unit Votes New Cancer Bill," *New York Times*, October 16, 1971.

13. Ibid.

14. "All-Out Struggle on Cancer Research Independence Enters Home Stretch; Rep. Rogers Expected to Prevail in Early Test, Refuses Compromise Offers," *The Blue Sheet*, October 13, 1971, p. 14.

15. "Cancer Society Ad 'Political,' Says Rep. Rogers," *American Medical News*, October 18, 1971, p. 1.

16. Sen. Gaylord Nelson, "The Conquest of Cancer Bill," *Congressional Record*, 92nd Cong., 1st sess., 1971, 117, pt. 28, pp. 36425-36426. Several Congressional staff developed the estimate of $53,000 on the basis of the published newspaper advertisement rate guide. Costs for 20 of the 21 local newspaper ads were calculated to be $36,927.73. Using the rough average of $1,800 per ad, the total for the 21 was estimated to be $38,700. The costs of the full-page advertisements in the *New York Times, Washington Post,* and *Washington Evening Star* were calculated at $16,148. The combined cost of both sets of ads, therefore, was estimated to be $54,800, or nearly $55,000. The American Cancer Society advertisement read at the bottom: "Note: This advertisement is not paid for from funds contributed by the public to the American Cancer Society. It has been paid for by a group of concerned individuals." The ACS, in response to an inquiry, said that it would be indiscreet to reveal the names of these individuals. The advertisements in the 21 local newspapers were sponsored by the Citizens Committee for the Conquest of Cancer, and Sidney Farber, Emerson Foote, and Howard Metzenbaum were listed as officers of the committee. Metzenbaum, Mary Lasker, and others contributed to the costs of these advertisements, which were handled through the New York public relations firm of Wunderman, Ricotta and Kline, Inc.

17. See "Statement by Dr. H. Marvin Pollard, President, American Cancer Society, in response to Congressman Paul Rogers' announcement to the press today," ACS press release, October 15, 1971. Also see Ronald Sarro, "Cancer Agency Issue: Society Denies It Lobbies," *Washington Evening Star*, October 16, 1971.

18. "Germ War Center Will Be Converted Into a Cancer Lab," *New York Times*, October 15, 1971.

19. See Robert B. Semple, Jr., "Nixon Counts on Conversion of Military Facilities," *New York Times*, October 19, 1971, for the comments by President Nixon.

20. Schmidt had not been consulted by either Mary Lasker and the Citizens Committee or by the American Cancer Society on their respective decisions to go public. He had had a number of lengthy, vigorous arguments with Paul Rogers on various provisions of the two cancer bills, and the two men had come to respect each other as determined and intelligent opponents. Though Schmidt was committed to keeping the pressure on Rogers as much as possible, he did not think the way to reach the congressman was through an advertising campaign.

21. Letter from Mike Gorman to The Honorable Paul G. Rogers, M.C., October 18, 1971. All quotations are from this letter.

22. Personal interview with Alan C. Davis.

23. "Cancer Politics," editorial, *New York Times*, October 14, 1971.

24. Benno C. Schmidt, "Conquest of Cancer Needs Independent Leadership," *New York Times*, October 25, 1971.

25. Samuel Hellman, M.D. et al., "Cancer Bill," *New York Times*, October 26, 1971. The other signers were: Henry I. Kohn, Ph.D.; George Nichols, Jr., M.D.; James L. Tullis, M.D.; Francis D. Moore, M.D., M.Ch., L.L.D., S.D.; William S. Moloney, M.D., D.Sc. (hon); Richard E. Wilson, M.D.; Alan C. Aisenberg, M.D., Ph.D.; John W. Raker, M.D.; Milford D. Schulz, M.D.; William Silen, M.D.; Chester B. Rosoff, M.D.; and Howard H. Hiatt, M.D.

26. Robert Martin, M.D. et al., "More on the Cancer Bill," *Washington Post*, October 26, 1971.

27. Letter from Paul G. Rogers, M.C., and members of the subcommittee of public health, to members of the Committee on Interstate and Foreign Commerce, November 1, 1971. Attached to the letter was a staff analysis of H.R. 10681 and a number of newspaper editorials, articles, and letters-to-the-editor critical of taking NCI out of NIH.

28. See "Cancer Agency Independence in Doubt," *The Blue Sheet*, October 20, 1971, p.1; "Rogers May Yet Win Cancer Stakes," *Nature*, October 22, 1971, pp. 516-517; Nicholas Wade, "Cancer Legislation: Pro-NIH Bill Advances in House," *Science*, October 22, 1971, pp. 388-389; "Big Debate: How to Organize 'Conquest of Cancer,' " *Congressional Quarterly*, October 23, 1971, pp. 2169-2173.

29. Letter from Rep. William L. Springer to Professor John W. Moore, Duke University Medical Center, March 19, 1971.

30. Letter from Rep. William L. Springer to Dr. H. Marvin Pollard, American Cancer Society, March 19, 1971.

31. Memorandum from Lewis E. Berry, Minority counsel, Committee on Interstate and Foreign Commerce, to Minority members of subcommittee on public health, regarding "Cancer legislation," April 26, 1971.

32. Memorandum from Lewis E. Berry, Minority counsel, to Minority members of Committee on Interstate and Foreign Commerce, regarding "Cancer research legislation," April 26, 1971.

33. See U.S., Congress, House, Committee on Interstate and Foreign Commerce, *The National Cancer Attack Act of 1971*, H. Report No. 92-659, 92nd Cong., 1st sess., November 10, 1971, pp. 63-64.

34. Six Democrats—Pickle, Eckhardt (Tex.), Van Deerlin (Calif.), Adams (Wash.), Tiernan (R. I.), and Podell (N. Y.), and only one Republican—Frey (Fla.) supported the proposal. Representative Tim Lee Carter, balancing loyalty to the subcommittee against sympathy with the amendment, voted "present" (ibid., p. 64).

35. See House cancer hearings, p. 625.

36. See House Report No. 92-659, op. cit., p. 43.

37. Ibid., p. 64.

38. See ibid., pp. 64-65.

39. Ibid., p. 65.
40. Harold M. Schmeck, Jr., "House Unit Votes Own Cancer Bill," *New York Times*, November 5, 1971. See also "House Committee Approves Rogers Cancer Bill Leaving NIH in Command; House-Senate Conference Outlook Clouded," *The Blue Sheet*, November 10, 1971, pp. 11-12.
41. Jude Wanniski, "Moon-Shot Plan for Whipping Cancer Hits Stiff Resistance," *National Observer*, November 6, 1971.
42. H. Report No. 92-659, op. cit.
43. Ibid., p. 19.
44. Ibid., p. 66.
45. Ibid., p. 67.
46. Representative Bob Eckhardt, also in "additional views" to the report, indicated support for his Texas colleague's views. His thinking, he wrote, had been greatly influenced by Dr. R. Lee Clark. Moreover, Eckhardt ventured, Dr. Carter had gone along with the subcommittee "to achieve an acceptable accommodation," even though Eckhardt believed he favored the Senate approach. For the House now to adopt the Pickle amendment need not be seen as an "embarrassment" to the health subcommittee. "It would simply be a recognition by the whole House," the congressman wrote, "that there are times when a given program like the conquest of cancer is pregnant with innovation and potentially a surge toward achievement" (ibid., p. 70).
47. Ibid., pp. 71-78.
48. See *Cancer News*, vol. 26 (Spring 1972), p. 10.
49. Ibid., pp. 10, 11.
50. "Report of the Legislative Committee to the Board of Directors, American Cancer Society," November 6, 1971.
51. In the past, conservative opposition within the Rules Committee had bottled up an extensive amount of legislation. The cancer bill faced no such problem, only the matter of scheduling time on a full legislative calendar.
52. "The National Cancer Attack Act of 1971," *Congressional Record*, 92nd Cong., 1st sess., 1971, 117, pt. 31, p. 41145 (hereafter referred to as House floor debate). Dulski, in remarks subsequently inserted in the *Congressional Record*, thanked the chairman of the subcommittee for relieving him of "the embarrassing predicament of being forced to raise doubts about a piece of legislation which I feel is far too important to be delayed on technical questions" (ibid., p. 41156).
53. Ibid., p. 41143. This gave Paul Rogers an additional advantage, since it meant that debate time would be controlled by Staggers and Springer, both favorable to passage of H.R. 11302.
54. House floor debate, p. 41145.
55. Ibid., p. 41146. The Buffalo visit had paid dividends in this regard also, since Rogers was able to announce that the director of Roswell Park "is enthusiastically endorsing the bill before you today."

56. Ibid.
57. See John K. Iglehart, "Administration Eases Pressure for New Cancer Research Agency," *National Journal*, November 27, 1971, pp. 2363-2364.
58. House floor debate, pp. 41148, 41149.
59. Ibid., p. 41150. Adams, distressed because Staggers had yielded him one minute for debate and Springer an additional two minutes, protested that "we cannot possibly have a debate under these circumstances."
60. Ibid., p. 41151.
61. Ibid., pp. 41161-41162. The five negative votes were cast by Adams, Tiernan, Rooney, Brasco (D., N. Y.), and Corman (D., Cal.). See Harold M. Schmeck, Jr., "Cancer Research is Voted by House," *New York Times*, November 16, 1971; "Rogers Cancer Bill Approved by Nixon," *The Blue Sheet*, November 17, 1971, p. 1. In an editorial, "Strengthening the War on Cancer," *New York Times*, November 26, 1971, Rogers was praised for daring "to stand up to the formidable pressure brought to bear on him by lobbyists for the Senate bill."
62. The date had initially been set for November 29, the first day after the Thanksgiving recess, but had been rescheduled because of a House vote on election reform legislation ("Sen. Kennedy Prepares Cancer Bill Compromise," *The Blue Sheet*, December 1, 1971, p. 4).
63. See Iglehart, op. cit., pp. 2363-2364; House floor debate, p. 41147.
64. Benno Schmidt had realized that Rogers and his colleagues were unwilling to concede anything on this question and had so advised Senators Kennedy and Javits. Schmidt had also concluded that this was a distinction without a difference and the senators had no difficulty accepting this interpretation.
65. Senator Kennedy played a rather passive role in the conference, unlike his behavior in the October conference on health manpower, and left much of the advocacy of the Senate position to Senator Javits.
66. See "Establishment of a Conquest of Cancer Agency Conference Report," *Congressional Record*, 92nd Cong., 1st sess., 1971, 117, pt. 35, p. 45837 for a copy of the letter (hereafter referred to as Senate ratification debate).
67. Ibid.
68. See U.S., Congress, House, Committee of Conference, *National Cancer Act of 1971*, H. Report No. 92-722, 92nd Cong., 1st sess., December 8, 1971, pp. 3-4, and 14 (hereafter referred to as Conference committee report).
69. See Senate ratification debate; Conference committee report, op. cit., pp. 6-7, 12, and 17.
70. Ibid., pp. 16-17.
71. Ibid., p. 17.
72. "Conference Report No. S. 1828, National Cancer Act of 1971,"

Congressional Record, 92nd Cong., 1st sess., 1971, 117, pt. 35, pp. 45850-45852.

73. Senate ratification debate, p. 45836.

74. See "The National Cancer Act of 1971," *Congressional Record*, 92nd Cong., 1st sess., 1971, 117, pt. 35, pp. 46102-46104, 46108-46109.

75. Ernest Borek, *The Sculpture of Life*, New York, Columbia University Press, 1973, pp. vii-viii.

76. Nixon's following remarks are from "The National Cancer Act of 1971," December 23, 1971, *Weekly Compilation of Presidential Documents*, vol. 7, No. 52 (December 27, 1971), pp. 1708-1712.

77. See, for example, Harold M. Schmeck, Jr., "Nixon Signs Cancer Bill; Cites Commitment to Cure," *New York Times*, December 24, 1971; Stuart Auerbach, "U.S. Campaign to Find Cancer Cure Is Launched With Nixon's Signature," *Washington Post*, December 24, 1971; and Jonathan Spivak, "Cancer Bill, Signed by Nixon, Seen Having Major Impact on Future Medical Research," *Wall Street Journal*, December 24, 1971.

78. See "137 Attend Cancer Bill Signing; Congress, Industrial & Academic Researchers Represented," *The Blue Sheet*, January 5, 1972, pp. S-40 to S-42 for the guest list. One prominent absentee was Dr. John Cooper, president of the AAMC; he had become persona non grata at the White House and was not invited.

79. Nixon, "The National Cancer Act of 1971," op. cit.

· 11 ·

1. J. Clarence Davies, 3rd, *Setting the National Agenda*, unpublished manuscript, Washington, D. C., 1976, p. I-7.

2. E. E. Schattschneider, *The Semi-Sovereign People*, New York, Holt, Rinehart, and Winston, 1961, p. 68.

3. Julius H. Comroe, Jr., and Robert D. Dripps, "Scientific Basis for the Support of Biomedical Science," *Science*, April 9, 1976, p. 105.

4. Comroe, "Lags Between Initial Discovery and Clinical Application to Cardiovascular Pulmonary Medicine and Surgery," Part I of *Report of the President's Biomedical Research Panel, Appendix B*, Washington, D. C., April 30, 1976.

5. The Rand Corporation, "Policy Analysis for Federal Biomedical Research," part III of ibid., also published as A. P. Williams, G. M. Carter, A. J. Harman, E. B. Keeler, W. G. Manning, Jr., C. R. Neu, M. L. Pearce, and R. A. Rettig, *Policy Analysis for Federal Biomedical Research*, Santa Monica, Calif., The Rand Corporation, R-1945-PBRP/RC, March 1976.

6. See Murray Edelman, *The Symbolic Uses of Politics*, Urbana, Il-

linois, University of Illinois Press, 1964, for a discussion of the importance of symbols in elite manipulation of mass political behavior.

7. "Cancer Institute Course Defended by Rauscher in Farewell Interview; Control Program Is 'Where the Emphasis Has Got to Be'; Future Good," *The Blue Sheet*, September 29, 1976, p. 5.

8. U.S. Department of Health, Education, and Welfare, *National Cancer Program: Digest of Scientific Recommendations for the National Cancer Program Plan*, Washington, D.C., GPO, 1974.

9. U.S. Department of Health, Education, and Welfare, *National Cancer Program: The Strategic Plan*, Washington, D. C., 1973.

10. U.S. Department of Health, Education, and Welfare, *National Cancer Program: Operational Plan, FY 1976-1980*, Washington, D. C., August 1974.

11. See Nicholas Wade, "Special Virus Cancer Program: Travails of a Biological Moonshot," *Science*, December 24, 1971, pp. 1306-1311.

12. Barbara J. Culliton, "Cancer: Select Committee Calls Virus Program a Closed Shop," *Science*, December 14, 1973, pp. 1110-1112; and Culliton, "Virus Cancer Program: Review Panel Stands By Criticism," *Science*, April 12, 1974, pp. 143-145; See also the *Report of the ad hoc Review Committee of the Virus Cancer Program*, submitted in draft to the National Cancer Advisory Board, November 26-28, 1973, final report submitted to NCAB, March 18-20, 1974.

13. In Public Law 93-352, 93rd Congress, July 23, 1974, the act that reauthorized the National Cancer Act of 1971, sec. 472 provided that biomedical and behavioral research and development contracts administered by NIH and the three institutes of the Alcohol, Drug Abuse, and Mental Health Administration should be subject to "appropriate scientific peer review."

14. Frank J. Rauscher, Jr., "Budget and the National Cancer Program," *Science*, May 24, 1974, p. 873.

15. See U.S. Department of Health, Education, and Welfare, *The Cancer Centers Program*, Washington, D. C., September 1974.

16. U.S. General Accounting Office, *Comprehensive Cancer Centers: Their Locations and Role in Demonstration*, Washington, D. C., March 17, 1976.

17. U.S. Department of Health, Education, and Welfare, *National Cancer Program: Cancer Control Program*, Washington, D. C., November 27, 1973.

18. In mid-1976, for example, a controversy arose over the relation of benefits to risk in the use of mammography—low-dosage X-ray tests—for breast cancer detection in women between 35 and 50. See, for instance, Stuart Auerbach, "X-Ray Test for Breast Cancer Disputed," *Washington Post*, July 15, 1976; Jane E. Brody, "Mammography Puzzle: Benefits Versus Risks," *New York Times*, July 26, 1976; and Auerbach, "Women Divided on Value of Breast Cancer X-Rays," *Washington Post*, July 31, 1976.

19. John Cairns, "The Cancer Problem," *Scientific American*, November 1975, pp. 64-78.
20. See Nicholas Wade "Cancer Institute: Expert Charges Neglect of Carcinogenesis Studies," *Science*, May 7, 1976, pp. 529-531.
21. See U.S. Department of Health, Education, and Welfare, *National Cancer Program: Report of the Director*, Washington, D. C., January 1973, *National Cancer Program: Report of the Director*, 1974, Washington, D. C., and *National Cancer Program: Report of the National Cancer Advisory Board*, Washington, D. C., January 1973.
22. See U.S., Congress, Senate, Committee on Labor and Public Welfare, *National Cancer Act of 1974*, Hearing, 93rd Cong., 2nd sess., January 30, 1974, and *National Cancer Act Amendments of 1974*, Report No. 93-736, 93rd Cong., 2nd sess., March 20, 1974; see also U.S., Congress, House, Committee on Interstate and Foreign Commerce, *National Cancer Act Amendments-1974*, Hearings, 93rd Cong., 2nd sess., February 5 and 6, 1974, and *National Cancer Amendments of 1974*, Report No. 93-954, 93rd Cong., 2nd sess., March 27, 1974. The conference committee report is U.S., Congress, House, Committee on Interstate and Foreign Commerce, *National Cancer Act Amendments of 1974*, Report No. 93-1164, 93rd Cong., 2nd sess., June 27, 1974.
23. In early 1975, Daniel S. Greenberg, a Washington, D. C. journalist, wrote a very critical article about the slow improvement in five-year survival rates for the more frequent forms of cancer ("A Critical Look at Cancer Coverage," *Columbia Journalism Review*, January/February 1975, pp. 40-44). This article (reprinted as "Cancer: Now, the Bad News," *Washington Post*, January 19, 1975), drew a response from NCI (see William S. Grey, "The Data Tell Only Part of the Story" in the same issue of the *Post*). In late 1975, it was reported that the cancer death rate jumped 5.2 percent in the first seven months of 1975 compared to the similar period of the year before; this was compared to an annual increase in the incidence of cancer of about 1 percent since 1933. See Morton Mintz, "Cancer Death Rate Up," *Washington Post*, November 7, 1975, and "Cancer Death Rate in U.S. Is Up 5.2% Over 1974 Figure," *New York Times*, November 9, 1975. The issue of what is happening to the survival rate is sufficiently critical to an appraisal of progress in the cancer crusade to warrant continued congressional attention.
24. The news first appeared in connection with White House confirmation that a "plumbers unit" had existed in 1972 to plug national security leaks; see "Nixon Had 2 Aides Fight News Leaks," *New York Times*, December 13, 1972. See also "Department of Health, Education, and Welfare, and Department of Transportation: Remarks of Press Secretary Ronald L. Ziegler Announcing the Resignation of Certain Officials in the Two Departments, December 15, 1972," *Weekly Compilation of Presidential Documents*, vol. 9, December 18, 1972, p. 1759. See also Barbara J. Culliton, "Health Hierarchy: Marston Fired and He's Not the

Only One," *Science,* December 22, 1972, pp. 1268-1270. Many senior officials across nearly all departments and agencies of the federal government were summarily dismissed in this manner.

25. See "National Institutes of Health: Announcement of Appointment of Dr. Robert S. Stone as Director, May 29, 1972," *Weekly Compilation of Presidential Documents*, vol. 9, June 1972, pp. 716-717; Harold M. Schmeck, Jr., "New Institutes of Health Chief: Robert Samuel Stone," *New York Times*, May 30, 1972; and Culliton, "NIH Director Stone: Another Manager on Nixon's Health Team," *Science* June 22, 1973, pp. 1258-1261. Senior NIH officials found it disappointing that Stone did not consult with anyone at NIH prior to accepting the position as director, nor did he apparently bargain for anything in return for accepting the appointment. Moreover, there were indications that Stone initially saw his tenure as concurrent with the second four-year term of President Nixon, thus binding the position of the NIH director to presidential politics in an entirely new and disturbing manner. The fact that Stone was strongly oriented toward management did little to allay NIH anxieties, though this orientation did coincide with the preferences of HEW and OMB.

26. Culliton, "Berliner Resigns from NIH," *Science*, vol. 180 (July 1972), p. 1344. Berliner also emphasized the personal reasons associated with the offer from his undergraduate institution. "The years are not likely to provide me another such opportunity to try my hand at the challenge of new professional activities." But the loss of Berliner was seen by many as one that the biomedical scientific community could ill afford.

27. See Stuart Auerbach, "NIH Aide Resigns, Hits Policy," *Washington Post*, January 18, 1974. Sherman's blast was amplified by an editorial, "Trouble at NIH," *Washington Post*, January 19, 1974, which spoke of the challenge "to restore NIH's world-wide reputation for excellence." Since Sherman's criticism and the *Post's* editorial were directed against the management of HEW Secretary Caspar Weinberger and Assistant Secretary Charles Edwards, the latter responded with a letter, "On the Question of Autonomy for NIH," *Washington Post*, January 24, 1974, in which he suggests that confidence in NIH "was diminished in the past through inadequate leadership and a misguided sense of the place of research in the nation's efforts to solve its health problems." "The real need," Edwards wrote, "is to establish effective methods for setting priorities among the institutes and programs of the NIH in order that the total biomedical research effort remain in balance. If, in fact, the NIH leadership had been more perceptive and responsive we might not have witnessed the removal of the cancer research effort from the administrative control of NIH, a move that threatens the further dissolution of biomedical research efforts."

28. See Harold Schmeck, Jr., "President Fills Top 2 Health Jobs," *New York Times*, April 22, 1975.

29. For a summary of these new laws see "Congressional Initiatives in Biomedical and Behavioral Research," in *Report of the President's Biomedical Research Panel*, Appendix D, Washington, D. C., April 30, 1976, pp. 33-41.

30. See Title II of op. cit., P.L. 93-352.

31. Panel members were: Franklin D. Murphy, Chairman, Times Mirror Corporation, Los Angeles, California; Ewald W. Busse, Duke University Medical Center, Durham, North Carolina; Robert H. Ebert, Harvard Medical School, Boston, Massachusetts; Albert L. Lehninger, Department of Physiological Chemistry, The Johns Hopkins University School of Medicine, Baltimore, Maryland; Paul A. Marks, Columbia University, College of Physicians and Surgeons, New York, New York; Benno C. Schmidt, J. H. Whitney and Company, New York, New York; and David B. Skinner, Department of Surgery, University of Chicago Hospitals and Clinics, Chicago, Illinois. See *Report of the President's Biomedical Research Panel*, Submitted to the President and the Congress of the United States, Washington, D. C., April 30, 1976. Also, ibid., *Appendix A: The Place of Biomedical Science in Medicine and the State of the Science; Appendix B: Approaches to Policy Development for Biomedical Research: Strategy for Budgeting and Movement from Invention to Clinical Application; Appendix C: Impact of Federal Health-Related Research Expenditures Upon Institutions of Higher Education;* and *Appendix D: Selected Staff Papers*. See also ibid., *Supplement 1. Analysis of Selected Biomedical Research Programs: Case Histories; Supplement 2. Impact of Federal Health-Related Research Expenditures Upon Institutions of Higher Education; Supplement 3. Written Statements Supplementing Verbal Testimonies of Witnesses;* and *Supplement 4. Statements of Professional, Scientific, and Voluntary Health Organizations.*

32. See Barbara J. Culliton, "Kennedy Hearings: Year-Long Probe of Biomedical Research Begins," *Science*, July 2, 1976, pp. 32-35.

33. *Report*, p. 5.

34. Culliton, "Kennedy Hearings," p. 33.

· AFTERWORD ·

1. Benno Schmidt, "Report to the President on the National Cancer Program," January 31, 1977, reprinted in *The Blue Sheet*, February 16, 1977, pp. S-2 through S-11.

2. Daniel S. Greenberg, "Cancer: Now the Bad News," *The Washington Post*, January 19, 1975.

3. William S. Gray, " 'The Data . . . Tell Only Part of the Story,' " *The Washington Post*, January 19, 1975.

4. James E. Enstrom and Donald F. Austin, "Interpreting Cancer Survival Rates," *Science*, March 4, 1977, pp. 847-851.

5. The three reports prepared for the National Cancer Institute on this matter are available in a single volume: National Cancer Institute, *The Final Reports of the National Cancer Institute Ad Hoc Working Groups on Mammography Screening for Breast Cancer and a Summary Report of Their Joint Findings and Recommendations*, Washington, D.C., Department of Health, Education, and Welfare, March 1977.

6. U.S., Congress, Senate, Committee on Labor and Public Welfare, *Breast Cancer, 1976*, Hearings, 94th Cong., 2nd sess., May 4, 1976.

7. See *Report of the President's Biomedical Research Panel, Appendix A: The Place of Biomedical Science in Medicine and the State of the Science*, Washington, D.C., Department of Health, Education, and Welfare, April 30, 1976.

8. The concern for carcinogenesis was strongly expressed in Comptroller General of the United States, *Federal Efforts to Protect the Public from Cancer-Causing Chemicals Are Not Very Effective*, Washington, D.C., General Accounting Office, June 16, 1976. The House subcommittee responsible for NCI appropriations has also indicated its deep concern with the environmental causes of cancer in its report accompanying the fiscal 1977 appropriation. See U.S., Congress, House, Committee on Appropriations, *Departments of Labor, and Health, Education, and Welfare, and Related Agencies Appropriation Bill, 1977*, Report No. 92-1219, 94th Cong., 2nd sess., June 8, 1976, pp. 24-27.

9. Perhaps the most impressive book of this kind is Stewart Alsop's *Stay of Execution: A Sort of Memoir*, Philadelphia and New York, J. B. Lippincott Company, 1973.

Index

Abel, I. W., 85, 341 n.15
Adams, Brock, 263-265, 266, 269, 271
agenda-setting, xiii-xvii; cancer as agenda item, 46, 282-284; role of Mary Lasker, 282-284; role of Panel of Consultants, 132, 223-224, 284, 288-291; role of press, 294
American Association for Cancer Research, 42, 55, 152, 183, 354 n.64
American Cancer Society (ACS), 21, 24, 67, 260; board of trustees takeover, 21-22; 58th annual meeting, 267; House testimony, 227-229; pressure on House, 201-202, 249-250, 258; resolution on H.R. 11302, 267; special citations, 267; support for President's Cancer Attack Panel, 260; support for S. 34, 159, 174, 183, 184
American College of Cardiology, 126
American College of Physicians, 172
American Heart Association, 159-161, 241-242
American Medical Association (AMA), 38-39, 171, 214
American Society for Clinical Investigation, 231-232
Amos, Harold, 301
Anderson, M. D., Hospital and Tumor Institute, 37-38, 55, 79
Association of American Cancer Institutes, 31-32, 230
Association of American Medical Colleges (AAMC), 127-130; opposes S. 34, 143-146; legislative priorities, 146; organizes opposition to S. 34, 171-173; opposes S. 1828, 192-194; helps Rogers draft H.R. 10681, 203; supports H.R. 10681, 240
Association of Professors of Medicine, 182-183

Baker, Carl G., 72-73, 95, 101, 137, 140, 159, 167, 197, 206, 223; research management interests, 70-71; relations with Panel, 98, 100
Baltimore, David, 235-236
Berliner, Robert W., 73-74, 105-106, 309
Blair, Deeda (Mrs. William McCormick Blair), 34
Blair, William McCormick, 85, 97, 280
Bobst, Elmer, 85-86, 98, 122; ACS board takeover, 21-22; contacts with Nixon on cancer initiative, 122-123, 177-178, 185; on directed research, 98; receives ACS first National Founders Award, 267
Bone, Hugh, 45
Brennan, Michael J., 232
Bridges, Styles, 26
Burchenal, Joseph, 84, 86, 90, 91, 92, 96, 100, 110, 152, 155, 279
Bush, George, 87, 122

cancer; fear of, 4; mortality, morbidity, 3, 5-6; nature of, 4-5; prevention of, 6; treatment of, 6-8
cancer centers, 253; comprehensive cancer centers, 107, 302-304; four major centers, 37; historical relations with medical schools, 37-38, 54-55; Roswell Park Memorial Institute, 244-247
cancer chemotherapy, 56-57. *See also* NCI cancer chemotherapy program
cancer control program; authorized in 1937, 50; "end game" provision, 295; program, 50-51, 303-305; reauthorized in 1971, 303; termination, 209-210; transfer from NCI to PHS, 51
cancer crusade; hyperbole of, 318;

representative, 78-79; cancer of liver diagnosed, 206; death, 315; instrumental in House Concurrent Resolution 675, 82; organizes support for S. 34, 134, 135; special citation from ACS, 267

Rauscher, Frank J., 297-298, 301, 305, 310
Regional Medical Programs, 40. *See also* Heart Disease, Cancer, and Stroke Amendments of 1965, President's Commission on Heart Disease, Cancer, and Stroke
research management, 10-15, 299-302. *See also* National Cancer Program, Panel of Consultants
research planning, 70-72, 299-300. *See also* National Cancer Institute, National Cancer Program, Panel of Consultants
Rhoads, Cornelius P., 56
Rhoads, Jonathan E., 84, 90, 159, 227-229, 279
Richardson, Elliot, 137, 165-166, 167-168, 169-170, 176, 178, 180, 184, 186-187, 206-210, 267
Rockefeller, Laurance, 85, 101, 122, 267
Rockefeller, Nelson A., 324
Rogers, Paul G., 199-200; accepts President's Cancer Attack Panel, 255; alternate cancer bill proposed, 202-203; attacks elimination of cancer control programs, 209-210; blasts cancer crusaders, 257; educates House Health Subcommittee, 204; floor position on H.R. 11302, 268-270; introduces H.R. 10681, 203; jurisdictional controversy with Dulski, 269; and miscalculation of cancer crusaders, 201; political problems with cancer legislation, 200-201; repays favor to Hastings, 244; secures unanimous subcommittee, 256-267; strategy of, 205, 226-

227, 251; support in House Commerce Committee, 262; visitors, 202
Rooney, John J., 82-83, 213
Roswell Park Memorial Institute, 30, 37, 55, 84, 244-247
Roy, William, R., 205, 208
Ruina Committee report, 68-69
Ruina, Jack P., 68-69
Rusch, Harold P., 84, 90

Saffiotti, Umberto, 305-306
Satterfield, David E., 204, 207, 228
Scheele, Leonard, 23
Schereschewsky, Joseph W., 43, 47
Schmidt, Benno C., 86-87, 88-89, 92, 157, 191, 219, 287, 310; appointed chairman of President's Cancer Panel, 278; chairman of NCI "executive committee," 297-298; chairman of Panel of Consultants, 86-88, 89; discusses Panel defections, 191-192; 5th report to the President, 319-320; managing partner, J. H. Whitney & Co., 88; negotiates between Senate and White House, 185-186; opposes President's Cancer Attack Panel, 263; position on independent cancer agency, 95-96; position on research management, 100-101; prepares for reception of report, 92-93, 122; relations with other Panel members, 90-91; preference for report format, 90; Senate testimony, 152-159; testimony in House, 215-224; view of Panel Consultants' mandate, 92, 152; visits Rogers, 202; Washington, D.C., experience, 88
Schmitz, John C., 205, 206, 209, 228-229
Schneider, Howard A., 146-147, 238
Schultz, George, 93, 122, 123-124, 159
Science magazine, 165, 173-174
scientists (biomedical) in politics, 9, 161-162, 289, 293. *See also*

scientists (*cont.*)
    medical-scientific community
Scott, Wendell, 84, 215, 217, 219, 267
Senate Health Subcommittee, 170,
    172, 177, 181; establishment of
    Panel of Consultants, 81-82; June 10
    hearing, 186-194; March hearings
    scheduled, 134
S. 34 (Conquest of Cancer Act), 134,
    147-149, 181-183; ACS support,
    159, 174; American Heart Associa-
    tion support, 159-161; American
    Medical Association opposition,
    171; AAMC opposition, 143-146;
    academic opposition, 171-173;
    Bobst urges support by Nixon, 178;
    contents, 134-135; FASEB tes-
    timony against, 146-147; effect of
    Nixon's proposal on, 180-181;
    HEW testimony on, 137-139; lack
    of medical-scientific support for,
    183; Panel of Consultants' testimony
    on, 151-159
S. 1828, 197-198, 209; adopted, 79 to
    1, 196; case for in House, 232-233;
    divergence between White House
    and HEW, 184-185; legislative his-
    tory, 194-196
Senate Resolution 376, 81-82
Sessoms, Stuart, 61
Shannon, James A., 21, 29, 51, 149-
    151
Sherman, John F., 106, 309
Skipper, Howard E., 62
space program analogy, 77-78, 79, 89,
    94, 97-98, 99-100, 105, 125-126,
    127, 128, 130, 136, 141-142, 144,
    154, 157, 158, 174, 176, 208, 211,
    295, 342 n.29, 357 n.96
Spratt, John S., 230-31

Springer, William, 262, 270, 271, 273
Staggers, Harley O., 198-199, 202,
    205, 207, 265, 270
Steinfeld, Jesse L., 137-139, 169
Stone, Robert S., 309
Surgeon General of U.S., 40, 43, 45,
    137-139
Sweek, Robert F., 89-90, 92, 94-95,
    96; preference for directed research,
    98-100
Symington, James W., 204-205, 208

Taft, Howard, 42-43
targeted research, 14. *See also* contract
    research, directed research
Temin, Howard, 235
Tiernan, Robert O., 264, 265, 266,
    269, 271

Virus Cancer Program, 69-70, 71, 112,
    300-301
Visscher, Maurice D., 240
Voegtlin, Carl, 43, 47

Wall Street Journal, 93, 175
Washington Post, 181, 209, 260-261
Wasserman, Lewis, 85, 285
wealth and cancer, 285-286
Weinberger, Caspar, 124, 313
Whitney, J. H., & Co., 88
Williams, Harrison, 119, 170, 176,
    182

Yarborough, Ralph W., 80-81, 102,
    103; introduces S. 4564, 113
Yarbro, John W., 231-232

Zinder Committee report, 300-301
Zinder, Norman, 300-301

LIBRARY OF CONGRESS CATALOGING
IN PUBLICATION DATA

Rettig, Richard A.
  Cancer crusade.

  Includes bibliographical references and index.
  1. Cancer—Law and legislation—United States.
2. United States.   Laws, statutes, etc.   National
cancer act of 1971.   I.   Title.
KF3803.C3R47   344′.73′043   77-72134
ISBN 0-691-07558-1